Microbial Biorefineries

Microbial-based biorefineries offer a promising, science-driven pathway toward sustainable development. By transforming waste streams into valuable bioproducts, these scalable technologies not only advance the global circular bioeconomy but also provide tangible solutions to some of the most pressing socioeconomic issues of our time. This book showcases the role of various microbial cell factories in waste beneficiation, highlighting both the opportunities and the barriers to the industrial implementation of these technologies.

Features:

- Expands the horizon of biomass valorization by presenting novel, scalable technologies that can be directly applied by scientists and engineers.
- Explores emerging R&D trends in microbial-based biorefineries, including the valorization of novel and underutilized biodegradable waste streams such as agri-waste, fruit waste, and fats/oils/greases, alongside traditional residues like agricultural biomass.
- Focuses on the upcycling and beneficiation of diverse organic waste streams into a broader portfolio of value-added products, including biochemicals, biomaterials, food additives, biofuels, and biostimulants.
- Contextualizes innovations through comprehensive techno-economic assessments and life cycle studies, evaluating the economic and environmental viability of these solutions.

By focusing on scalable, economically viable innovations and positioning microbial cell factories as key enablers of resource valorization, this book bridges policy aspirations and practical applications, offering readers in chemical, biological, and environmental engineering and related disciplines concrete pathways to translate circular economy concepts into industrial reality.

Patrick Sekoai is a senior lecturer at North-West University, Mahikeng, South Africa. He obtained his PhD in Bioprocess Engineering from the University of the Witwatersrand, South Africa.

Sizwe Ndlovu is an associate professor at North-West University, Potchefstroom, South Africa. He obtained his PhD in Microbiology from Stellenbosch University, South Africa.

Olivier Habimana is an associate professor at the Guangdong Technion-Israel Institute of Technology, Shantou, China. Prof. Habimana holds a PhD in Environmental Microbiology from the Université Paris-Sud XI-INRA, France.

Microbial Biorefineries

Global Innovations in Waste Beneficiation and Circular Economy Solutions

Edited by
Patrick Sekoai, Sizwe Ndlovu,
and Olivier Habimana

CRC Press
Taylor & Francis Group
Boca Raton London New York

CRC Press is an imprint of the
Taylor & Francis Group, an **informa** business

Designed cover image: © 2027 Shutterstock

First edition published 2027
by CRC Press
2385 NW Executive Center Drive, Suite 320, Boca Raton FL 33431

and by CRC Press
4 Park Square, Milton Park, Abingdon, Oxon, OX14 4RN

CRC Press is an imprint of Taylor & Francis Group, LLC

ISBN: 978-1-041-25108-8 (hbk)
ISBN: 978-1-041-25109-5 (pbk)
ISBN: 978-1-003-74515-0 (ebk)

DOI: 10.1201/9781003745150

Typeset in Times
by codeMantra

Contents

Preface

The circular bioeconomy is a burgeoning economic strategy focused on the sustainable use of our natural resources. Unlike the traditional linear model, the circular bioeconomy fosters the development of new value chains that integrate dimensions of economic growth, environmental sustainability, social equity, and ethics. Strong scientific evidence corroborates the use of different microorganisms in promoting the tenets of the circular bioeconomy. These biocatalysts can upcycle large amounts of by-products generated by various biomass sectors, producing diverse bio-based products. Microbial-based biorefineries have been successfully demonstrated at higher technology readiness levels (TRL $\geq$ 6), and some have even reached an industrial scale, highlighting their significance in this fast-evolving technological field.

This book showcases the role of microbial cell factories toward waste beneficiation and the role they can play in advancing global circular bioeconomic trends, highlighting both the opportunities and the barriers limiting the industrial implementation of these technologies. Furthermore, emerging R&D trends in microbial-based biorefineries are broadly covered in this book, including the valorization of novel and underutilized biodegradable waste streams such as agri-waste, fruit waste, and fats/oils/greases (FOG), alongside traditional residues like agricultural biomass. The explored biorefinery innovations are also scrutinized by techno-economic assessments (TEA) to investigate their economic potential for producing biofuels, biochemicals, biomaterials, and biostimulants in line with market and societal needs.

By aligning closely with the United Nations Sustainable Development Goals (SDGs)—notably SDG 7 (Affordable and Clean Energy), SDG 9 (Industry, Innovation, and Infrastructure), SDG 12 (Responsible Consumption and Production), and SDG 13 (Climate Action)—this book provides a holistic and timely scientific contribution to the global transition toward a circular and inclusive bioeconomy framework.

Patrick Sekoai, Sizwe Ndlovu, and Olivier Habimana

Contributors

Ayodeji Adedoyin
Department of Biotechnology
and Food Technology
Doornfontein Campus
University of Johannesburg
Johannesburg, South Africa

Adebowale Aladejana
Department of Biotechnology
and Food Technology
Doornfontein Campus
University of Johannesburg
Johannesburg, South Africa

Itumeleng Baloyi
Division of Biochemistry
Department of Biochemistry,
Genetics & Microbiology
University of Pretoria
Hatfield, South Africa

Ryan Bosch
Division of Biochemistry
Department of Biochemistry,
Genetics & Microbiology
University of Pretoria
Hatfield, South Africa

Tumelo Bosena
Department of Biological Sciences
Faculty of Science
University of Botswana
Gaborone, Botswana

Tidimalo Coetzee
Department of Biological Sciences
Faculty of Science
University of Botswana
Gaborone, Botswana

Eunice Sefakor Dogbe
Department of Chemical Engineering
Stellenbosch University
Matieland, South Africa

Obinna Ezeokoli
Unit for Environmental Sciences
and Management
North-West University
Potchefstroom, South Africa

Johann F. Görgens
Department of Chemical Engineering
Stellenbosch University
Matieland, South Africa

Olivier Habimana
Department of Biotechnology
and Food Engineering
Guangdong Technion-Israel Institute
of Technology
Shantou, China

Lebang Kgosietsile
Department of Biological Sciences
Faculty of Science
University of Botswana
Gaborone, Botswana

Janus Louw
Department of Chemical Engineering
Stellenbosch University
Matieland, South Africa

Jeanne Louw
Department of Biological Sciences
Faculty of Science
University of Botswana
Gaborone, Botswana

Samkelo Malgas
Division of Biochemistry
Department of Biochemistry, Genetics & Microbiology
University of Pretoria
Hatfield, South Africa

Pako Manyala
Department of Biological Sciences
Faculty of Science
University of Botswana
Gaborone, Botswana

Olga Mashandule
Division of Biochemistry
Department of Biochemistry, Genetics & Microbiology
University of Pretoria
Hatfield, South Africa

Lebohang Moloi
Department of Biological Sciences
Faculty of Science
University of Botswana
Gaborone, Botswana

Pako Motshwarakgole
Department of Biological Sciences
Faculty of Science
University of Botswana
Gaborone, Botswana

Sizwe Ndlovu
Unit for Environmental Sciences and Management
North-West University
Potchefstroom, South Africa

Thando Ndlovu
Department of Biological Sciences
Faculty of Science
University of Botswana
Gaborone, Botswana

Marni Oberholzer
Division of Biochemistry
Department of Biochemistry, Genetics & Microbiology
University of Pretoria
Hatfield, South Africa

Sana Samson
Division of Biochemistry
Department of Biochemistry, Genetics & Microbiology
University of Pretoria
Hatfield, South Africa

Patrick Sekoai
Department of Microbiology
Faculty of Natural and Agricultural Sciences
North-West University
Mafikeng, South Africa

Freedom Tshabuse
Department of Biochemistry and Microbiology
University of Zululand
KwaDlangezwa, South Africa

Kadima Tshiyoyo
Division of Biochemistry
Department of Biochemistry, Genetics & Microbiology
University of Pretoria
Hatfield, South Africa

1 A Review of Green Extraction Technologies for Phytochemicals from Agricultural Processing Waste Residues

Kadima Tshiyoyo, Lebohang Moloi, Marni Oberholzer, Sana Samson, Ryan Bosch, Olga Mashandule, Itumeleng Baloyi, and Samkelo Malgas

1.1 INTRODUCTION

The agricultural sector generates a substantial amount of waste that is not effectively used, impacting human health and the environment (Bala et al., 2023). Globally, approximately 1.3 billion tons of food are lost and wasted annually, leading to a high environmental footprint. According to the Food and Agriculture Organization (FAO), this waste represents one-third of the food produced for human consumption (Capanoglu et al., 2022; Gustavsson et al., 2011). Further, the United Nations Sustainable Development Goals (SDGs), which guide the global effort towards sustainable living, have set food waste reduction targets (SDG 12) for the 2030 agenda (Bala et al., 2023). Additionally, numerous countries have implemented policies that move toward a circular economy. In addition to minimising waste, converting waste food and agricultural by-products into valuable or useful products can contribute to the SDGs. Biomass, such as agricultural waste, can be valorised into several novel bioactive compounds and other bioproducts to contribute towards a global bio-economy (Bala et al., 2023; Capanoglu et al., 2022). The innovation of biorefinery development on natural biomass to find a wide range of by-products is valuable in various production fields (agricultural and pharmaceutical); when combined with energy generation, this system can reduce or eliminate problematic waste streams (Herrero et al., 2015). Valorising non-edible crop residues such as leaves, peels, pits, pomace, pulp, and seeds is essential (Torres-Valenzuela et al., 2020). These agricultural wastes are a valuable source of bioactive phytochemicals, such as carotenoids,

DOI: 10.1201/9781003745150-1

essential oils, polyphenols, and tocopherols, which have antioxidant, antidiabetic, anticarcinogenic, and antimicrobial properties (Yadav et al., 2024). The assessment of these by-products as bioactive compound sources has gained interest over the last decade (Torres-Valenzuela et al., 2020).

Extracting phytochemicals from agricultural waste improves the sustainability of farming practices and offers an economically viable means of producing high-value compounds. Selective solvents are used to extract important phytochemicals present in biological material from their inactive components (Rasul, 2018).

Against this background, the purpose of this chapter is to analyse and offer an up-to-date synopsis of the potential benefits, advantages, and disadvantages of applying green extraction technologies for the recovery of phytochemicals from agricultural waste materials.

1.2 METHODOLOGY

Several reviews and experimental articles related to this topic were searched using various electronic databases, including Google Scholar, PubMed, and ScienceDirect. Most of the research articles considered in this study were published after 2015. However, some important articles published before 2015 were also added, such as articles on extraction methods with significant results that could not be replaced. The literature search focused on the agricultural waste sources, conventional and green extraction technologies used, and the extracted phytochemical classes obtained from the specific extraction technologies. Figure 1.1 shows the word cloud generated from keyword occurrences in different publications.

FIGURE 1.1 Word cloud of keywords associated with agricultural wastes and green technologies from various publications (https://www.wordclouds.com/).

1.3 EXTRACTION METHODS

Extraction is important for separating the desired natural products from the raw materials (Zhang et al., 2018). These extraction methods are classified into two types: conventional and non-conventional, which are usually sustainable.

Conventional extraction methods include percolation (plant material exposed to flowing solvent), maceration (soaking plant material in a solvent at room temperature), decoction and hydrodistillation (use of water or steam to facilitate the extraction of essential oils and other volatile compounds from plant materials) (Zhang et al., 2018). These extraction methods are important for valorising agricultural waste by recovering valuable phytochemicals applicable in various industrial sectors. Table 1.1 summarises the recent applications of conventional extraction methods in agricultural and food waste, highlighting waste sources, the class of phytochemical extracted, and the optimal extraction conditions.

Despite their widespread use, these conventional methods require large amounts of solvents and prolonged extraction time to generate appreciable yields of phytochemicals (Rasul, 2018). These methods also usually involve high temperatures, which can

TABLE 1.1
Recent applications of some conventional extraction methods of phytochemicals from agricultural waste feedstocks

Extraction method	Plant	Plant parts	Phytochemical extracted	Extraction conditions	References
Soxhlet extraction	Grape	Pomace	(Poly)phenols	Solvent: Ethanol Temperature: 60°C Duration: 6 h	Abbasi-Parizad et al. (2022) and Waseem et al. (2023)
	Sweet potato	Leaves	Carotenoids	Solvent: Hexane Temperature: 70°C Duration: 5 h	Waseem et al. (2023)
	Tomato	Skin	Carotenoids, flavonoids	Solvent: 80% methanol Duration: 25 min Temperature: 45°C	Yadav et al. (2024) and Capanoglu et al. (2022)
	Rice	Grains	Phenolic acids, flavonoids		
	Blueberry	Pomace	Anthocyanins		
Hydrodistillation	Orange	Peels	Essential oils	Direct steam injection/water at 100°C	Capanoglu et al. (2022) and Azeez et al. (2017)
	Ginger	Root			
Maceration	Sugarcane	Bagasse	Cellulose	Cold extraction with acetone	Rasheed et al. (2024) and Chaka (2022)
	Corn	Cob			
Decoction	Ginger	Root	Gingerol	Boil in water for 30 min	Wang et al. (2016)
	Turmeric	Rhizome	Curcumin		

TABLE 1.2
Advantages and disadvantages of conventional and non-conventional phytochemical extraction techniques

Extraction technique	Advantages	Disadvantages
Conventional	• Simple methods • Simple equipment • A wide range of compounds • Established protocols	• Long extraction times • High energy usage • Slow process • Usage of organic solvents • Large volumes of solvents • High temperature and pressure • Environmental concerns • Less sensitive
Non-conventional (green)	• Faster extraction times • Less organic solvents • Moderate temperature • Environmentally friendly • High selectivity • High efficiency	• Less established • Optimisation challenges • Reproducibility challenges • Complex equipment • Costly equipment

degrade any thermolabile phytochemicals to be extracted, resulting in lower bioactivity and yields if not considered during the procedure (Bitwell et al., 2023). Table 1.2 highlights the advantages and disadvantages of conventional and non-conventional methods observed in previous studies (Zhang et al., 2018; Mohammad Azmin et al., 2016; Osorio-Tobón, 2020; Bitwell et al., 2023).

Due to the drawbacks outlined in Table 1.2, particularly the environmental concerns associated with conventional extraction methods, the pursuit of more efficient and sustainable (green) extraction technologies has intensified. In recent years, several non-conventional green extraction techniques have gained prominence because of their distinct advantages over traditional solvent-based approaches. These include enzyme-assisted extraction (EAE), supercritical fluid extraction (SFE), ultrasound-assisted extraction (UAE), microwave-assisted extraction (MAE), pressurised liquid extraction (PLE), and pulsed electric field-assisted extraction (PEAE) (Bitwell et al., 2023). Collectively, these innovative techniques enable the recovery of target bioactive compounds under milder processing conditions, using moderate temperatures, shorter extraction times, and environmentally benign solvents (Osorio-Tobón, 2020). Beyond their ecological benefits, these green technologies are economically viable, offering enhanced extraction efficiency, improved preservation of biological activities, and reduced energy requirements (Bitwell et al., 2023). Moreover, growing evidence suggests that integrating multiple techniques can produce synergistic effects, further improving the yield and selectivity of phytochemical extraction. Consequently, combined or hybrid extraction strategies are increasingly being explored as an emerging frontier in sustainable bioprocessing (Kumar et al., 2023).

1.4 GREEN EXTRACTION TECHNIQUES

1.4.1 Enzyme-Assisted Extraction

EAE relies on the use of enzymes to dismantle the polymeric structure of the plant material for the release of specific phytochemicals. The efficiency of EAE depends on the biochemical and physicochemical properties of the enzyme(s) used to extract a particular phytochemical (Puri et al., 2012). A short extraction time and reduced quantity of the extraction solvent are required, rendering this method convenient and environmentally friendly and yielding a good quality product (Streimikyte et al., 2022). Moreover, extractions are carried out under controlled temperatures, qualifying this technique as ideal for extracting heat-sensitive biomolecules such as flavonoids, lipids, natural pigments, and oils. Enzymes commonly used in this method include cellulases, hemicellulases, ligninases, pectinases, and proteases (Puri et al., 2012; Nadar et al., 2018), as the plant cell wall comprises cellulose, hemicellulose, lignin, pectin, and protein (Gligor et al., 2019). The enzymes target specific macromolecules within the bulk biomass complex, releasing intracellular components within the cytosolic spaces and those that are cell wall-bound into the external environment (Nadar et al., 2018; Gligor et al., 2019). Therefore, it is crucial to choose the appropriate enzyme for high productivity (Puri et al., 2012).

A study by Heemann et al. (2019) evaluated the use of EAE to recover (poly)phenols from green yerba mate. Mate is a tea made from the infusion of the dry leaves of *Ilex paraguariensis* and is one of the most consumed non-alcoholic beverages in Latin America (Heck and De Mejia, 2007). The authors applied a response surface methodology to compare the extraction with and without enzymes, where a commercial blend of enzymes, viscozyme was used, consisting of a range of carbohydrases including arabanase, beta-glucanase, cellulase, hemicellulase, and xylanase. The study reported an increase in yield from 38.67% to 52.08% in the presence of the enzyme operating at 50.0°C, with an enzyme load of 168 fungal beta-glucanase units (FGBU)/100 g dry matter, 120 min incubation, and pH 4.50 (Heemann et al., 2019). Some studies have investigated the effect of temperature, enzyme loading, and incubation time on the yield of desired compounds. Cortes-Ferre et al. (2022) investigated the extraction of phenolics from habanero seeds using cellulase-assisted extractions at different temperatures (30°C–60°C), protein loadings (250 and 2,500 UI/L), and extraction times (0–150 min). The authors reported that the highest phenolic content (337.96 mg GAE/L) was observed at 30°C, enzyme loading of 2,500 UI/L, and 150 min incubation period (Cortes-Ferre et al., 2022).

Commercial enzymes currently used for EAE are highly priced; therefore, processing large amounts of raw plant material tends to be costly (Cheng et al., 2015). Moreover, the enzymes have poor reusability and recyclability. EAE cannot meet the high industrial demands of the products due to enzyme inactivation under industrial operational conditions (high temperatures and extreme pH) (Das et al., 2021). To overcome these challenges, enzyme immobilisation has been used to improve the efficiency of enzymes against unfavourable industrial operational conditions (Maghraby et al., 2023). During immobilisation, enzymes or their cocktail are attached to an inert material to increase the rigidity and stability of enzymes. This technology also enables the enzyme to be reused multiple times over a long period, saving on the costs of buying new enzymes. Immobilised enzymes can be separated from unreacted substrate and soluble reaction products (Bilal and Iqbal, 2019). Various studies have explored using EAE with immobilised enzymes on different agricultural residues, highlighting its versatility and effectiveness across different plant matrices (Tizón Alba et al., 2023; Vardakas et al., 2024). Vardakas et al. (2024) evaluated the extraction of phenolics from olive leaves using EAE with different immobilised enzyme mixtures and doses (Vardakas et al., 2024). Optimum phenolic content (605.55 mg GAE/L of extract) was recovered when using a combination of pectinase and viscozyme at a 1:1 volume ratio, and the authors concluded that the reported phenolic content was higher than that recovered by conventional methods (Vardakas et al., 2024).

EAE is an excellent technique with several advantages, such as high yields of high-quality phytochemicals and increased efficiency and specificity in the extraction of bioactive compounds from agricultural wastes. However, the method also has limitations, including long processing time, complexity, incomplete cell wall breakdown, and co-extraction of undesirable compounds (Cheng et al., 2015; Nadar et al., 2018). This drawback can be counteracted by combining this method with other green extraction technologies, such as MAE (Khruengsai et al., 2023), PLE (Domínguez-Rodríguez et al., 2021), SFE (Mushtaq et al., 2017), and UAE (Bosch and Malgas, 2023). In combination, EAE can enhance the extraction process by breaking down complex matrices and hydrolysing cell walls and other barriers in the agricultural waste, thereby facilitating the release and availability of the target compounds for the benefit of the second method, thus increasing the extraction of phytochemicals of interest in high yields in their pure form (Mushtaq et al., 2017; Das et al., 2021).

1.4.2 Assisted Extraction

MAE is an eco-friendly technique that requires the oscillation of microwave wavelengths between 1 mm and 1 m, with frequencies ranging from 0.3 GHz (1 m) to 300 GHz (1 mm) (Pinto et al., 2021). Under these conditions, the microwaves can penetrate the sample and interact with polar components. The heating procedure includes two distinct processes, dipole rotation and ionic conduction, encouraging ion and dipole collisions and friction that permit quick heating of the solvent and excellent extraction efficiency (Pinto et al., 2021). According to Alupului et al. (2012), the extraction mechanism for MAE entails the following three ordered steps: solute separation from the active sites of the sample matrix because of elevated pressure and temperature; solvent diffusion through the sample matrix; and finally, solute from the sample matrix is released into the solvent. It is a method for extracting more intact organic and organometallic compounds. The extraction conditions, such as power, irradiation time, type of solvents, solvent ratio, and composition, need to be optimised to achieve the highest yield of compounds. The extraction process generates excessive heat, destroying the desired phytochemicals (Murugesan et al., 2024). Several studies have investigated phytochemical extraction from different waste sources. Table 1.3 shows waste sources, phytochemicals extracted, optimised conditions, and extraction yield using MAE.

TABLE 1.3
Application of non-conventional (green) technologies for the extraction of phytochemicals from agro-industrial waste

Food waste	Phytochemical extracted	Optimal conditions	Yield (%)/extract quantity	References
		Enzyme-assisted extraction		
Green yerba mate leaves	(Poly)phenols	Temperature: 66°C Carbohydrase: 168 FGBU/100 g Time: 120 min pH: 4.50	TPC: 3.40 mg GAE/g dry matter	Heemann et al. (2019)
Habanero chilli pepper seeds	Phenolics	Temperature: 30°C Cellulase: 2,500 UI/L Time: 150 min	TPC: 337.96 mg GAE/L of extract	Cortes-Ferre et al. (2022)
Olive leaves	Phenolics	Temperature: 50°C Pectinex and viscozyme: 0.1 mL enzyme/100 g substrate (50–50 v/v) Time: 30 min pH: 4.0	TPC: 605.55 mg GAE/L extract	Vardakas et al. (2024)
Raspberry pomace	(Poly)phenols Phytosterols	Temperature: 60°C Enzyme: 1.2 units of thermostable alkaline protease/100 g pomace press-cake (PPC) Time: 2 h pH: 9	(Poly)phenols: 3.7 g/100 g PPC Phytosterols: 259.7 mg/100 g PPC	Saad et al. (2019)
		Microwave-assisted extraction		
Fruits hulls of tea-oil tree	(Poly)phenols	Liquid-to-solid ratio: 15.33:1 (mL/g) Time: 35 min Temperature: 76°C	15.05% ± 0.04% (w/w) TFC: 140.06 mg/g dry matter Gallic acid: 66.18 mg/g dry matter	Zhang et al. (2011)
Cocoa shell waste	Phytosterols	Time: 10 min Power: 500 W Temperature: 70°C	Phytosterols: 3546.1 mg/100 g dry matter	Ibrahim et al. (2020)
Coffee pulp waste	Phenolics Flavonoids	Time: 85 min Sample to solvent ratio: 1:100 g/100 ml Solvent: Ethanol – 42.5% (v/v) Power: 1,000 W	TPC: 38.68 mg/g dry matter TFC: 27.00 mg/g dry matter	Tran et al. (2022)

(Continued)

TABLE 1.3
(Continued)

Food waste	Phytochemical extracted	Optimal conditions	Yield (%)/extract quantity	References
Tomato waste	Carotenoids	Time: 90 s Power: 300 W Temperature: 35°C–75°C	5.74 mg lycopene/100 g dry matter 4.83 mg beta-carotene/100 g dry matter	Lasunon et al. (2021)
	Phenolics	Power: 450 W Time: 30 s Temperature: ~46°C	TPC: 280.10 mg GAE/100 g dry matter	
	Flavonoids	Power: 180 W Time: 90 s Temperature: ~47°C	TFC: 9832.52 mg catechin/100 g dry matter	
		Pulsed electrical field-assisted extraction		
Banana peels	(Poly)phenols	Field strength: 4 kV/cm Time: 2 min	1.664 ± 0.226 mg GAE/mg dry matter	Hendrawan et al. (2019)
Pomelo peels	Naringin	Field strength: 4 kV/cm Pulse: 30	(15.80–16.19 mg/ml) 20% increase in yield	Niu et al. (2021)
Lemon peels	(Poly)phenols	Field strength: 7 kV/cm Time and pulse: 30 pulses of 30 µsec	300% increase in yield (84 mg of hesperidin in 100 g fresh weight (FW) and 176 mg of eriocitrin in 100 g FW	Peiró et al. 2019)
Orange peels	(Poly)phenols	Field strength: 7 kV/cm Time and pulses: 60 µsec (20 pulses of 3 µsec)	159% increase in yield	Luengo et al. (2013)
Potato peels	(Poly)phenols	Field strength: 1 kV/cm Total specific energy input: 5 kJ/kg	10% increase in yield	Frontuto et al. (2019)
Blueberry pomace	(Poly)phenols	Field strength: 20 kV/cm Pulses: 100 Total specific energy input: 41.03 kJ/kg	10.52 mg of gallic acid equivalent (GAE) per g of dry weight (dw)	Lončarić et al. (2020)
Tomato peel	Lycopene	Field strength: 5 kV/cm Total specific energy input: 5 kJ/kg	12%–18% increase in yield	Pataro et al. (2020)
		Pressurised liquid extraction		
Blueberries	Anthocyanins	Temperature: 40°C Pressure: 20 MPa	3.9%–8% (w/w)	Paes et al. (2014)
Pomegranate	Dietary fibre	Pressure: 1,500 psi Time: 20 min	30% (w/w) from peels 18% (w/w) from fruit	García et al. (2021)

(Continued)

TABLE 1.3
(Continued)

Food waste	Phytochemical extracted	Optimal conditions	Yield (%)/extract quantity	References
Palm fibre	Carotenoids	Solvent: Ethanol Pressure: 2–6 MPa Temperature: 35°C–55°C Time: 5 min	305 ± 18 μg α-carotene/g extract 713 ± 46 μg β-carotene/g extract	Cardenas-Toro et al. (2015)
Pomegranate peel	Phenolics	Solvent: Water Temperature: 70°C Ultrasound power: 480 W	61.72 ± 7.70 mg/g dry matter	Sumere et al. (2018)
Beetroot leaves	Phenolics	Solvents: Ethanol:water (70/30) Temperature: 40°C Pressures: 7.5, 10 and 12.5 MPa	36.03% ± 0.04% (w/w) for leaves TPC: 252 ± 2 mg GAE/g extract	Lasta et al. (2019)
		Supercritical fluid extraction		
Asparagus	Phenolics	Temperature: 65°C Pressure: 30 MPa Solvent: CO_2 Co-solvent: Water and ethanol (1:1) Time: 60 min	2.78 mg/g lyophilised mass	Solana et al. (2015)
Blackberries	Phenolics Flavonoids	Temperature: 40°C Pressure: 300 bar Solvent: CO_2 Co-solvent: Ethanol Time: 60 min	4.42% (w/w) dry matter	Nastić et al. (2020)
Sesame husks	Essential oils	Temperature: 35°C Pressure: 20 MPa Solvent: CO_2 Time: 210 min	26.47 g oil/100 g dry sample	Carvalho et al. (2012)
Orange peel	Essential oils	Temperature: 40 Pressure: 35 MPa Solvent: CO_2 Time: 90 min	0.48% ± 0.02% w/w peels	Barrales et al. (2018)
Avocado peels, pulps and seeds	Essential oils	Temperature: 50 Pressure: 250 bar Solvent: CO_2 Time: 40 min	Peels: 16.89% ± 0.56% (w/w) dry weight Pulps: 56.10% ± 0.66% (w/w) dry weight Seeds: 1.61% ± 0.31% (w/w) dry weight	Wang et al. (2020)

(Continued)

TABLE 1.3
(Continued)

Food waste	Phytochemical extracted	Optimal conditions	Yield (%)/extract quantity	References
Papaya seeds	Phenolics	Temperature: 50 Pressure: 20 MPa Solvent: CO_2 Co-solvent: Ethanol Time: 180 min	24.52% ± 0.06% yield (w/w) dry weight TPC: 34.23±2.75 mg GAE/g extract	Castro-Vargas et al. (2019)
Grape pomace	(Poly)phenols Phytosterols Vitamins	Temperature: 60 Pressure: 250 bar Solvent: CO_2 Co-solvent: Ethanol Time: 15 min	TPC: 570 ± 10 μg GAE/g dry matter β-Sitosterol: 1.55 ± 0.28 mg/g dry matter α-Tocopherol: 156.6 ± 41.1 μg GAE/g dry matter Vitamin C: 2612.0 ± 440.0 μg GAE/g dry matter	Aresta et al. (2020)
Orange peel	Phenolics	Temperature: 90 Pressure: 170 bar Solvent: CO_2 Co-solvent: Ethanol Time: 120 min	0.78%–1.24% (w/w) extract/dry matter	Trabelsi et al. (2016)
Pomegranate seed	Phenolics Fatty acids Carotenoids	Temperature: 40–60 Pressure: 200 bar Solvent: CO_2 Co-solvent: Ethanol Time: 20 min	TPC: 1.11±0.02 mg/100 g seed oil TFC: 0.14±0.01 mg/100 g seed oil β-Carotene: 0.02 mg/100 g seed oil Phytosterols: 3142.4 mg/100 g seed oil	Silva et al. (2019)
		Ultrasound-assisted extraction		
Apple peel	Pectin	Solvent: Acidified water pH: 2.3 Frequency: 20 kHz Temperature: 63.22°C Time: 18.04 min	8.93% (w/w)	Shivamathi et al. (2019)
Apricot kernel	Fatty acids	Solvent: n-Hexane Frequency: 40 kHz Power: 700 W Temperature: 51.72°C Time: 43.95 min	40.86%–46.01% (w/w)	Gayas et al. (2017)

(Continued)

TABLE 1.3
(Continued)

Food waste	Phytochemical extracted	Optimal conditions	Yield (%)/extract quantity	References
Apricot pomace	ρ-Coumaric acid Ferulic acid Rutin	Solvent: 50% ethanol Temperature: 50°C Time: 90 min	TPC: 1.206 mg GAE/g DM TFC: 1.015 mg CE/g DM	Dinçel Kasapoğlu et al. (2023)
Eggplant peel	High methoxy pectin	Solvent: Acidified water pH: 1.5 Power: 50 W Time: 30 min	33.64 g ± 1.12/100 g DM	Kazemi et al. (2019)
Garlic husk	Polyphenols	Solvents: Water, methanol and ethanol Temperature: 40°C Time: 30 min	TPC: 473.73 ± 23.6 GAE/100 g TFC: 126.39 ± 9.54 CE/100 g	Naeem et al. (2022)
Grape marc and seeds	Polyphenols	Solvent: 80% ethanol Power: 253 W Time: 27 min	TPC: 27 ± 1 mg/g TFC: 3.6 ± 0.3 mg/g	Ueda et al. (2023)
Jackfruit waste	Caffeic acid	Solvent: ChCl-LA Power: 750 W Time: 7 min	TPC: 53.16 ± 1.23 mg GAE/g DW	Ly and Sothornvit (2024)
Mandarin and lime peels	Hesperidin Naringin	Solvent: 80% ethanol Frequency: 38.5 kHz Power: 50.93 W Temperature: 40°C Time: 30 min	TPC: 3,083.61 mg GAE/100 g DW TFC: 2,539.82 mg QE/100 g DW Hesperidin: 1,374.20 mg/100 g DW	Singanusong et al. (2015)
Onion peel	Polyphenols	Solvents: Water, methanol, and ethanol Temperature: 40°C Time: 30 min	TPC: 914.44 ± 39.4 mg/100 g DW TFC: 178.19 ± 14.5 mg/100 g DW	Naeem et al. (2022)
Orange peel	Caffeic acid Hesperidin Carotenoids	Solvent: 50% ethanol Power: 400 W Time: 30 min	TPC: 105.96 mg GAE/100 g Carotenoids: 0.63 mg β-carotene/100 g Hesperidin: 113.03 ± 0.08 mg/100 g	Montero-Calderon et al. (2019)
Orange peel	Carotenoids	Solvent : Menthol:eucalyptol (1:20 w/v), Frequency: 20 kHz Power: 200 W Time: 20 min	Carotenoids: 359.3 ± 3.5 mg/100 g DW	Viñas-Ospino et al. (2023)

(Continued)

TABLE 1.3
(Continued)

Food waste	Phytochemical extracted	Optimal conditions	Yield (%)/extract quantity	References
Pineapple pulp and peel	Bromelain	Solvent: Water Amplitude: 99.96% Power: 200 W Time: 26.8 min	Bromelain: 196.46 ± 3.29 U/ml	Mala et al. (2021)
Tomato waste	Pectin	Solvent: Ammonium oxalate/oxalic acid Frequency: 37 kHz Time: 15 min	16.3%–18.5% (w/w)	Grassino et al. (2016)
		Twin-screw extrusion		
Wheat straw and bran	Wheat bran hemicellulose	Solvent: Dilute NaOH solution Liquid/solid ratio: 7 Bran/alkali ratio: 6.6 Dry flow rate: 13.8 kg/h Temperature: 50°C	Hemicellulose: 24% (of total) Xylan: 44.6% (w/w) dry weight	Zeitoun et al. (2010)
Alfalfa	Protein	Solvent: Water Liquid/solid ratio: 7.7 Temperature: 31°C	Protein: 31% (w/w) dry weight	Colas et al. (2013)
Industrial hemp waste (hurds and dust)	Hydroxy cinnamic acids	Solvent: Dilute NaOH in water/ethanol Hurds: Dry flow rate: 2.2–2.5 kg/h Liquid/solid ratio: 6.8 Temperature: 50°C Dust: Dry flow rate: 1.5 kg/h Liquid/solid ratio: 9.5 Temperature: 50°C	Hurds: p-Cinnamic acid: 1814 mg/kg of raw organic matter (ROM) Ferulic acid: 99 mg/kg ROM Dust: p-Cinnamic acid: 1150 mg/kg ROM Ferulic acid: 95 mg/kg ROM	Candy et al. (2017)

A study focused on (poly)phenols extraction from the fruit hulls of the tea-oil tree used a central composite design to optimise MAE conditions. The optimal parameters were a liquid-to-solid ratio of 15.33:1 (mL/g), 35 min residence time, and 76°C. The (poly)phenols yield was high, with gallic acid being the predominant compound found in the extracts (Zhang et al., 2011). In another study, MAE was used to extract (poly)phenols from blackcurrant by-products. The extraction process was optimised using a design of experiments (DOE) approach, finding that higher microwave power and shorter extraction time led to the best results. The extracts were then used in active packaging, showing antimicrobial and antioxidant properties that could extend the shelf-life of fruits (Alchera et al., 2022).

In a study by Ibrahim et al. (2020), MAE was utilised to extract phytosterols from cocoa shell waste (CSW), and ethanol was used as a solvent, with a conventional maceration extraction also used for comparison purposes. From the study, the optimum condition for MAE of β-sitosterol was 10 min residence time using 500 W of microwave power at 70°C, with a greater yield of β-sitosterol obtained from MAE compared to the traditional maceration method (Ibrahim et al., 2020). Similar results were obtained from a study by López-Salazar et al. (2019), where the MAE of β-sitosterol β-D-glucoside (BSSG) from *Agave angustifolia* achieved a significantly higher yield of BSSG in just 5 s of extraction time in comparison to the 48 h maceration process (Ibrahim et al., 2020; López-Salazar et al., 2019).

Due to its localised and rapid heating capability, MAE can be completed within minutes, making it a highly reliable technique characterised by excellent stability and reproducibility. This efficiency makes MAE particularly suitable for analytical chemistry applications, where precision and repeatability are paramount (Chan et al., 2011). The superior performance of MAE is primarily attributed to the unique mechanism of microwave heating, which promotes enhanced mass transfer through direct interaction between microwave energy and the extraction medium. Scanning electron microscopy (SEM) analyses have revealed that microwave irradiation induces structural disruption of the sample matrix, facilitating rapid extraction as intracellular components are released and dissolved into the solvent upon cell rupture. Notably, MAE preserves the therapeutic integrity of the extracted bioactive compounds, minimising degradation during processing (Gao et al., 2007; Zhang et al., 2008; Chan et al., 2011).

Economically, MAE is considered a viable technique due to its relatively low equipment setup costs and affordability compared to other non-conventional technologies. It also presents minimal operational risks, as the process is typically conducted under atmospheric conditions, thereby avoiding high-pressure or high-temperature hazards. Despite these advantages, MAE is not without limitations (Chan et al., 2011). The use of non-polar solvents is generally discouraged because of their poor microwave absorbance, which reduces heating efficiency. However, certain target compounds exhibit high solubility only in non-polar solvents, posing challenges in solvent selection and extraction optimisation (López-Salazar et al., 2023). Another limitation of MAE is its restricted selectivity, which is strongly influenced by both the solvent system and extraction temperature (Osorio-Tobón, 2020). As a result, post-extraction purification steps such as liquid–liquid extraction are often required to isolate and concentrate the active compounds, particularly for pharmaceutical or nutraceutical applications (Chan et al., 2011).

1.4.3 Pulse Electric Field-Assisted Extraction

PEAE is a highly feasible green extraction technology that produces high-intensity pulsed electric fields from a high current flow (Zhang et al., 2019). Cell envelopes (membrane/cell wall) are electroporated by these intense pulsed electric fields, which make it easier for the solvent to penetrate the cytoplasm and to extract desired intracellular compounds (Guerrero-Beltran and Welti-Chanes, 2016). Thus, PEAE permits a greater mass transfer of intracellular components by enhancing the permeability of the cells (Barba et al., 2016). The factors that impact the efficiency of PEAE include field strength, specific energy input, pulse number, and temperature (Shi et al., 2022). Citrus fruit waste is a rich source of phytochemicals such as naringin, abundant in pomelo peel waste. Investigations were conducted on the effects of different pulsed electric field treatments (2–10 kV/cm, 10–50 pulses) on the naringin extraction yield from the pomelo peel's white capsule. The extraction yield of naringin was greatly enhanced by PEAE treatment, with a 20% increase observed at 4 kV/cm electric field intensity and 30 pulses compared to non-PEAE treatment. The initially applied input of electric field strength (2–4 kV/cm) and pulse numbers (0–30) boosted naringin yield; however, no significant increase in naringin yield was detected after that (Niu et al., 2021).

PEAE has some limitations, such as the reversibility of the membrane changes because of the strength of the electric field being applied during electroporation, making the efficiency of this method highly dependent on electric field strength and electrode gap. Moreover, air bubbles in the treatment chamber may affect the dielectric breakdown, making the process less uniform (Bozinou et al., 2019). Every year, tons of fruit pomace are wasted globally. PEAE can be used as a pretreatment to improve the extraction of bioactive compounds from fruit pomace. A study on peach pomace examined the impact of PEAE versus conventional thermal treatment on the extraction of several chemicals, including vitamin C, flavonoids, and anthocyanins. The authors reported that PEAE considerably enhanced the extraction rate of several bioactive compounds from peach pomace waste (Plazzotta et al., 2021). Table 1.3 shows waste feedstock, phytochemicals extracted, optimised conditions, and increase in yield using PEAE.

PEAE is cost-effective, with the additional benefit of reduced electrical energy consumption compared to other extraction methods (Ranjha et al., 2021). The solvent requirement is also less, and it is also time efficient compared to conventional extraction methods. Therefore, it is considered an environment-friendly or green extraction technology (Bozinou et al., 2019). Phytochemicals extracted by PEAE are not degraded, and several literature reported that using PEAE as a green extraction technology also preserves the antioxidant properties of bioactive compounds (Ranjha et al., 2021; Wang et al., 2021). It can be concluded that PEAE has the potential to process agricultural and industrial waste in an eco-friendly manner with improved yields of phytochemicals.

1.4.4 Pressurised Liquid Extraction

PLE is a green technique that uses a solvent that remains liquid at high temperature (~200°C) and pressure (~200 bar), just below its critical point (Chang et al., 2022). This technique is also called high-pressure high-temperature solvent extraction (HPHTSE), pressurised hot solvent extraction (PHSE), subcritical solvent extraction (SSE), accelerated solvent extraction (ASE), high-pressure solvent extraction (HPSE), and pressurised fluid extraction (PFE) (Barp et al., 2023). The instrument distributes the sample with an inactive material, that is, sand or drying agent, then inserts the combined sample into a steel extraction vessel and pumps the solvent, heating the vessel around 75°C–200°C with a rising pressure of around 100 atm

(Ebrahimi and Lante, 2022). In terms of solvent extraction, it can be applied as static, dynamic, or both. Yield recoveries could be improved by increasing the number of extraction cycles. Afterwards, the compressed gas purges the sample extract from the cell into a collection vessel (Barp et al., 2023).

Temperature is considered the main parameter for PLE, and when the temperature is increased, the solvent properties are altered (Castellanos-Gallo et al., 2022). The most suitable temperature parameters for thermolabile and thermostable compounds are 40°C–60°C and 75°C–220°C, respectively, whereas pressure is 3.5–20 MPa. Choosing a solvent with similar polar characteristics as the target analytes is important. The solvent's role is to solubilise the target analytes while minimising the co-extraction of other biomass components (Perez-Vazquez et al., 2023). The suggested green solvent for PLE is water:ethanol mixtures, as the solvent can decrease or reduce toxic solvents and recover sample yield through decreased extraction time.

Elevated temperatures reduce the liquid solvent's viscosity and aid the spread and dissolution of substances of interest in the matrix. Additionally, higher temperatures promote the breakdown of analyte-matrix bonds, facilitating substance diffusion on the matrix surface (Perez-Vazquez et al., 2023). Other parameters, such as the adjustment of viscosity, diffusivity, and surface tension of the solvent, should be considered, as they could affect the mass transfer (Castellanos-Gallo et al., 2022). Altering the pressure modifies the physical and chemical properties of the solution, thus lowering the viscosity and surface tension of the solvent. Then, the solubility of the solute in the solvent increases (Chang et al., 2022). Thus, optimisation of parameters like the number of cycles, extraction time, solvent, pressure, and temperature can improve effectiveness.

PLE is the most used extraction technology to obtain bioactive compounds from natural sources, offering beneficial advantages and scalable extraction yields (Herrero et al., 2015). For instance, anthocyanins were extracted from fresh and freeze-dried blueberries using PLE at a temperature and pressure of 40°C and 20 MPa, respectively. Yields ranged from 3.9% to 8%, with higher yields observed from the freeze-dried samples (Paes et al., 2014). Another study used PLE to extract dietary fibre from pomegranate peel and fruit in a pressurised liquid extractor at 1,500 psi for 20 min; the authors reported a yield of 30% and 18% from the peel and fruit, respectively (García et al., 2021). A comparative study by Cardenas-Toro et al. (2015) extracted carotenoids from pressed palm fibre and husk using PLE and Soxhlet extraction methods (Cardenas-Toro et al., 2015). The authors performed PLE in ethanol at a pressure range of 2–6 MPa and temperature range of 35°C–55°C for 5 min, while Soxhlet was conducted for 6 h using ethanol and hexane at 78.4°C and 69.1°C, respectively. Despite the long extraction time and high temperature used in Soxhlet extraction, the study reported a higher yield (675 μg/g) for PLE extraction, while the extraction yields for Soxhlet were observed to be 317 and 465 μg/g for the ethanol and hexane, respectively (Cardenas-Toro et al., 2015).

Some of the studies combined two extraction techniques to improve extraction yields. Sumere et al. (2018) applied UAE and PLE to evaluate whether this would increase the yield of phenolic compounds from pomegranate peel (Sumere et al., 2018). The authors demonstrated that ultrasound power (0–800 W at the generator) and the sample's particle size (1.05 and 0.68 mm) influenced the yield of the phenolic compounds. Chemat et al. (2019a) and Sumere et al. (2018) mentioned that the rise in total phenolic yield is mostly due to the mechanical effect of the ultrasound treatment and cavitation. The combination of these two factors enhanced the surface area of interaction among the solid and liquid phases, thus improving the solvent's penetration in the matrix, and fast-tracked the analytes' mass transfer into the extraction solvent (Dias et al., 2021; Chemat et al., 2019a). PLE achieves rapid, effective, and selective extraction with varied compound polarities, presenting less extraction time and solvent utilisation than conventional maceration extraction (Sumere et al., 2018). Hence, PLE is one of the unorthodox methods to conventional methods, as it presents improved extract quality and good yields (Alexandre et al., 2017). Table 1.3 shows some applications of the PLE method in extracting bioactive compounds from different waste sources.

1.4.5 Supercritical Fluid Extraction

Supercritical fluid (SCF) technology is another green extraction method that uses non-toxic solvents to minimise energy utilisation and environmental pollution (Bruno et al., 2019). This technology uses substances that, when subjected to temperatures and pressures above their critical point, exhibit properties of both gases and liquids as the solvent, facilitating the extraction of phytochemicals from biomass (Fraguela-Meissimilly et al., 2023; Braga et al., 2023). This solvent system significantly enhances the extraction of the desired compounds compared to traditional solvents such as organic solvents (Al Khawli et al., 2019).

Carbon dioxide is the most used solvent in SFE because it is non-toxic, readily available at high purity, non-flammable, and recyclable; has a low critical temperature (31°C) and a low pressure (74 bar); and is generally recognised as safe (GRAS) (Braga et al., 2023). Carbon dioxide can reach a supercritical state at relatively low temperatures and pressures, facilitating the clean extraction of solvent-free compounds as the solvent is converted to its gaseous phase after extraction (Salami et al., 2021). However, its non-polar nature limits its use in the extraction of polar compounds. To overcome this, a small quantity of a polar organic solvent is used to act as a co-solvent to carbon dioxide for the extraction of polar compounds (De Melo et al., 2014; Fraguela-Meissimilly et al., 2023). The nature of the co-solvent firmly depends on the nature of the compounds to be extracted. Essential oils have been extracted from different agricultural waste sources, such as fruit peels and seeds, using SFE; due to the non-polar nature of essential oils, CO_2 was the solvent used, and no co-solvents were needed (Barrales et al., 2018; Wang et al., 2020). However, several studies have investigated the extraction of phenolics from different agricultural waste sources; due to the polar nature of phenolic compounds, the use of co-solvents such as water and ethanol was necessary in the extraction process (Solana et al., 2015; Nastić et al., 2020; Castro-Vargas et al., 2019; Aresta et al., 2020). Several studies have investigated using SCF technology to extract phytochemicals from agricultural waste sources, including vegetables, fruits, herbs/spices, grains, peels, seeds, and pomaces. Table 1.3 illustrates the applications of SCF technology in extracting phytochemicals from different agricultural wastes and the conditions used for the extraction process. The reflux extraction method is another conventional method similar to Soxhlet; it is associated with disadvantages such as thermal degradation, long extraction times, and complex equipment (Luksta and Spalvins, 2023). A comparative study by Yaqoob et al. (2023) investigated the extraction of carotenoids from dried *Citrus reticulata* fruit peel using reflux extraction and SFE with different solvents (Yaqoob et al., 2023). The authors performed reflux extraction using solvents such as methanol, ethanol, and acetone for 4 h at 30°C, while SFE was conducted with CO_2 at 400 bar and 60°C with acetone, ethanol, and methanol as co-solvents. The authors reported the highest yield (5.17 mg/100 g) of carotenoids using SFE (Yaqoob et al., 2023). There are disadvantages associated with SFE, including high equipment and maintenance costs, complex equipment, high power consumption, and restrictions in the extraction of polar compounds; thus, the use of a co-solvent may impact extraction yield and parameters (Pattnaik et al., 2021; Al Khawli et al., 2019; Fraguela-Meissimilly et al., 2023; De Melo et al., 2014).

1.4.6 Ultrasound-Assisted Extraction

Ultrasound is defined as sound waves with frequencies above 20 kHz and below 10 MHz, and its use in extractions disrupts the biomass composed of the extracellular matrix, cell walls, and membranes (McClements, 1995). UAE is the process in which ultrasound is used to release wanted compounds from biomass into a specific solvent. Like conventional extraction methods, UAE is dependent on the use of solvents. Polar solvents, such as water, methanol, ethanol, and deep eutectic solvents, are used to extract polar compounds such as (poly)phenols (flavonoids and phenolic acids) (Pal and Jadeja, 2019). Non-polar solvents such as hexane and chloroform extract non-soluble compounds, such as carotenoids (Viñas-Ospino et al., 2023) and fatty acids (Gayas et al., 2017). Table 1.3 illustrates the application of UAE for the extraction of phytochemicals from different agricultural wastes and the conditions used during the extraction process. During UAE use, Montero-Calderon et al. (2019) found that an increase in time, ethanol concentration, and power increased the yield of carotenoids, polyphenols, and antioxidants in the extracts. UAE has shown promise in bioactive peptide extraction, as it conserves the proteolytic activity compared to other techniques (Mala et al., 2021). UAE was used to allow penetration of enzyme cocktails for cellulase in the ultrasound-EAE of pectin from orange peels (Bosch and Malgas, 2023). Ultrasound can also be used in conjunction with other non-conventional techniques, such as MAE. An example is using UAE as a pretreatment to extract essential oils from citrus peel waste (CPW). CPW was subjected to ultrasonication before the conventional hydrodistillation, increasing the essential oil yield (Sandhu et al., 2021).

The UAE has many advantages over conventional and other green technologies; for example, the UAE decreases the reaction time and power required (Shivamathi et al., 2019). UAE is also simple to use and does not require complicated or sophisticated apparatus or consumables. A study showed that UAE pretreatment of plant materials decreased the extraction time while maintaining or improving the quality and yield of the extracted compounds (Nora and Borges, 2017). The UAE requires the use of solvents and an increased solvent-to-mass ratio (SMR). It has been proven that an increase in the SMR causes an increase in extraction efficiency (Dzah and Dzigbor, 2023). Sonication of phytochemical feedstocks can also lead to the degradation of molecules; this has been illustrated by alterations in the colour and antioxidant capacity of phytochemicals and lipid degradation due to cavitation post-UAE application (Pingret et al., 2013). This should be considered if biomass with a high lipid content is used.

1.4.7 Twin-Screw Extrusion

Extrusion is a processing technique in which solid feed material is inserted into a long chamber that contains one or more rotating screws. The screws convey the material across the length of the chamber and press it through a narrow gap, die-mould, or filter screen, mixing, cooking, or otherwise altering the properties of the material (Martin, 2016). Twin-screw extrusion (TSE), on the other hand, uses two meshed screws, though some details may apply to extrusion in general. In TSE, the material is both homogenised by the screws and has considerable shear force applied to it as it passes through the narrow gaps between them (Vergnes, 2021). This shear force releases intrinsic liquid and soluble compounds from the material, separating them from the now-refined insoluble remains (Karama, 2010).

Plant proteins, seed and essential oils, and other phytochemicals can be separated from insoluble lignocellulose when using plant matter as a feed material (Colas et al., 2013; Uitterhaegen and Evon, 2017; Candy et al., 2017; Trujillo-Juárez et al., 2021). These components can be extracted from agricultural lignocellulose waste by extrusion and then used to manufacture nutraceutical products (Sachdeva et al., 2020). The shear force applied to the agricultural waste also modifies the biomolecular and structural features of the lignocellulose itself, which can enhance downstream valorisation efforts or be used to create new products (Duque et al., 2017).

Soluble compounds can be extracted during TSE by adding an appropriate solvent, allowing admixture with the material, and then filtering the solvent with dissolved compounds from the system (Kartika et al., 2010). Solvents can be added at various points, potentiating sequential extraction and filtration steps with different solvents to garner multiple compounds with different physicochemical properties in a single extrusion process. With a well-optimised system, TSE can provide a sequential extraction of various classes of compounds from a feed material in a single, continuous procedure. Organic solvents or acidified/alkalised water is a cheap and widely available medium to process or extract phytochemicals from lignocellulose waste using TSE (Chemat et al., 2019b). The use of these conventional solvents can render green extraction technologies redundant, as they generate environmentally harmful waste products. While solvents can be used in TSE to improve the extraction of specific compounds, the technology offers interesting alternatives (Uitterhaegen and Evon, 2017). Bioactive phytochemicals and oils can be pressed and collected from feed material without the addition of water or solvent (Trujillo-Juárez et al., 2021). Alternatively, the pressure and temperature within the TSE system can be adjusted to transition water into a subcritical state, which modifies its solvent ability to be more like some organic solvents (Evon et al., 2018; Ferreira et al., 2023). This technique could obviate the need for harmful or non-renewable solvents when using TSE for extraction.

Bioextrusion is an application of TSE in which enzymes are added to a heat- and shear-stress-controlled region of the chamber to bioprocess the material before extrusion (Gatt et al., 2018). During bioextrusion, temperature control, moisture content, screw profile, and speed are modified to limit enzyme denaturation and optimise biocatalysis (Xu et al., 2017; Gatt et al., 2018). Because TSE can mix highly viscous materials, it can operate under low moisture conditions, minimising water consumption and wastewater production (Gatt et al., 2018).

While the modularity and adjustability of TSE are among its main advantages, they also present challenges. The heat generated during extrusion can degrade or alter heat-sensitive nutrients through processes such as the Maillard reaction, reducing the product's nutrient density and bioactivity (Singh et al., 2007). TSE can both fractionate and pre-treat lignocellulose from agricultural waste to produce a clean and refined material for the synthesis of prebiotics and other nutraceuticals (Duque et al., 2017). Functional protein hydrolysates with antioxidant activities have been produced from plant material using bioextrusion and may be used to generate proteinaceous nutraceuticals or other products to address the issues of malnutrition (Zheng et al., 2006; Montoya-Rodríguez et al., 2014).

In the context of green extraction technologies, TSE offers a continuous, single-step mechano-thermo-chemical extraction and preprocessing system capable of performing sequential extraction. The extraction process is modular and can be used to optimise the production of certain products from various agricultural waste products and improve their quality. The system is energy efficient and high-throughput and may be used without environmentally harmful or non-renewable solvents, making it shine as a green technology.

1.5 REPORTS ON THE USE OF GREEN EXTRACTION TECHNOLOGIES FOR PHYTOCHEMICAL RECOVERY FROM AGRICULTURAL PROCESSING WASTE

Implementing green extraction technologies to recover phytochemicals from agricultural processing waste represents a rapidly advancing sector that prioritises sustainable practices to minimise environmental impact. These methodologies employ innovative techniques and environmentally friendly solvents to extract valuable phytochemicals from plant materials, specifically focusing on waste products generated by agricultural processes. Table 1.3 illustrates literature reports detailing the application of these non-conventional extraction methods on agricultural wastes over the past 15 years.

1.6 CONCLUSION

Emerging non-conventional green extraction technologies such as EAE, MAE, SFE, UAE, and TSE hold significant promise for sustainable valorisation of agricultural wastes. Each technique offers tunable parameters that can be optimised according to specific feedstocks and targeted phytochemicals. Compared with conventional solvent-based methods, these technologies enhance extraction efficiency while reducing solvent consumption, energy demand, and environmental impact. Their adaptability enables the transformation of diverse agricultural residues into high-value bioactive compounds, positioning them as integral components of microbial biorefineries. The wider adoption of these techniques could strengthen agro-industrial processing chains and promote a circular, resource-responsible economy. Future research should prioritise optimisation, process integration, and techno-economic feasibility assessments to advance the scalable application of these green extraction methods for phytochemical production within biorefinery systems.

REFERENCES

Abbasi-Parizad, P., Scarafoni, A., Pilu, R., Scaglia, B., De Nisi, P. & Adani, F. (2022). The recovery from agro-industrial wastes provides different profiles of anti-inflammatory polyphenols for tailored applications. *Frontiers in Sustainable Food Systems*, 6, 996562.

Al Khawli, F., Pateiro, M., Domínguez, R., Lorenzo, J.M., Gullón, P., Kousoulaki, K., Ferrer, E., Berrada, H. & Barba, F.J. (2019). Innovative green technologies of intensification for valorisation of seafood and their by-products. *Marine Drugs*, 17, 689.

Alchera, F., Ginepro, M. & Giacalone, G. (2022). Microwave-assisted extraction of polyphenols from blackcurrant by-products and possible uses of the extracts in active packaging. *Foods*, 11, 2727.

Alexandre, E.M., Araújo, P., Duarte, M.F., de Freitas, V., Pintado, M. & Saraiva, J.A. (2017). Experimental design, modeling, and optimisation of high-pressure-assisted extraction of bioactive compounds from pomegranate peel. *Food and Bioprocess Technology*, 10, 886–900.

Alupului, A., Călinescu, I., & Lavric, V. (2012). Microwave extraction of active principles from medicinal plants. *U.P.B. Scientific Bulletin, Series B: Chemistry and Materials Science*, 74, 129–142.

Aresta, A., Cotugno, P., De Vietro, N., Massari, F. & Zambonin, C. (2020). Determination of polyphenols and vitamins in wine-making by-products by supercritical fluid extraction (SFE). *Analytical Letters*, 53, 2585–2595.

Azeez, S., Narayana, C., & Oberoi, H. (2017). Extraction and utilisation of bioactive compounds from agricultural waste. In Q.V. Vuong (Eds.), *Utilisation of bioactive compounds from agricultural and food production waste* (pp. 26–32). CRC Press.

Bala, S., Garg, D., Sridhar, K., Inbaraj, B.S., Singh, R., Kamma, S., Tripathi, M. & Sharma, M. (2023). Transformation of agro-waste into value-added bioproducts and bioactive compounds: Micro/nano formulations and application in the agri-food-pharma sector. *Bioengineering*, 10, 152.

Barba, F.J., Zhu, Z., Koubaa, M., Sant'Ana, A.S. & Orlien, V. (2016). Green alternative methods for the extraction of antioxidant bioactive compounds from winery wastes and by-products: A review. *Trends in Food Science & Technology*, 49, 96–109.

Barp, L., Višnjevec, A.M. & Moret, S. (2023). Pressurised Liquid Extraction: A powerful tool to implement extraction and purification of food contaminants. *Foods*, 12, 2017.

Barrales, F.M., Silveira, P., Barbosa, P.d.P.M., Ruviaro, A.R., Paulino, B.N., Pastore, G.M., Macedo, G.A. & Martinez, J. (2018). Recovery of phenolic compounds from citrus by-products using pressurised liquids—An application to orange peel. *Food and Bioproducts Processing*, 112, 9–21.

Bilal, M. & Iqbal, H.M. (2019). Sustainable bioconversion of food waste into high-value products by immobilised enzymes to meet bio-economy challenges and opportunities–A review. *Food Research International*, 123, 226–240.

Bitwell, C., Indra, S.S., Luke, C. & Kakoma, M.K. (2023). A review of modern and conventional extraction techniques and their applications for extracting phytochemicals from plants. *Scientific African*, 19, e01585.

Bosch, R. & Malgas, S. (2023). Ultrasound-assisted enzymatic extraction of orange peel pectin and its characterisation. *International Journal of Food Science & Technology*, 58, 6784–6793.

Bozinou, E., Karageorgou, I., Batra, G., Dourtoglou, G.V. & Lalas, I.S. (2019). Pulsed electric field extraction and antioxidant activity determination of Moringa oleifera dry leaves: A comparative study with other extraction techniques. *Beverages*, 5, 8.

Braga, M.E., Gaspar, M.C. & de Sousa, H.C. (2023). Supercritical fluid technology for agri-food materials processing. *Current Opinion in Food Science*, 50, 100983.

Bruno, S.F., Ekorong, F.J.A.A., Karkal, S.S., Cathrine, M. & Kudre, T.G. (2019). Green and innovative techniques for recovery of valuable compounds from seafood by-products and discards: A review. *Trends in Food Science & Technology*, 85, 10–22.

Candy, L., Bassil, S., Rigal, L., Simon, V. & Raynaud, C. (2017). Thermo-mechano-chemical extraction of hydroxycinnamic acids from industrial hemp by-products using a twin-screw extruder. *Industrial Crops and Products*, 109, 335–345.

Capanoglu, E., Nemli, E. & Tomas-Barberan, F. (2022). Novel approaches in the valorisation of agricultural wastes and their applications. *Journal of Agricultural and Food Chemistry*, 70, 6787–6804.

Cardenas-Toro, F.P., Alcázar-Alay, S.C., Coutinho, J.P., Godoy, H.T., Forster-Carneiro, T. & Meireles, M.A.A. (2015). Pressurised liquid extraction and low-pressure solvent extraction of carotenoids from pressed palm fiber: Experimental and economical evaluation. *Food and Bioproducts Processing*, 94, 90–100.

Carvalho, R., Galvão, E., Barros, J., Conceição, M. & Sousa, E. (2012). Extraction, fatty acid profile and antioxidant activity of sesame extract (*Sesamum Indicum* L.). *Brazilian Journal of Chemical Engineering*, 29, 409–420.

Castellanos-Gallo, L., Ballinas-Casarrubias, L., Espinoza-Hicks, J.C., Hernández-Ochoa, L.R., Muñoz-Castellanos, L.N., Zermeño-Ortega, M.R., Borrego-Loya, A. & Salas, E. (2022). Grape pomace valorisation by extraction of phenolic polymeric pigments: A review. *Processes*, 10, 469.

Castro-Vargas, H.I., Baumann, W., Ferreira, S.R. & Parada-Alfonso, F. (2019). Valorisation of papaya (*Carica papaya* L.) agroindustrial waste through the recovery of phenolic antioxidants by supercritical fluid extraction. *Journal of Food Science and Technology*, 56, 3055–3066.

Chaka, K.T. (2022). Extraction of cellulose nanocrystals from agricultural by-products: A review. *Green Chemistry Letters and Reviews*, 15, 582–597.

Chan, C.-H., Yusoff, R., Ngoh, G.-C. & Kung, F.W.-L. (2011). Microwave-assisted extractions of active ingredients from plants. *Journal of Chromatography A*, 1218, 6213–6225.

Chang, Y., Shi, X., He, F., Wu, T., Jiang, L., Normakhamatov, N., Sharipov, A., Wang, T., Wen, M. & Aisa, H.A. (2022). Valorisation of food processing waste to produce valuable polyphenolics. *Journal of Agricultural and Food Chemistry*, 70, 8855–8870.

Chemat, F., Abert-Vian, M., Fabiano-Tixier, A.S., Strube, J., Uhlenbrock, L., Gunjevic, V. & Cravotto, G. (2019a). Green extraction of natural products. Origins, current status, and future challenges. *TrAC Trends in Analytical Chemistry*, 118, 248–263.

Chemat, F., Abert Vian, M., Ravi, H.K., Khadhraoui, B., Hilali, S., Perino, S. & Fabiano Tixier, A.-S. (2019b). Review of alternative solvents for green extraction of food and natural products: Panorama, principles, applications and prospects. *Molecules*, 24, 3007.

Cheng, X., Bi, L., Zhao, Z., & Chen, Y. (2015). Advances in enzyme-assisted extraction of natural products. In *Proceedings of the 3rd International Conference on Material, Mechanical and Manufacturing Engineering (IC3ME 2015)* (pp. 371–375). Atlantis Press.

Colas, D., Doumeng, C., Pontalier, P.Y. & Rigal, L. (2013). Twin-screw extrusion technology, an original solution for the extraction of proteins from alfalfa (Medicago sativa). *Food and Bioproducts Processing*, 91, 175–182.

Cortes-Ferre, H.E., Antunes-Ricardo, M. & Gutiérrez-Uribe, J.A. (2022). Enzyme-assisted extraction of anti-inflammatory compounds from habanero chili pepper (*Capsicum chinense*) seeds. *Frontiers in Nutrition*, 9, 942805.

Das, S., Nadar, S.S. & Rathod, V.K. (2021). Integrated strategies for enzyme assisted extraction of bioactive molecules: A review. *International Journal of Biological Macromolecules*, 191, 899–917.

De Melo, M., Silvestre, A. & Silva, C. (2014). Supercritical fluid extraction of vegetable matrices: Applications, trends and future perspectives of a convincing green technology. *The Journal of Supercritical Fluids*, 92, 115–176.

Dias, A.L.B., de Aguiar, A.C. & Rostagno, M.A. (2021). Extraction of natural products using supercritical fluids and pressurised liquids assisted by ultrasound: Current status and trends. *Ultrasonics Sonochemistry*, 74, 105584.

Dinçel Kasapoğlu, E., Kahraman, S. & Tornuk, F. 2023. Extraction optimisation and characterisation of cellulose nanocrystals from apricot pomace. *Foods*, 12, 746.

Domínguez-Rodríguez, G., García, M.C., Marina, M.L. & Plaza, M. (2021). Pressurised liquid extraction combined with enzymatic-assisted extraction to obtain bioactive non-extractable polyphenols from sweet cherry (*Prunus avium* L.) pomace. *Nutrients*, 13, 3242.

Duque, A., Manzanares, P. & Ballesteros, M. (2017). Extrusion as a pretreatment for lignocellulosic biomass: Fundamentals and applications. *Renewable Energy*, 114, 1427–1441.

Dzah, C.S. & Dzigbor, A. (2023). Ultrasound assisted extraction: A relook at solvent to material ratio, its effects on process efficiency and how it can be exploited for different uses. *Journal of Food Process Engineering*, 46, e14339.

Ebrahimi, P. & Lante, A. (2022). Environmentally friendly techniques for the recovery of polyphenols from food by-products and their impact on polyphenol oxidase: A critical review. *Applied Sciences*, 12, 1923.

Evon, P., Vandenbossche, V., Candy, L., Pontalier, P.Y., & Rouilly, A. (2018). Twin-screw extrusion: A key technology for the biorefinery. In S.K. Rakshit & S.K. Ghosh (Eds.), *Biomass extrusion and reaction technologies: Principles to practices and future potential* (Chapter 2, pp. 25–44). American Chemical Society.

Ferreira, C., Moreira, M.M., Delerue-Matos, C. & Sarraguça, M. (2023). Subcritical water extraction to valorize grape biomass — A step closer to circular economy. *Molecules*, 28(22), 7538.

Fraguela-Meissimilly, H., Bastías-Monte, J.M., Vergara, C., Ortiz-Viedma, J., Lemus-Mondaca, R., Flores, M., Toledo-Merma, P., Alcázar-Alay, S. & Gallón-Bedoya, M. (2023). New trends in supercritical fluid technology and pressurised liquids for the extraction and recovery of bioactive compounds from agro-industrial and marine food waste. *Molecules*, 28, 4421.

Frontuto, D., Carullo, D., Harrison, S., Brunton, N., Ferrari, G., Lyng, J. & Pataro, G. (2019). Optimisation of pulsed electric fields-assisted extraction of polyphenols from potato peels using response surface methodology. *Food and Bioprocess Technology*, 12, 1708–1720.

Gao, M., Huang, W., RoyChowdhury, M. & Liu, C. (2007). Microwave-assisted extraction of scutellarin from Erigeron breviscapus Hand-Mazz and its determination by high-performance liquid chromatography. *Analytica Chimica Acta*, 591, 161–166.

García, P., Fredes, C., Cea, I., Lozano-Sánchez, J., Leyva-Jiménez, F.J., Robert, P., Vergara, C. & Jimenez, P. (2021). Recovery of bioactive compounds from pomegranate (*Punica granatum* L.) peel using pressurised liquid extraction. *Foods*, 10, 203.

Gatt, E., Rigal, L. & Vandenbossche, V. (2018). Biomass pretreatment with reactive extrusion using enzymes: A review. *Industrial Crops and Products*, 122, 329–339.

Gayas, B., Kaur, G. & Gul, K. (2017). Ultrasound-assisted extraction of apricot kernel oil: Effects on functional and rheological properties. *Journal of Food Process Engineering*, 40, e12439.

Gligor, O., Mocan, A., Moldovan, C., Locatelli, M., Crişan, G. & Ferreira, I.C. (2019). Enzyme-assisted extractions of polyphenols–A comprehensive review. *Trends in Food Science & Technology*, 88, 302–315.

Grassino, A.N., Halambek, J., Djaković, S., Brnčić, S.R., Dent, M. & Grabarić, Z. (2016). Utilisation of tomato peel waste from canning factory as a potential source for pectin production and application as tin corrosion inhibitor. *Food Hydrocolloids*, 52, 265–274.

Guerrero-Beltran, J.A. & Welti-Chanes, J. (2016). Pulsed electric fields. In Caballero, B., Finglas, P.M., Toldrá, F. (Eds.), *Encyclopedia of food and health* (pp. 561–565). Amsterdam, Netherlands.

Gustavsson, J., Cederberg, C., Sonesson, U., van Otterdijk, R., & Meybeck, A. (2011). *Global food losses and food waste: Extent, causes and prevention.* Food and Agriculture Organization of the United Nations (FAO).

Heck, C.I. & De Mejia, E.G. (2007). Yerba Mate Tea (*Ilex paraguariensis*): A comprehensive review on chemistry, health implications, and technological considerations. *Journal of Food Science*, 72, R138–R151.

Heemann, A.C.W., Heemann, R., Kalegari, P., Spier, M.R. & Santin, E. (2019). Enzyme-assisted extraction of polyphenols from green yerba mate. *Brazilian Journal of Food Technology*, 22, e2017222.

Hendrawan, Y., Larasati, R., Wibisono, Y., Umam, C., Sutan, S.M., & Choviya Hawa, L. (2019). Extraction of phenol and antioxidant compounds from kepok banana skin with PEF pretreatment. *IOP Conference Series: Earth and Environmental Science*, 230, 012065.

Herrero, M., del Pilar Sánchez-Camargo, A., Cifuentes, A. & Ibáñez, E. (2015). Plants, seaweeds, microalgae and food by-products as natural sources of functional ingredients obtained using pressurised liquid extraction and supercritical fluid extraction. *TrAC Trends in Analytical Chemistry*, 71, 26–38.

Ibrahim, N., Mahmud, M., & Nurdin, S. (2020). Microwave-assisted extraction of β-sitosterol from cocoa shell waste. *IOP Conference Series: Materials Science and Engineering*, 736, 012106.

Karama, M. (2010). *Advances in structural analysis of advanced materials.* Trans Tech Publications.

Kartika, I.A., Pontalier, P.Y. & Rigal, L. (2010). Twin-screw extruder for oil processing of sunflower seeds: Thermo-mechanical pressing and solvent extraction in a single step. *Industrial Crops and Products*, 32, 297–304.

Kazemi, M., Khodaiyan, F. & Hosseini, S.S. (2019). Utilisation of food processing wastes of eggplant as a high potential pectin source and characterisation of extracted pectin. *Food Chemistry*, 294, 339–346.

Khruengsai, S., Promhom, N., Sripahco, T., Siriwat, P. & Pripdeevech, P. (2023). Optimisation of enzyme-assisted microwave extraction of Zanthoxylum limonella essential oil using response surface methodology. *Scientific Reports*, 13, 12872.

Kumar, M., Barbhai, M.D., Puranik, S., Natta, S., Senapathy, M., Dhumal, S., Singh, S., Kumar, S., Deshmukh, V.P. & Anitha, T. (2023). Combination of green extraction techniques and smart solvents for bioactives recovery. *TrAC Trends in Analytical Chemistry*, 169, 117286.

Lasta, H.F.B., Lentz, L., Rodrigues, L.G.G., Mezzomo, N., Vitali, L. & Ferreira, S.R.S. (2019). Pressurised liquid extraction applied for the recovery of phenolic compounds from beetroot waste. *Biocatalysis and Agricultural Biotechnology*, 21, 101353.

Lasunon, P., Phonkerd, N., Tettawong, P. & Sengkhamparn, N. (2021). Effect of microwave-assisted extraction on bioactive compounds from industrial tomato waste and its antioxidant activity. *Food Research*, 5, 468–474.

Lončarić, A., Celeiro, M., Jozinović, A., Jelinić, J., Kovač, T., Jokić, S., Babić, J., Moslavac, T., Zavadlav, S. & Lores, M. (2020). Green extraction methods for extraction of polyphenolic compounds from blueberry pomace. *Foods*, 9, 1521.

López-Salazar, H., Camacho-Díaz, B.H., Ávila-Reyes, S.V., Pérez-García, M.D., González-Cortazar, M., Arenas Ocampo, M.L. & Jiménez-Aparicio, A.R. (2019). Identification and quantification of β-sitosterol β-D-glucoside of an ethanolic extract obtained by microwave-assisted extraction from Agave angustifolia Haw. *Molecules*, 24, 3926.

López-Salazar, H., Camacho-Díaz, B.H., Ocampo, M.A. & Jiménez-Aparicio, A.R. (2023). Microwave-assisted extraction of functional compounds from plants: A review. *Bioresources*, 18, 6614.

Luengo, E., Álvarez, I. & Raso, J. (2013). Improving the pressing extraction of polyphenols of orange peel by pulsed electric fields. *Innovative Food Science & Emerging Technologies*, 17, 79–84.

Luksta, I. & Spalvins, K. (2023). Methods for extraction of bioactive compounds from products: A review. *Environmental and Climate Technologies*, 27, 422–437.

Ly, L. & Sothornvit, R. (2024). Novel sequential microwave-ultrasonic-assisted extraction of phenolic compounds from jackfruit (*Artocarpus heterophyllus* Lam.) peel waste using choline chloride-lactic acid deep eutectic solvent. *Food and Bioprocess Technology*, 17, 4249–4261.

Maghraby, Y.R., El-Shabasy, R.M., Ibrahim, A.H. & Azzazy, H.M.E.-S. (2023). Enzyme immobilisation technologies and industrial applications. *ACS Omega*, 8, 5184–5196.

Mala, T., Sadiq, M.B. & Anal, A.K. (2021). Comparative extraction of bromelain and bioactive peptides from pineapple byproducts by ultrasonic-and microwave-assisted extractions. *Journal of Food Process Engineering*, 44, e13709.

Martin, C. (2016). Twin screw extruders as continuous mixers for thermal processing: A technical and historical perspective. *AAPS PharmSciTech*, 17, 3–19.

McClements, D.J. (1995). Advances in the application of ultrasound in food analysis and processing. *Trends in Food Science & Technology*, 6, 293–299.

Mohammad Azmin, S.N.H., Abdul Manan, Z., Wan Alwi, S.R., Chua, L.S., Mustaffa, A.A. & Yunus, N.A. (2016). Herbal processing and extraction technologies. *Separation & Purification Reviews*, 45, 305–320.

Montero-Calderon, A., Cortes, C., Zulueta, A., Frigola, A. & Esteve, M.J. (2019). Green solvents and ultrasound-assisted extraction of bioactive orange (*Citrus sinensis*) peel compounds. *Scientific Reports*, 9, 16120.

Montoya-Rodríguez, A., de Mejía, E.G., Dia, V.P., Reyes-Moreno, C. & Milán-Carrillo, J. (2014). Extrusion improved the anti-inflammatory effect of amaranth (maranthus hypochondriacus) hydrolysates in LPS-induced human THP-1 macrophage-like and mouse RAW 264.7 macrophages by preventing activation of NF-κB signaling. *Molecular Nutrition & Food Research*, 58, 1028–1041.

Murugesan, S., Maran, P., Venkatesan, M. & Alexander, R.A. (2024). Microwave assisted extraction of polyphenols from Pithecellobium dulce benth fruit peels and evaluation of its anticancer and antioxidant activity. *Waste and Biomass Valorization*, 15, 841–855.

Mushtaq, M., Sultana, B., Akram, S., Anwar, F., Adnan, A. & Rizvi, S.S. (2017). Enzyme-assisted supercritical fluid extraction: An alternative and green technology for non-extractable polyphenols. *Analytical and Bioanalytical Chemistry*, 409, 3645–3655.

Nadar, S.S., Rao, P. & Rathod, V.K. (2018). Enzyme assisted extraction of biomolecules as an approach to novel extraction technology: A review. *Food Research International*, 108, 309–330.

Naeem, U., Arshad, M.U., Saeed, F. & Imran, A. (2022). Extraction and characterisation of polyphenols from fruits and vegetable waste through green extraction technologies with special reference to antioxidant profile. *Journal of Food Processing and Preservation*, 46, e16807.

Nastić, N., Lozano-Sánchez, J., Borrás-Linares, I., Švarc-Gajić, J. & Segura-Carretero, A. (2020). New technological approaches for recovering bioactive food constituents from sweet cherry (Prunus avium L.) stems. *Phytochemical Analysis*, 31, 119–130.

Niu, D., Ren, E.-F., Li, J., Zeng, X.-A. & Li, S.-L. (2021). Effects of pulsed electric field-assisted treatment on the extraction, antioxidant activity and structure of naringin. *Separation and Purification Technology*, 265, 118480.

Nora, F.M.D. & Borges, C.D. (2017). Ultrasound pretreatment as an alternative to improve essential oils extraction. *Ciência Rural*, 47, e20170173.

Osorio-Tobón, J.F. (2020). Recent advances and comparisons of conventional and alternative extraction techniques of phenolic compounds. *Journal of Food Science and Technology*, 57, 4299–4315.

Paes, J., Dotta, R., Barbero, G.F. & Martínez, J. (2014). Extraction of phenolic compounds and anthocyanins from blueberry (*Vaccinium myrtillus* L.) residues using supercritical CO_2 and pressurised liquids. *The Journal of Supercritical Fluids*, 95, 8–16.

Pal, C.B.T. & Jadeja, G.C. (2019). Deep eutectic solvent-based extraction of polyphenolic antioxidants from onion (*Allium cepa* L.) peel. *Journal of the Science of Food and Agriculture*, 99, 1969–1979.

Pataro, G., Carullo, D., Falcone, M. & Ferrari, G. (2020). Recovery of lycopene from industrially derived tomato processing by-products by pulsed electric fields-assisted extraction. *Innovative Food Science & Emerging Technologies*, 63, 102369.

Pattnaik, M., Pandey, P., Martin, G.J., Mishra, H.N. & Ashokkumar, M. (2021). Innovative technologies for extraction and microencapsulation of bioactives from plant-based food waste and their applications in functional food development. *Foods*, 10, 279.

Peiró, S., Luengo, E., Segovia, F., Raso, J. & Almajano, M.P. (2019). Improving polyphenol extraction from lemon residues by pulsed electric fields. *Waste and Biomass Valorization*, 10, 889–897.

Perez-Vazquez, A., Carpena, M., Barciela, P., Cassani, L., Simal-Gandara, J. & Prieto, M.A. (2023). Pressurised liquid extraction for the recovery of bioactive compounds from seaweeds for food industry application: A review. *Antioxidants*, 12, 612.

Pingret, D., Fabiano-Tixier, A.-S. & Chemat, F. (2013). Degradation during application of ultrasound in food processing: A review. *Food Control*, 31, 593–606.

Pinto, D., Silva, A.M., Freitas, V., Vallverdú-Queralt, A., Delerue-Matos, C. & Rodrigues, F. (2021). Microwave-assisted extraction as a green technology approach to recover polyphenols from Castanea sativa shells. *ACS Food Science & Technology*, 1, 229–241.

Plazzotta, S., Ibarz, R., Manzocco, L. & Martín-Belloso, O. (2021). Modelling the recovery of biocompounds from peach waste assisted by pulsed electric fields or thermal treatment. *Journal of Food Engineering*, 290, 110196.

Puri, M., Sharma, D. & Barrow, C.J. (2012). Enzyme-assisted extraction of bioactives from plants. *Trends in Biotechnology*, 30, 37–44.

Ranjha, M.M.A., Kanwal, R., Shafique, B., Arshad, R.N., Irfan, S., Kieliszek, M., Kowalczewski, P.Ł., Irfan, M., Khalid, M.Z. & Roobab, U. (2021). A critical review on pulsed electric field: A novel technology for the extraction of phytoconstituents. *Molecules*, 26, 4893.

Rasheed, H.A., Adeleke, A.A., Nzerem, P., Olosho, A.I., Ogedengbe, T.S. & Jesuloluwa, S. (2024). Isolation, characterisation and response surface method optimisation of cellulose from hybridised agricultural wastes. *Scientific Reports*, 14, 14310.

Rasul, M.G. (2018). Conventional extraction methods use in medicinal plants, their advantages and disadvantages. *International Journal of Basic Sciences and Applied Computing*, 2, 10–14.

Saad, N., Louvet, F., Tarrade, S., Meudec, E., Grenier, K., Landolt, C., Ouk, T.-S. & Bressollier, P. (2019). Enzyme-assisted extraction of bioactive compounds from raspberry (*Rubus idaeus* L.) pomace. *Journal of Food Science*, 84, 1371–1381.

Sachdeva, V., Roy, A. & Bharadvaja, N. (2020). Current prospects of nutraceuticals: A review. *Current Pharmaceutical Biotechnology*, 21, 884–896.

Salami, A., Asefi, N., Kenari, R.E. & Gharekhani, M. (2021). Extraction of pumpkin peel extract using supercritical CO_2 and subcritical water technology: Enhancing oxidative stability of canola oil. *Journal of Food Science and Technology*, 58, 1101–1109.

Sandhu, H.K., Sinha, P., Emanuel, N., Kumar, N., Sami, R., Khojah, E. & Al-Mushhin, A.A. (2021). Effect of ultrasound-assisted pretreatment on extraction efficiency of essential oil and bioactive compounds from citrus waste by-products. *Separations*, 8, 244.

Shi, L., Zhao, W., Yang, Z., Subbiah, V. & Suleria, H.A.R. (2022). Extraction and characterisation of phenolic compounds and their potential antioxidant activities. *Environmental Science and Pollution Research*, 29, 81112–81129.

Shivamathi, C., Moorthy, I.G., Kumar, R.V., Soosai, M.R., Maran, J.P., Kumar, R.S. & Varalakshmi, P. (2019). Optimisation of ultrasound assisted extraction of pectin from custard apple peel: Potential and new source. *Carbohydrate Polymers*, 225, 115240.

Silva, L.d.O., Ranquine, L.G., Monteiro, M. & Torres, A.G. (2019). Pomegranate (Punica granatum L.) seed oil enriched with conjugated linolenic acid (cLnA), phenolic compounds and tocopherols: Improved extraction of a specialty oil by supercritical CO_2. *The Journal of Supercritical Fluids*, 147, 126–137.

Singanusong, R., Nipornram, S., Tochampa, W. & Rattanatraiwong, P. (2015). Low power ultrasound-assisted extraction of phenolic compounds from mandarin (Citrus reticulata Blanco cv. Sainampueng) and lime (*Citrus aurantifolia*) peels and the antioxidant. *Food Analytical Methods*, 8, 1112–1123.

Singh, S., Gamlath, S. & Wakeling, L. (2007). Nutritional aspects of food extrusion: A review. *International Journal of Food Science & Technology*, 42, 916–929.

Solana, M., Boschiero, I., Dall'Acqua, S. & Bertucco, A. (2015). A comparison between supercritical fluid and pressurised liquid extraction methods for obtaining phenolic compounds from Asparagus officinalis L. *The Journal of Supercritical Fluids*, 100, 201–208.

Streimikyte, P., Viskelis, P. & Viskelis, J. (2022). Enzymes-assisted extraction of plants for sustainable and functional applications. *International Journal of Molecular Sciences*, 23, 2359.

Sumere, B.R., de Souza, M.C., Dos Santos, M.P., Bezerra, R.M.N., da Cunha, D.T., Martinez, J. & Rostagno, M.A. (2018). Combining pressurised liquids with ultrasound to improve the extraction of phenolic compounds from pomegranate peel (*Punica granatum* L.). *Ultrasonics Sonochemistry*, 48, 151–162.

Tizón Alba, A., Aliaño-González, M.J., Palma, M., Fernández Barbero, G. & Carrera, C. (2023). Enhancing efficiency of enzymatic-assisted extraction method for evaluating bioactive compound analysis in mulberry: an optimisation approach. *Agronomy*, 13, 2548.

Torres-Valenzuela, L.S., Ballesteros-Gómez, A. & Rubio, S. (2020). Green solvents for the extraction of high added-value compounds from agri-food waste. *Food Engineering Reviews*, 12, 83–100.

Trabelsi, D., Aydi, A., Zibetti, A.W., Della Porta, G., Scognamiglio, M., Cricchio, V., Langa, E., Abderrabba, M. & Mainar, A.M. (2016). Supercritical extraction from Citrus aurantium amara peels using CO_2 with ethanol as co-solvent. *The Journal of Supercritical Fluids*, 117, 33–39.

Tran, T.M.K., Akanbi, T., Kirkman, T., Nguyen, M.H. & Van Vuong, Q. (2022). Maximising recovery of bioactive compounds from coffee pulp waste using microwave-assisted extraction. *European Journal of Engineering and Technology Research*, 7, 1–6.

Trujillo-Juárez, L.G., Hernández-Meléndez, Ó., Gimeno, M., Gracia-Fadrique, J. & Bárzana, E. (2021). Extraction of essential oil from waste grapefruit peel using a pilot-scale twin-screw extruder. *ACS Food Science & Technology*, 1, 1198–1205.

Ueda, J.M., Griebler, K.R., Finimundy, T.C., Rodrigues, D.B., Veríssimo, L., Pires, T.C., Gonçalves, J., Fernandes, I.P., Pereira, E. & Barros, L. (2023). Polyphenol composition by HPLC-DAD-(ESI-) MS/MS and bioactivities of extracts from grape agri-food wastes. *Molecules*, 28, 7368.

Uitterhaegen, E. & Evon, P. (2017). Twin-screw extrusion technology for vegetable oil extraction: A review. *Journal of Food Engineering*, 212, 190–200.

Vardakas, A., Kechagias, A., Penov, N. & Giannakas, A.E. (2024). Optimisation of enzymatic assisted extraction of bioactive compounds from Olea europaea leaves. *Biomass*, 4, 647–657.

Vergnes, B. (2021). Average shear rates in the screw elements of a corotating twin-screw extruder. *Polymers*, 13, 304.

Viñas-Ospino, A., López-Malo, D., Esteve, M.J., Frígola, A. & Blesa, J. (2023). Improving carotenoid extraction, stability, and antioxidant activity from Citrus sinensis peels using green solvents. *European Food Research and Technology*, 249, 2349–2361.

Wang, M., Yu, P., Chittiboyina, A.G., Chen, D., Zhao, J., Avula, B., Wang, Y.-H. & Khan, I.A. (2020). Characterisation, quantification and quality assessment of avocado (*Persea americana* Mill.) oils. *Molecules*, 25, 1453.

Wang, M., Zhou, J., Collado, M.C. & Barba, F.J. (2021). Accelerated solvent extraction and pulsed electric fields for valorisation of rainbow trout (Oncorhynchus mykiss) and sole (Dover sole) by-products: Protein content, molecular weight distribution and antioxidant potential of the extracts. *Marine Drugs*, 19, 207.

Wang, Y., Wang, L., Zhu, X., Wang, D. & Li, X. (2016). Choleretic activity of turmeric and its active ingredients. *Journal of Food Science*, 81, H1800–H1806.

Waseem, M., Majeed, Y., Nadeem, T., Naqvi, L.H., Khalid, M.A., Sajjad, M.M., Sultan, M., Khan, M.U., Khayrullin, M. & Shariati, M.A. (2023). Conventional and advanced extraction methods of some bioactive compounds with health benefits of food and plant waste: A comprehensive review. *Food Frontiers*, 4, 1681–1701.

Xu, E., Wu, Z., Jiao, A., Long, J., Li, J. & Jin, Z. (2017). Dynamics of rapid starch gelatinisation and total phenolic thermomechanical destruction moderated via rice bio-extrusion with alpha-amylase activation. *RSC Advances*, 7, 19464–19478.

Yadav, S., Malik, K., Moore, J.M., Kamboj, B.R., Malik, S., Malik, V.K., Arya, S., Singh, K., Mahanta, S. & Bishnoi, D.K. (2024). Valorisation of agri-food waste for bioactive compounds: Recent trends and future sustainable challenges. *Molecules*, 29, 2055.

Yaqoob, M., Aggarwal, P. & Babbar, N. 2023. Extraction and screening of kinnow (*Citrus reticulata* L.) peel phytochemicals, grown in Punjab, India. *Biomass Conversion and Biorefinery*, 13, 11631–11643.

Zeitoun, R., Pontalier, P.Y., Marechal, P. & Rigal, L. (2010). Twin-screw extrusion for hemicellulose recovery: Influence on extract purity and purification performance. *Bioresource Technology*, 101, 9348–9354.

Zhang, B., Yang, R. & Liu, C.-Z. (2008). Microwave-assisted extraction of chlorogenic acid from flower buds of Lonicera japonica Thunb. *Separation and Purification Technology*, 62, 480–483.

Zhang, L., Wang, Y., Wu, D., Xu, M. & Chen, J. (2011). Microwave-assisted extraction of polyphenols from Camellia oleifera fruit hull. *Molecules*, 16, 4428–4437.

Zhang, Q.-W., Lin, L.-G. & Ye, W.-C. (2018). Techniques for extraction and isolation of natural products: A comprehensive review. *Chinese Medicine*, 13, 1–26.

Zhang, Z.H., Wang, L.H., Zeng, X.A., Han, Z. & Brennan, C.S. (2019). Non-thermal technologies and its current and future application in the food industry: A review. *International Journal of Food Science & Technology*, 54, 1–13.

Zheng, X.-q., Li, L.-t., Liu, X.-l., Wang, X.-j., Lin, J. & Li, D. (2006). Production of hydrolysate with antioxidative activity by enzymatic hydrolysis of extruded corn gluten. *Applied Microbiology and Biotechnology*, 73, 763–770.

2 Harnessing the Physico-Chemical Properties of Biochar for the Augmentation of Dark Fermentation Processes

An Overview

Patrick Sekoai, Obinna Ezeokoli, and Sizwe Ndlovu

2.1 INTRODUCTION

The transition from a fossil-based linear economy to a circular bioeconomy is no longer an option as previously thought by many people but an indispensable strategy that will enable the utilization of our resources in a sustainable manner while alleviating the socioeconomic burdens derived from fossil fuels (Hughes and Jones, 2020). Biomass valorization focusing on renewable energy production forms part of the circular bioeconomy principles, and these initiatives are accelerated throughout the world (including developing nations) to mitigate various global issues such as environmental pollution, energy crisis, and depletion of fossil reserves (Sekoai et al., 2018, 2021, 2026). Biomass materials are converted/valorized into clean fuels and valuable chemicals using biological and/or chemical processes. From an environmental and economic standpoint, biological methods are more desirable due to their eco-friendliness and affordability. Among the sustainable strategies that are being explored by the scientific community, dark fermentation (DF) is viewed as the technology with the most benefits for producing renewable H_2 due to its outstanding properties such as (i) its carbon-sequestration abilities, (ii) its high energy content (122 kJ/g) that is 2.75 times greater than that of fossil fuels, (iii) its ability to use diverse biomasses including organic wastes, and (iv) its ability to use mild fermentation conditions. The world market for biotechnological products, including renewable

 DOI: 10.1201/9781003745150-2

fuels and biopolymers, was estimated at $64.2 billion in 2016 and is expected to reach $113 billion in 2030 (Bio-era, 2009).

Despite the excellent properties of DF, this process produces undesirable H_2 yields, and this limits its scalability because the DF experiments generate yields that are 20%–50% of the theoretical yield due to the proliferation of H_2-limiting reactions, which shift the metabolic reactions from acidogenesis (DF-producing reactions) to solventogenesis (DF-inhibiting reactions). The feedstocks must also undergo extensive biomass fractionation steps before the acidogenic H_2-producing process, and this escalates the overall DF costs. In addition, the majority of published studies in this field are carried out at bench-scale conditions, which means that the actual DF conditions are not well understood in the literature (Sekoai et al., 2020; Sekoai et al., 2018). Over the past two decades, several optimization methods (e.g., pretreatment methods, metabolic engineering, two-stage processes, inert gas-sparging, parametric optimization) have been studied in DF processes in attempts to overcome the thermodynamic limitations facing this technology. However, these different optimization methods have not been successful in attaining the desired H_2 yields so that this technology can be scaled up (Banu et al., 2021; Mahmod et al., 2021). Hence, novel R&D approaches are constantly being explored to reinvigorate interest in DF technology and advance its development as a future clean fuel.

Biochar, a carbonaceous by-product that originates from the incomplete combustion of organic materials, has garnered enormous attention in environmental science and engineering in recent years due to its outstanding properties and may be a useful bio-additive in DF. Biochar exhibits diverse features that make it a favorable additive in DF processes. It reduces the lag phase during microbial growth, increases the surface-area-to-volume ratio, increases the digestibility of feedstocks during the fermentation process, provides a balanced carbon-to-nitrogen (C/N) ratio, aids in the inhibition of toxins, helps in the supplementation of trace elements, and provides high ion exchange capacity and adsorption ability (Oliveira et al., 2017).

2.2 SIGNIFICANCE OF STUDY AND CONTRIBUTION TO KNOWLEDGE

The use of biochar as a bio-additive in anaerobic digestion and environmental remediation is extensively covered in the literature (Oliveira et al., 2017; Luo et al., 2021; Sun et al., 2021; Zhao et al., 2021), but little attention has been devoted to exploring the role of biochar in the DF process and the possible mechanisms that are triggered when this additive is coupled with the dominant H_2-producing species as a result of its intrinsic properties. Several scientific hypotheses have been proposed regarding its role in bio-H_2 reactions, but these reports have not provided deeper insights as there is little information that covers this topic. Therefore, the feasibility of using biochar as a way of enriching the acidogenic H_2-producing processes will be discussed herein. This chapter examines the characteristics of biochar that make it a suitable additive in DF processes. The possible mechanisms that are induced when biochar is coupled with DF-producing reactions will also be discussed. Furthermore, a short section that discusses the environmental benefits of using anaerobic digestates that are derived from biochar-integrated processes is included. Finally, the chapter provides some recommendations that could fast-track the use of biochar in acidogenic H_2-producing processes. This scholarly work aims to provide a basis upon which further scientific studies can be undertaken toward the use of biochar as a sustainable bioaugmentation additive in circular bio-based processes such as DF to fast-track the United Nations' Sustainable Development Goals.

2.3 AN OVERVIEW OF THE SYNTHESIS OF BIOCHAR AND ITS CHARACTERIZATION

Biochar is derived from the incomplete combustion of organic materials such as agricultural waste (Lee et al., 2017), cattle manure (Kiran et al., 2017), waste-activated sludge (Wu et al., 2019), aquatic plants (Roberts et al., 2015), and digestates (Wang et al., 2020) as shown in Table 2.1. It is synthesized within a temperature range of 350°C–700°C in low- or no-oxygen environments (Tomczyk et al., 2020) using thermochemical methods such as pyrolysis (Shen et al., 2021; Zhang et al., 2020), hydrothermal carbonization (Álvarez et al., 2020), and gasification (Kraisornkachit et al., 2018). These thermochemical methods usually employ feedstocks with low moisture, while hydrothermal liquefaction (HTL) is applied to feedstocks with high moisture (≥70% w/w) (Kumar et al., 2021). Biochar has numerous physico-chemical properties such as functional groups (C–O, C=O, COOH, OH, and $-NH_x$) and several micro- and macro-elements (calcium, copper, magnesium, potassium, sodium, zinc, and iron), which all contribute to its porosity and reactivity (Wan et al., 2020). It is also characterized by pore size (PS), pore volume (PV), specific surface area (SSA), pH, electrical conductivity (EC), and cation exchange capacity (CEC) (Leng et al., 2021). Among these parameters, PS, PV, and SSA are regarded as the most crucial parameters that determine the physical properties of the synthesized biochar (Chen et al., 2015; You et al., 2017). It is noteworthy to highlight that the PS is highly dependent on the SSA and is categorized into three groups, namely, micropores (<2 nm), mesopores (2–50 nm), and macropores (>50 nm) (Hu et al., 2008). The micropore size has been shown to provide suitable microenvironments for the proliferation of microbial communities (Chiappero et al., 2020). Furthermore, it has been reported that the SSA and PV of biochar usually fall within the range of 8–132 m^2/g and 0.016–0.083 cm^3/g, respectively (Leng et al., 2021). But when suitable precursors alongside the optimal parameters are used in pyrolysis, the SSA and PV can reach up to 490.8 m^2/g (Chen et al., 2008) and 0.25 cm^3/g (Yang et al., 2016), respectively.

The pH values of biochar are usually in the alkaline range (7.0–10.0) due to the ash constituents and the loss of electrons in the acidic functional groups. Moreover, the pH tends to increase when the pyrolysis temperature is increased (Yin et al., 2019). The type of feedstocks used to produce the biochar also contributes to its pH values. For example, corn straw, soybean, sugarcane, rice husk, wheat straw, peanut shell, and corn straw have been reported to produce higher pH values of around 8.0–10.0 (Yuan et al., 2011), while feedstocks like waste-activated sludge, canola straw, manures, food waste, and sludge produce neutral pH values in the range of 6.5–7.0 (Ambaye et al., 2020).

TABLE 2.1
Synthesis and characterization of biochars that are applicable in various processes

Feedstock	Method	T (°C)	HR (°C/min)	SSA (m^2/g)	PS (cm^3/g)	pH	Ash content (%)	C (%)	H (%)	O (%)	N (%)	References
Sewage sludge	Pyrolysis	300	7.0	4.5	0.010	6.8	56.5	30.7	3.11	11.2	4.11	Ahmad et al. (2014)
Corn stover	Fast pyrolysis	500	–	109.10, 216.08	–	2.0	20.42, 33.92	59.4, 42.71	1.71, 2.74	1.25, 1.27	16.05, 18.31	Zhou et al. (2020)
Wood pellets	Pyrolysis	700, 800	–	–	–	7.41	5.49	83.16	3.24	7.24	0.40	Zhang et al. (2014)
Forest waste	Pyrolysis	450, 550	–	–	–	8.1	11.38	–	–	–	–	Cimon et al. (2020)
Municipal sludge	Pyrolysis	500	10.0	25.424	0.05608	8.58	74.21	17.46	0.70	10.449	1.54	Chen et al. (2014b)
Municipal sludge	Pyrolysis	600	10.0	20.27	0.05268	9.04	77.90	18.40	0.34	7.35	1.38	Chen et al. (2014b)
Municipal sludge	Pyrolysis	700	10.0	32.17	0.06842	9.91	81.53	16.92	0.21	6.86	0.95	Chen et al. (2014b)
Municipal sludge	Pyrolysis	800	10.0	48.50	0.08989	10.18	83.93	16.20	0.03	3.64	0.50	Chen et al. (2014b)
Municipal sludge	Pyrolysis	900	10.0	67.60	0.09855	10.17	88.07	15.92	0.11	2.44	0.53	Chen et al. (2014b)
Pine waste	Pyrolysis	700	–	15	–	6.5–7.0	–	–	–	–	–	Chen et al. (2014a)
Activated carbon	Hydrothermal liquefaction	300	–	971.33	–	6.46	–	78.12	2.65	16.92	0.34	Ren et al. (2020)

(Continued)

TABLE 2.1
(Continued)

Feedstock	Method	T (°C)	HR (°C/min)	SSA (m^2/g)	PS (cm^3/g)	pH	Ash content (%)	C (%)	H (%)	O (%)	N (%)	References
Corn straw	Hydrothermal liquefaction	300	–	49.54	–	5.71	–	59.54	4.25	9.76	2.22	Ren et al. (2020)
Food waste	Gasification	550	7.0	294.7	0.05	–	2.5	47.71	7.07	39.99	2.27	Igalavithana et al. (2020)
Rice straw and $1Fe_3$	Carbonization	500	–	16.95	0.0166	8.53	–	–	–	–	–	Qin et al. (2017)
Switchgrass	Slow pyrolysis	500	–	50.2	–		52.5	39.4	1.3	58.6	0.70	Brewer et al. (2009)
Cassava slag	Hydrothermal carbonization	180	–	–	–	2.0–12.0	–	–	–	–	–	Wu et al. (2019)

2.4 AN OVERVIEW OF THE PARAMETERS GOVERNING THE PHYSICO-CHEMICAL PROPERTIES OF BIOCHAR

Although this section is broadly covered in scientific literature, this information is still relevant as it will enable scientists specializing in DF to acquire in-depth knowledge regarding the optimal setpoint parameters that produce biochar with desirable properties. It has been documented in several studies that the physico-chemical properties of biochars are governed by several parameters, which contribute to their overall structure/properties. The notable parameters include the biomass types, methods of synthesis, and the operational conditions used during the synthesis of biochars (e.g., heating rate, pH, temperature, residence time, carrier gas). For this reason, it is important to fully understand these process parameters to successfully synthesize biochars that are effective in the enhancement of acidogenic H_2 processes.

The properties of biochar are highly attributed to the type of feedstocks used during its synthesis (Nzediegwu et al., 2021). Ippolito et al. (2020) indicated that lignocellulosic-derived biochars have a high SSA and PV compared to other feedstocks such as animal manures. The higher SSA and PV values in woody biochars are due to the breakdown of their cellulosic structures into smaller pores, resulting in a high SSA (Weber and Quicker, 2018). On the other hand, biochars derived from livestock manures have low SSA due to structural deformations and have less distinct porous structures (Ahmad et al., 2014). Huang et al. (2021) observed that hyacinth biochar had a higher carbon content (83.29%) than chicken manure biochar (32.46%), while Liu et al. (2019) observed that biochar from livestock manure had a low carbon content and a higher nitrogen content than biochar produced from corn stalk. Pan et al. (2019) revealed that high lignin and carbon contents produce biochar with high SSA and high functional groups. Hence, these results underscore the importance of using suitable feedstocks for the synthesis of biochars with optimal physico-chemical properties.

Temperature is an important parameter that affects the overall structure of biochar due to its impact on the elemental analysis and surface functional groups, which in turn affect the SSA, CEC, and PV, among other properties (Sbizzaro et al., 2021; Low and Yee, 2021). A study by Igalavithana et al. (2018) showed that biochars produced at low temperatures (300°C–500°C) exhibited higher oxygen content while reducing the hydrogen and nitrogen content at elevated temperatures (500°C–1,000°C). Luo et al. (2015) observed that increasing the temperature (500°C–600°C) increased the carbon content in biochars while reducing the hydrogen and oxygen content. Xie et al. (2021) demonstrated that high-temperature pyrolysis biochars (600°C and 800°C) had high stability and were rich in hydroxyl and amino groups. However, between 300°C and 700°C, the hydrogen content decreased while the nitrogen content remained relatively stable at ~400°C (Xie et al., 2021). Chatterjee et al. (2020) revealed that elevated pyrolysis temperatures (600°C and 700°C) resulted in biochar that had a large surface area (303–325 m^2/g), high carbon content (82%–84%), and low ash content (4%–5%), respectively. Elsewhere, Mumme et al. (2014) observed that pyrochar (produced at 500°C) resulted in high carbon content (~30%) and high pH (pH 9.3) compared to the hydrochar that was produced through the hydrothermal method (at 230°C), which produced low carbon content (~7%) and low pH (pH 4.8). These results are consistent with the literature as it has been reported that hydrochars possess a low carbon content, lower pH, and less aromatic structure, but yield higher hydrogen content than pyrochars (Arauzo et al., 2020). In summary, the thermochemical methods used to produce the biochars should be configured based on the properties of the chosen feedstocks, affordability, and their intended applications. The heating rate is also considered an important parameter that impacts the SSA and PV of biochar (Liang et al., 2021; Li et al., 2020). At a low heating rate, the breakdown of biomass is significantly reduced during the secondary decomposition process, and this increases the biochar yield (Leng and Huang, 2018). It has been reported that a low heating rate stabilizes the biochar and also contributes to the formation of aromatic structures by retaining the structural complexity of biomass than a high heating rate (Leng and Huang, 2018; Calvelo Pereira et al., 2011). On the contrary, a high heating rate usually decreases the biochar yield due to the disruption of the biomass structure, as evidenced by Noor et al. (2012) and Li et al. (2016). Similarly, Angin (2013) observed that the yield decreased (34.18% to 27.70%) upon increasing the heating rate (10 to 50°C/min). In a study investigating the pyrolysis properties of beech wood under varying pyrolysis temperatures (600°C–2,000°C) and heating rates (5°C–450°C/s), Zeng et al. (2015) observed an increase in SSA and PV at a low heating rate and high temperature. The authors observed that increasing the heating rate adversely affected the PV, fixed carbon, and H/C and O/C ratios. Contradicting results were also reported in the literature. For example, studies from Zhao et al. (2018), Li et al. (2016), and Mohanty et al. (2013) reported an increase in the SSA upon increasing the heating rate. These conflicting results may be attributed to the variations in the feedstock types used during the synthesis of biochar.

The residence time affects the transformation and the chemical properties of biochars (Wang et al., 2019). Wang et al. (2019) demonstrated that the biochar yields were significantly decreased when elevating the residence time from 30 to 90 min. Likewise, Hodgson et al. (2016) observed that increasing the residence time from 120 to 360 min led to a dramatic decrease in the total carbon content. This is likely attributed to the cleavage of biomass molecules at a prolonged residence time. In addition, Zhao et al. (2018) indicated that the SSA and the morphology of biochars were mainly impacted by the residence time. Herein, the SSA increased with an increase in residence time; however, it decreased thereafter due to the longer residence times, which destroy the biomass structure and the morphology of biochars (Zhao et al., 2018). More studies should focus on examining the optimal residence times for various biomasses as these differ due to structural and morphological variations in feedstocks used during pyrolysis.

Lee et al. (2017) demonstrated the significance of carrier gas and its effects on the physico-chemical properties of biochar during the pyrolytic conversion of agricultural waste. The authors showed that biochar produced under CO_2-enriched conditions exhibited a high SSA (109.15 m^2/g) compared with the biochar that was produced under N_2-enriched conditions (32.46 m^2/g). Nevertheless, the presence of CO_2 reduces the production of biochar because CO_2 typically accelerates the breakdown of volatile organic carbons (Lee et al., 2017). Crombie and Mašek (2015) also observed a decrease in biochar yield upon increasing the carrier gas (N_2).

2.5 THE ROLE OF BIOCHAR AND ITS POSSIBLE AUGMENTATION MECHANISMS ON DF PROCESSES

Studies that have investigated the role of biochar on anaerobic microorganisms (e.g., acidogens, acetogens, and methanogens) demonstrate that biochar provides optimal microenvironments for the proliferation of biofilm-forming microbial communities. This is primarily achieved by (i) improving the buffering capacity of the fermentation medium; (ii) aiding in the inhibition of ammonia; (iii) shortening the lag phase of microbial growth; (iv) providing the key micro- and macro-nutrients to anaerobic microorganisms; (v) removing the inhibitory by-products like HMF (hydroxymethylfurfural), furfural, and acidic compounds that affect the DF process; and (vi) promoting the direct interspecies electron transfer (DIET) as depicted in Figure 2.1 (Kumar et al., 2021; Luo and Angelidaki, 2012). In recent acidogenic H_2 production studies, it was shown that the inherent properties of biochar were effective in maintaining the neutral pH range during acidogenesis, thereby promoting the growth of predominant H_2-producing species like *Clostridium* sp. (Li et al., 2020).

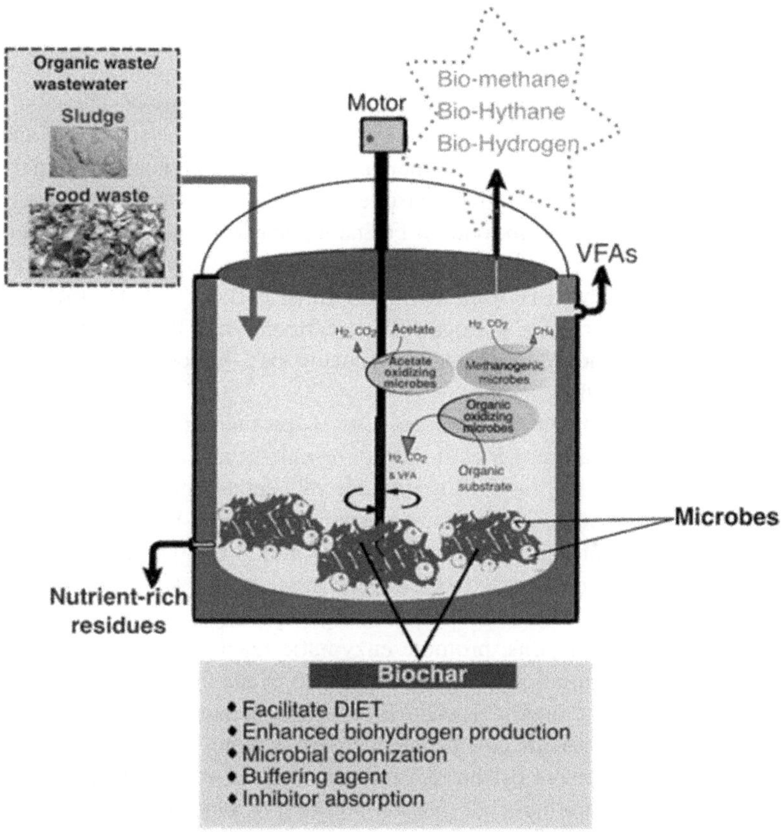

FIGURE 2.1 Schematic diagram indicating the roles of biochar in the DF process. Adapted and modified from Kumar et al. (2021).

Zhu et al. (2021) demonstrated that the use of biochar as a support material in DF enhanced the H_2 yield from 0 to 286.83 mL H_2, reduced the redox potential, and enhanced the utilization of acetate and butyrate, which are the main metabolites generated during DF-producing reactions. Sharma and Melkania (2017) reported that the use of biochar was effective in the inhibition of ammonia from co-cultures of *Enterobacter aerogenes* and *E. coli*. Herein, biochar promoted the formation of different acidogenic biofilms within the acidogenic reactor (Sharma and Melkania, 2017). Similarly, Sunyoto et al. (2017) studied the impact of adding biochar alongside the impact of initial pH on H_2 generation from food waste at varying biochar concentrations of 0–18.6 g/L. The authors observed a significant improvement of 26% in H_2 production rate and 41% in H_2 yield at 10 g/L substrate and initial pH of 6.0, compared to the control. Elsewhere, it was revealed that the incorporation of biochar altered the microbial community structure and favored the proliferation of H_2-producing species (*Bacillus* sp. and *Clostridium* sp.) while suppressing the non-acidogenic species (*Methylobacterium* sp. and *Sphingomonas* sp.) (Liao et al., 2019). These results are also consistent with the findings of Jindo et al. (2012) who observed a notable increase in the growth of Gram-positive bacteria, particularly the cultivation of *Bacillus* sp., which are among the dominant H_2 producers during their investigation of the role of biochar on microbial community dynamics during the biocomposting of organic manures.

Yang and Wang (2019) reported that the use of co-additives (biochar + Fe^0 nanoparticles) increased the H_2 yield by improving the hydrolysis of the substrate (grass). The co-addition of these enrichment additives resulted in an H_2 yield that was 89%, 53%, and 15% higher than the control experiment (the experiment that did not include the biochar), and the individual Fe^0 nanoparticles experiment, respectively. In addition, these additives reduced the lag phase and accelerated the bioconversion of grass, resulting in a high H_2 yield, as highlighted earlier. The synergistic effects observed for the biochar + Fe^0 nanoparticles treatment were likely attributed to the complementary properties of these additives during the DF process (Yang and Wang, 2019; Sekoai et al., 2019).

The effects of biochar on the generation of H_2-producing metabolites have also been studied in the literature. Lu et al. (2020) revealed that coupling the acidogenic reactors with biochars led to the enrichment of acetate (primary DF-producing metabolite) alongside the development of microbial biofilms. Biochars derived from coconut waste and longan shell generated a 109.6% and 71.8% increase in acetate production, respectively, and exhibited excess functional groups and high bulk internal crystallinity (Lu et al., 2020). These functional groups were reported to enhance the microbial activity of acidogens, promote enzymatic reactions, increase adsorption, and decrease the accumulation of by-products (Lu et al., 2020). In another related study, it was reported that the incorporation of biochar led to an increase in the accumulation of volatile fatty acids (VFAs) (acetate, propionate, and butyrate), and these were later used as precursors by the H_2 producers (Khoei et al., 2021). In summary, the aforementioned results indicate that the addition of biochars in acidogenic processes is also effective in enriching the DF precursors, which in turn enhances the H_2 process yields. The mechanisms of biochar on acidogenic species have not been fully elucidated in the literature. However, it has been proposed that biochar facilitates

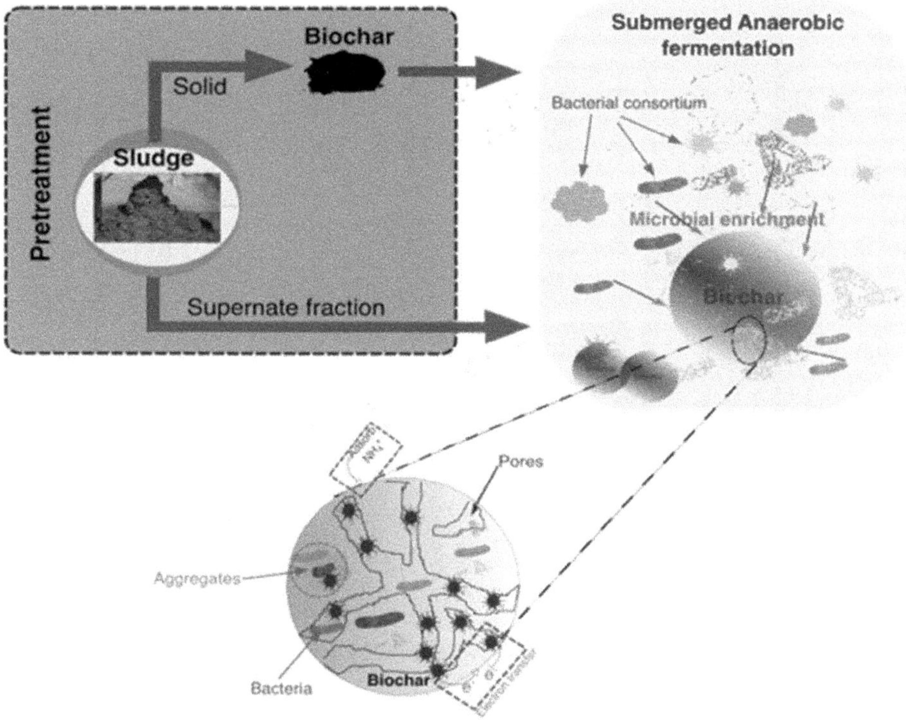

FIGURE 2.2 Schematic diagram of DIET mechanism on acidogenic species via biochar. Adapted and modified from Zhai et al. (2020).

microbial biofilm formation within the acidogenic reactors due to its large surface area (125 m^2/g) and PS, which usually attracts a wide range of microbial species (Cooney et al., 2016) (Figure 2.2). The functional groups in biochars also aid in the development of biofilms, and the high crystallinity accelerates the transfer of electrons (Lu et al., 2020). It also promotes synergistic interactions within the biofilms by enhancing DIET between microorganisms, which leads to the fast conversion of substrates. It provides the essential micro- and macro-nutrients to acidogenic species, as stated above (Lehmann and Joseph, 2012). Biochar also maximizes the activity of the hydrogenases and dehydrogenases, which are the essential enzymes that regulate the acidogenic H_2-producing reactions (Yang and Wang, 2019). Studies that have explored the use of biochar toward the enrichment of acidogenic processes are shown in Table 2.2.

2.6 THE PRODUCTION AND APPLICATION OF BIOCHAR IN DEVELOPING COUNTRIES – SOUTH AFRICA

As a nation that is rapidly expanding in economic and population growth, coupled with severe climatic variations, South Africa is faced with a major challenge of producing sufficient crops to meet the country's ever-increasing food demands. In

TABLE 2.2
Effect of biochar as a bio-additive on acidogenic H_2-producing processes

Carbon source	Inoculum source	Type of biochar	Acidogenic conditions (pH, T)	Role of biochar on acidogenic process	References
Growth medium	*Clostridium tyrobutyricum* ATCC 25755, *E. harbinense* Yuan-3	Rice straw	Initial pH = 7.0, T = 37°C	There was a dramatic increase in H_2 yield upon the addition of biochar on ethanol-type and butyrate-type fermentations.	Li et al. (2020)
Glucose	Mixed cultures	Corn-bran residue	Initial pH = 7.1, T = 37°C	The synergist effect of biochar and Fe^{2+} reduced the lag phase from 4.25 to 2.21 h, and the H_2 increased from 158 to 234 mL H_2/glucose.	Zhang et al. (2017)
Food waste	Mixed cultures	Pine sawdust	Initial pH = 7.0, T = 37°C	The addition of biochar increased the H_2 production rate by 32.5% and shortened the lag phase by 36%.	Sunyoto et al. (2016)
Grass	Mixed cultures	Sawdust	Initial pH = 7.0, T = 37°C	Biochar alongside Fe^0 NPs synergistically enhanced the growth of *Clostridium* species and substrate utilization during DF process.	Yang and Wang (2019)
Growth medium	*Clostridium beijerinckii* F-6	Rice straw	Initial pH = 7.0, T = 37°C	The addition of biochar significantly promoted the co-production of H_2 and butanol during the acidogenic process.	Wu et al. (2019)
Food waste	*Clostridium acetobutulicum*	Sewage sludge	Initial pH = 7.0, T = 37°C	The synergist effect of biochar and resin increased the H_2 yield by 94% and reduced the lag phase by 41.5%.	Rezaeitavabe et al. (2020)
Growth medium	Photosynthetic bacteria HAU-M1	Corn stover	Initial pH = 5.0–7.0, T = 30°C	Biochar and nano-scale Fe^0 improved the acid tolerance of acidogenic processes.	Zhu et al. (2021)
Organic waste	*Enterobacter aerogenes* and *E. coli*	Woody biomass	Initial pH = 5.5, T = 37°C	Biochar addition shortened the lag phase from 12.5 to 8.1 h and enhanced the H_2 production rate.	Sharma and Melkania (2017)
Growth medium	Mixed cultures	Corncob	Initial pH = 7.0, T = 37°C	The microbial community structure shifted from non-acidogenic species to acidogenic species like *Clostridium* sp.	Zhang et al. (2021)

(Continued)

TABLE 2.2
(Continued)

Carbon source	Inoculum source	Type of biochar	Acidogenic conditions (pH, T)	Role of biochar on acidogenic process	References
Growth medium	*Thermoanaerobacterium thermosaccharolyticum* W16	Cornstalk	Initial pH = 5.5–7.0, T = 60°C	An increase in biochar concentration (0–10 g/L) improved the H_2 yield from 24 to 154 mL H_2/g substrate.	Zhao et al. (2021)
Food waste	Mixed cultures	Pine sawdust	Initial pH = 7.0, T = 35°C	The use of biochar increased the H_2 production rates by 32.5% and increased the production of VFAs.	Sunyoto et al. (2016)
Growth medium	Photosynthetic bacteria HAU-M1	Corn stover	Initial pH = 5.0–7.0, T = 30°C	Biochar stabilized the overall photosynthetic H_2 process and promoted the consumption of VFAs.	Zhu et al. (2021)
Glucose	Mixed cultures	Rice husk, white popinee, bamboo, coconut	Initial pH = 7.0, T = 36°C	The biochars acted as biofilm carriers and promoted the growth of acidogenic microbes.	Lu et al. (2020)
Food waste	Mixed cultures	Pinewood	Initial pH = 6.4, T = 32°C	The elements in biochar were shown to contribute to the increase of VFAs and *Clostridium butyricum*.	Sugiarto et al. (2021)
Glucose	Mixed cultures	Calcium lignosulfonate	Initial pH = 7.0, T = 54°C	The biochar enhanced the microbial activity and process stability.	Zhao et al. (2019)
Food waste	Mixed cultures	Sawdust, wheat bran, peanut shell	Initial pH = 5.5, T = 35°C	Biochar addition shortened the lag phase by 18%–62% and increased the H_2 production rate by 18%–110%.	Wang et al. (2018)
Sugarcane bagasse	Mixed cultures	Fallen leaves	Initial pH = 7.0, 37°C	The H_2 yield increased by 317.1% through the stimulation of bacterial growth, enhancing the enzymatic activities, manipulating the $NADH/NAD^+$ ratio, and boosting electron transfer efficiency.	Bu et al. (2021)

addition, the indiscriminate use of chemical fertilizers has led to several problems such as the depletion of soil nutrients, alteration of the indigenous soil microbiota, salinization, environmental pollution, eutrophication, and health problems caused by the excessive use of chemical fertilizers (Kumar et al., 2021).

Biochar can play a pivotal role in addressing these issues as it is capable of enhancing soil and crop productivity by improving soil moisture, restoring soil properties, and stimulating the soil microbiome. The production and application of biochar have been practiced for several decades in developed nations like North America, Europe, and Asia (Agegnehu et al., 2017). In these regions, it is mainly used for promoting soil fertility, crop improvement, and environmental remediation (Agegnehu et al., 2017). However, developing countries such as South Africa have not explored the full economic potential of biochar technologies as few local scientists and industries have pursued this technological route, implying a shortage of knowledge regarding the socioeconomic impact of biochar in South Africa. New studies are now being conducted to obtain an in-depth understanding of the potential impact of this sustainable additive in South Africa. A study that investigated the properties of biochar derived from the country's bagasse showed that it is well suited for soil amendment applications due to its high water-holding capacity, CEC, high surface area, and high composition of micro- and macro-nutrients (Carrier et al., 2012). Dzvene et al. (2019) observed that biochar produced from livestock manure enhances maize germination. This was attributable to the biochar's properties such as high pH, EC, and high composition of nutrient content (Dzvene et al., 2019). Some authors have conducted techno-economic assessments to delve into the economic implications of pursuing this technology and see if it is commercially implementable. It was shown that the technology of biochar becomes profitable when it is produced via a biorefinery platform leading to a myriad of value-added bio-based products such as biochar, wood vinegar, and bio-coke (van Coller and Marx, 2022). Besides, biochar presents a lot of opportunities in South Africa as the country produces more than 30 million tons of biowaste each year, and the continuous production of biowaste poses serious environmental concerns as 90% of this waste is either landfilled and/or incinerated (Oelofse et al., 2021). The production of biochar is also supported by the country's strategic policy frameworks such as the National Development Plan, the Department of Science and Innovation's Decadal Plan, the Bioeconomy Strategy, and the Waste Economy Master Plan – these are aimed at accelerating the country's circular economy practices (Sekoai et al., 2023). To examine the large-scale implementation of this technology in South Africa, several local biochar companies have been recently established, and these include Phambili Energy, Hive Carbon, Organic Matter, NewCarbon, Green Houston, Carbon Gold, and Biochar Kilns International.

2.7 BENEFITS OF BIOCHAR-DERIVED DIGESTATES FROM THE ACIDOGENIC PROCESS

The sustainable management of DF residues (acidogenic digestates) is important due to their environmental effects (Kumar et al., 2021; Enaime et al., 2020). The direct discharge of acidogenic residues can adversely affect the ecosystem and cause various environmental concerns such as nutrient imbalance and eutrophication (Guo et al.,

2020). On the contrary, the use of biochar-derived acidogenic digestate has been shown to contribute to the growth of microorganisms by supplying key nutrients when used as biofertilizers (Wang et al., 2020; Kumar et al., 2021). Biochar digestates improve soil fertility by augmenting the soil microbiome and also provide the key micro- and macro-nutrients (e.g., calcium, copper, magnesium, potassium, sodium, zinc, and iron) that are required by plants for growth (Luo et al., 2015; Wang et al., 2017). The biochar constituents in these digestates also act as chelating agents and absorb excess compounds such as ammonia, nitrates, phosphates, and sulfates (Sarkhot et al., 2013).

Furthermore, these constituents have an impact on the stabilization of the soils' pH due to their composition of calcium and magnesium (Abdelhafez and Abbas, 2020). Hung et al. (2017) showed that biochar-derived digestate has a high composition of calcium (26.3%), phosphorus (11.5%), magnesium (0.86%), and silica (0.68%). The use of biochar-derived digestates could offer sustainable agricultural practices by

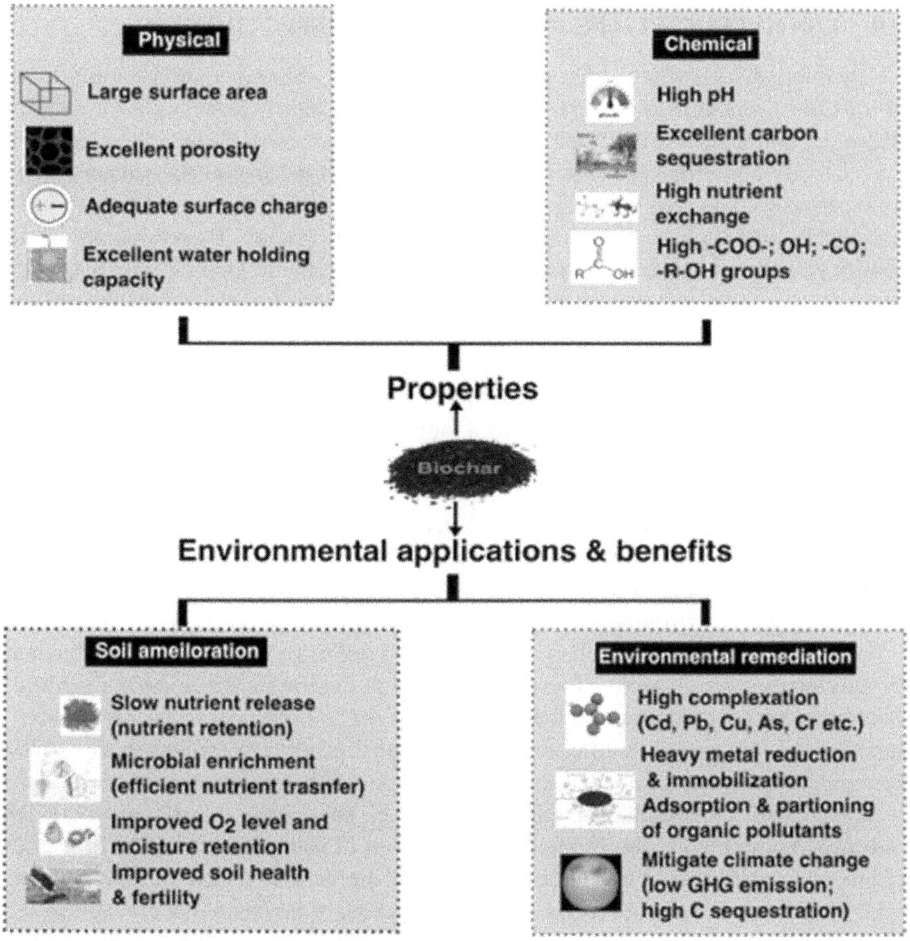

FIGURE 2.3 Characteristics of biochar and its applications. Adapted and modified from Oliveira et al. (2017).

reducing the heavy reliance on chemical fertilizers (Kumar et al., 2021). These results were substantiated by a study by Holatko et al. (2021), who examined the effect of carbon-enriched (with biochar) digestate on soil microflora. The authors observed that the activity of the soil microflora was significantly enhanced by the carbon-enriched digestate (Holatko et al., 2021). Wang et al. (2020) demonstrated that the biochar-derived digestate could be used in garden soils as an efficient phosphorous fertilizer. Fu et al. (2017) applied a digestate derived from a biochar-copper composite for tetracycline removal and observed a tetracycline removal efficiency of up to 97.8%, which implies that the residual biochars could be effective in the treatment of antibiotic pollutants. Therefore, these results highlight the benefits of residual biochars in environmental applications. The physico-chemical properties of biochar that make it a suitable bio-additive/support material in various applications are summarized in Figure 2.3.

2.8 KNOWLEDGE GAPS AND FUTURE PERSPECTIVES

It is therefore evident that the use of biochar presents a wide range of opportunities in DF due to its remarkable properties as shown in this chapter. However, a lot remains unclear regarding the use of this bio-additive in DF because most studies in the literature are relatively new. Hence, further research is required to unravel the various metabolic pathways triggered when biochar interacts with the predominant H_2-producing microorganisms such as *Clostridium* sp. Although the optimal H_2-producing conditions (e.g., pH, temperature, substrate concentration, organic loading rate, hydraulic retention time) are well documented in the literature, no studies have examined the impact of these setpoint conditions when biochar is integrated into acidogenic processes. Consequently, there might be some variations in these optimal conditions when biochar is incorporated in the acidogenic processes due to its composition of complex molecules/elements that have a significant effect on the acidogenic reactions. The use of biochar has been shown to have a profound effect on the formation of diverse biofilm communities (Sunyoto et al., 2016); however, the microbial assemblages (at phylum level and genus level) associated with these biofilms are not well elucidated in the literature. Therefore, understanding the key biofilm-forming species when biochar is coupled in these DF processes will be instrumental in identifying the microenvironments that favor the synergistic interactions of acidogenic biofilm formers. More research is also needed to optimize the properties of biochar for maximizing the H_2 yields. It is also essential to determine the scientific methods that optimize the synthesis of biochar alongside the suitable characterization procedures for a better understanding of its bio/chemical profile (Yaashikaa et al., 2020). Research efforts are also needed to understand the combined synergistic interactions of biochar with other enhancement additives (e.g., nanoparticles) on the overall DF pathways. Furthermore, the economic implications of using biochar in DF processes and its circularity should be studied to get a holistic view of its use in this scientific field. Finally, studies should focus on the engineering of biochars that are well suited for the DF process while taking into consideration the engineered biochars' affinity to acidogenic species, hydrolytic enzymes, metabolite concentration, and other process variables discussed above.

2.9 CONCLUSION

Biochar is an economical and environmentally friendly bio-additive/support material that could be instrumental in advancing DF technology owing to its outstanding characteristics. This chapter demonstrated the feasibility of augmenting the DF yields using biochar as a sustainable bio-additive. The work also discussed the possible mechanisms that are triggered when biochar is used in DF processes, as this remains an untapped scientific area. Furthermore, the physico-chemical parameters that affect the biochar properties were elucidated. The production and application of biochar in developing nations like South Africa were also explored. Despite its numerous benefits, many research gaps need to be explored before this carbonaceous material can be fully implemented in DF, as shown in this work.

REFERENCES

Abdelhafez, A. A., & Abbas, M. H. H. (2020). *Applications of biochar for environmental safety*. IntechOpen. https://doi.org/10.5772/intechopen.87828

Agegnehu, G., Srivastava, A. K., & Bird, M. I. (2017). The role of biochar and biochar-compost in improving soil quality and crop performance: A review. *Applied Soil Ecology, 119*, 156–170. https://doi.org/10.1016/j.apsoil.2017.06.008

Ahmad, M., Rajapaksha, A. U., Lim, J. E., Zhang, M., Bolan, N., Mohan, D., et al. (2014). Biochar as a sorbent for contaminant management in soil and water: A review. *Chemosphere, 99*, 19–33. https://doi.org/10.1016/j.chemosphere.2013.10.071

Álvarez, M. L., Gascó, G., Palacios, T., Paz-Ferreiro, J., & Méndez, A. (2020). Fe oxides–biochar composites produced by hydrothermal carbonization and pyrolysis of biomass waste. *Journal of Analytical and Applied Pyrolysis, 151*, 104893. https://doi.org/10.1016/j.jaap.2020.104893

Ambaye, T. G., Rene, E. R., Dupont, C., Wongrod, S., & van Hullebusch, E. D. (2020). Anaerobic digestion of fruit waste mixed with sewage sludge digestate biochar: Influence on biomethane production. *Frontiers in Energy Research, 8*, 31. https://doi.org/10.3389/fenrg.2020.00031

Angin, D. (2013). Effect of pyrolysis temperature and heating rate on biochar obtained from pyrolysis of safflower seed press cake. *Bioresource Technology, 128*, 593–597. https://doi.org/10.1016/j.biortech.2012.10.150

Arauzo, P. J., Atienza-Martínez, M., Ábrego, J., Olszewski, M. P., Cao, Z., & Kruse, A. (2020). Combustion characteristics of hydrochar and pyrochar derived from digested sewage sludge. *Energies, 13*(16), 4164. https://doi.org/10.3390/en13164164

Banu, J. R., Usman, M. T. M., Kavitha, S., Yukesh, K. R., Yogalakshmi, K. N., Sivashanmugam, P., et al. (2021). A critical review on limitations and enhancement strategies associated with biohydrogen production. *International Journal of Hydrogen Energy, 46*(34), 16565–16590. https://doi.org/10.1016/j.ijhydene.2021.01.075

Bio-era. (2009). *United States economic impact of advanced biofuels production: Perspectives to 2030.*

Brewer, C. E., Schmidt-Rohr, K., Satrio, J. A., & Brown, R. C. (2009). Characterization of biochar from fast pyrolysis and gasification systems. *Environmental Progress & Sustainable Energy, 28*(3), 386–396. https://doi.org/10.1002/ep.10378

Bu, J., Wei, H. L., Wang, Y. T., Cheng, J. R., & Zhu, M. J. (2021). Biochar boosts dark fermentative H_2 production from sugarcane bagasse by selective enrichment/colonization of functional bacteria and enhancing extracellular electron transfer. *Water Research, 202*, 117440. https://doi.org/10.1016/j.watres.2021.117440

Calvelo Pereira, R., Kaal, J., Camps Arbestain, M., Pardo Lorenzo, R., Aitkenhead, W., & Hedley, M. (2011). Contribution to characterisation of biochar to estimate the labile fraction of carbon. *Organic Geochemistry, 42*(11), 1331–1342. https://doi.org/10.1016/j.orggeochem.2011.09.002
Carrier, M. C., Hardie, A. G., Uras, Ü., Görgens, J., & Knoetze, J. H. (2012). Production of char from vacuum pyrolysis of South African sugar cane bagasse and its characterization as activated carbon and biochar. *Journal of Analytical and Applied Pyrolysis, 96*, 24–32.
Chatterjee, R., Sajjadi, B., Chen, W.-Y., Mattern, D. L., Hammer, N., Raman, V., et al. (2020). Effect of pyrolysis temperature on physicochemical properties and acoustic-based amination of biochar for efficient CO_2 adsorption. *Frontiers in Energy Research, 8*, 85. https://doi.org/10.3389/fenrg.2020.00085
Chen, B., Zhou, D., & Zhu, L. (2008). Transitional adsorption and partition of nonpolar and polar aromatic contaminants by biochars of pine needles with different pyrolytic temperatures. *Environmental Science & Technology, 42*(14), 5137–5143. https://doi.org/10.1021/es8002684
Chen, H., Liu, D., Shen, Z., Bao, B., Zhao, S., & Wu, L. (2015). Functional biomass carbons with hierarchical porous structure for supercapacitor electrode materials. *Electrochimica Acta, 180*, 241–251. https://doi.org/10.1016/j.electacta.2015.08.133
Chen, S., Rotaru, A. E., Shrestha, P. M., Malvankar, N. S., Liu, F., Fan, W., et al. (2014a). Promoting interspecies electron transfer with biochar. *Scientific Reports, 4*, 5019. https://doi.org/10.1038/srep05019
Chen, T., Zhang, Y., Wang, H., Lu, W., Zhou, Z., Zhang, Y., et al. (2014b). Influence of pyrolysis temperature on characteristics and heavy metal adsorptive performance of biochar derived from municipal sewage sludge. *Bioresource Technology, 164*, 47–54. https://doi.org/10.1016/j.biortech.2014.04.048
Chiappero, M., Norouzi, O., Hu, M., Demichelis, F., Berruti, F., Di Maria, F., et al. (2020). Review of biochar role as additive in anaerobic digestion processes. *Renewable and Sustainable Energy Reviews, 131*, 110037. https://doi.org/10.1016/j.rser.2020.110037
Cimon, C., Kadota, P., & Eskicioglu, C. (2020). Effect of biochar and wood ash amendment on biochemical methane production of wastewater sludge from a temperature phase anaerobic digestion process. *Bioresource Technology, 297*, 122440. https://doi.org/10.1016/j.biortech.2019.122440
Cooney, M. J., Lewis, K., Harris, K., Zhang, Q., & Yan, T. (2016). Start up performance of biochar packed bed anaerobic digesters. *Journal of Water Process Engineering, 9*, e7–e13. https://doi.org/10.1016/j.jwpe.2014.12.004
Crombie, K., & Mašek, O. (2015). Pyrolysis biochar systems: Balance between bioenergy and carbon sequestration. *GCB Bioenergy, 7*(2), 349–361. https://doi.org/10.1111/gcbb.12137
Dzvene, A. R., Chiduza, C., Mnkeni, P. N. S., & Peter, P. C. (2019). Characterisation of livestock biochars and their effect on selected soil properties and maize early growth stage in soils of Eastern Cape province, South Africa. *South African Journal of Plant and Soil, 36*, 199–209.
Enaime, G., Baçaoui, A., Yaacoubi, A., & Lübken, M. (2020). Biochar for wastewater treatment—Conversion technologies and applications. *Applied Sciences, 10,* 3492. https://doi.org/10.3390/app10103492
Fu, D., Chen, Z., Xia, D., Shen, L., Wang, Y., & Li, Q. (2017). A novel solid digestate-derived biochar-Cu NP composite activating H_2O_2 system for simultaneous adsorption and degradation of tetracycline. *Environmental Pollution, 221*, 301–310. https://doi.org/10.1016/j.envpol.2016.11.078
Guo, M., Song, W., & Tian, J. (2020). Biochar-facilitated soil remediation: Mechanisms and efficacy variations. *Frontiers in Environmental Science, 8*, 183. https://doi.org/10.3389/fenvs.2020.521512

Hodgson, E., Lewys-James, A., Rao Ravella, S., Thomas-Jones, S., Perkins, W., & Gallagher, J. (2016). Optimisation of slow-pyrolysis process conditions to maximise char yield and heavy metal adsorption of biochar produced from different feedstocks. *Bioresource Technology, 214*, 574–581. https://doi.org/10.1016/j.biortech.2016.05.009

Holatko, J., Hammerschmiedt, T., Kintl, A., Danish, S., Skarpa, P., Latal, O., et al. (2021). Effect of carbon-enriched digestate on the microbial soil activity. *PLoS ONE, 16*, e0252262. https://doi.org/10.1371/journal.pone.0252262

Hu, S., Xiang, J., Sun, L., Xu, M., Qiu, J., & Fu, P. (2008). Characterization of char from rapid pyrolysis of rice husk. *Fuel Processing Technology, 89*, 1096–1105. https://doi.org/10.1016/j.fuproc.2008.05.001

Huang, H., Reddy, N. G., Huang, X., Chen, P., Wang, P., Zhang, Y., et al. (2021). Effects of pyrolysis temperature, feedstock type and compaction on water retention of biochar-amended soil. *Scientific Reports, 11*, 1–19. https://doi.org/10.1038/s41598-021-86701-5

Hughes, S. R., & Jones, M. A. (2020). The global demand for biofuels and biotechnology-derived commodity chemicals. In A. A. Vertès, N. Qureshi, H. P. Blaschek, & H. Yukawa (Eds.), *Green energy to sustain: Strategies for global industries* (pp. 137–156). John Wiley & Sons. https://doi.org/10.1002/9781119152057.ch7

Hung, C. Y., Tsai, W. T., Chen, J. W., Lin, Y. Q., & Chang, Y. M. (2017). Characterization of biochar prepared from biogas digestate. *Waste Management, 66*, 53–60. https://doi.org/10.1016/j.wasman.2017.04.034

Igalavithana, A. D., Choi, S. W., Dissanayake, P. D., Shang, J., Wang, C. H., Yang, X., et al. (2020). Gasification biochar from biowaste (food waste and wood waste) for effective CO_2 adsorption. *Journal of Hazardous Materials, 391*, 121147. https://doi.org/10.1016/j.jhazmat.2019.121147

Igalavithana, A. D., Mandal, S., Niazi, N. K., Vithanage, M., Parikh, S. J., Mukome, F. N. D., et al. (2018). Advances and future directions of biochar characterization methods and applications. *Environmental Reviews, 47*, 2275–2330. https://doi.org/10.1080/10643389.2017.1421844

Ippolito, J. A., Cui, L., Kammann, C., Wrage-Mönnig, N., Estavillo, J. M., Fuertes-Mendizabal, T., et al. (2020). Feedstock choice, pyrolysis temperature and type influence biochar characteristics: A comprehensive meta-data analysis review. *Biochar, 2*, 421–438. https://doi.org/10.1007/s42773-020-00067-x

Jindo, K., Sánchez-Monedero, M. A., Hernández, T., García, C., Furukawa, T., Matsumoto, K., et al. (2012). Biochar influences the microbial community structure during manure composting with agricultural wastes. *Science of the Total Environment, 416*, 476–481. https://doi.org/10.1016/j.scitotenv.2011.12.009

Khoei, S., Stokes, A., Kieft, B., Kadota, P., Hallam, S. J., & Eskicioglu, C. (2021). Biochar amendment rapidly shifts microbial community structure with enhanced thermophilic digestion activity. *Bioresource Technology, 341*, 125864. https://doi.org/10.1016/j.biortech.2021.125864

Kiran, Y. K., Barkat, A., Cui, X. Q., Feng, Y., Pan, F. S., Tang, L., et al. (2017). Cow manure and cow manure-derived biochar application as a soil amendment for reducing cadmium availability and accumulation by *Brassica chinensis* L. in acidic red soil. *Journal of Integrative Agriculture, 16*, 725–734. https://doi.org/10.1016/S2095-3119(16)61488-0

Kraisornkachit, P., Vivanpatarakij, S., Powell, J., & Assabumrungrat, S. (2018). Experimental study of dual fixed bed biochar-catalytic gasification with simultaneous feed of O_2-steam-CO_2 for synthesis gas or hydrogen production. *International Journal of Hydrogen Energy, 43*, 14974–14986. https://doi.org/10.1016/j.ijhydene.2018.06.054

Kumar, M., Dutta, S., You, S., Luo, G., Zhang, S., Show, P. L., et al. (2021). A critical review on biochar for enhancing biogas production from anaerobic digestion of food waste and sludge. *Journal of Cleaner Production, 305*, 127143. https://doi.org/10.1016/j.jclepro.2021.127143

Lee, J., Yang, X., Cho, S. H., Kim, J. K., Lee, S. S., Tsang, D. C. W., et al. (2017). Pyrolysis process of agricultural waste using CO_2 for waste management, energy recovery, and biochar fabrication. *Applied Energy, 185*, 214–222. https://doi.org/10.1016/j.apenergy.2016.10.092

Lehmann, J., & Joseph, S. (2012). *Biochar for environmental management: Science and technology*. Routledge. https://doi.org/10.4324/9781849770552

Leng, L., & Huang, H. (2018). An overview of the effect of pyrolysis process parameters on biochar stability. *Bioresource Technology, 270*, 627–642. https://doi.org/10.1016/j.biortech.2018.09.030

Leng, L., Xiong, Q., Yang, L., Li, H. H., Zhou, Y., Zhang, W., et al. (2021). An overview on engineering the surface area and porosity of biochar. *Science of the Total Environment, 763*, 144204. https://doi.org/10.1016/j.scitotenv.2020.144204

Li, A., Liu, H. L., Wang, H., Xu, H. B., Jin, L. F., Liu, J. L., et al. (2016). Effects of temperature and heating rate on the characteristics of molded bio-char. *BioResources, 11*, 3259–3274. https://doi.org/10.15376/biores.11.2.3259-3274

Li, Y., Xing, B., Ding, Y., Han, X., & Wang, S. (2020). A critical review of the production and advanced utilization of biochar via selective pyrolysis of lignocellulosic biomass. *Bioresource Technology, 312*, 123614. https://doi.org/10.1016/j.biortech.2020.123614

Liang, L., Xi, F., Tan, W., Meng, X., Hu, B., & Wang, X. (2021). Review of organic and inorganic pollutants removal by biochar and biochar-based composites. *Biochar, 3*, 255–281. https://doi.org/10.1007/s42773-021-00101-6

Liao, H., Li, Y., & Yao, H. (2019). Biochar amendment stimulates utilization of plant-derived carbon by soil bacteria in an intercropping system. *Frontiers in Microbiology, 10*, 1361.

Liu, Z., Zhang, Y., & Liu, Z. (2019). Comparative production of biochars from corn stalk and cow manure. *Bioresource Technology, 291*, 121855. https://doi.org/10.1016/j.biortech.2019.121855

Low, Y. W., & Yee, K. F. (2021). A review on lignocellulosic biomass waste into biochar-derived catalyst: Current conversion techniques, sustainable applications and challenges. *Biomass and Bioenergy, 154*, 106245. https://doi.org/10.1016/j.biombioe.2021.106245

Lu, J. H., Chen, C., Huang, C., Zhuang, H., Leu, S. Y., & Lee, D. J. (2020). Dark fermentation production of volatile fatty acids from glucose with biochar-amended biological consortium. *Bioresource Technology, 303*, 122921. https://doi.org/10.1016/j.biortech.2020.122921

Luo, G., & Angelidaki, I. (2012). Integrated biogas upgrading and hydrogen utilization in an anaerobic reactor containing enriched hydrogenotrophic methanogenic culture. *Biotechnology and Bioengineering, 109*, 2729–2736. https://doi.org/10.1002/bit.24557

Luo, K., Pang, Y., Wang, D., Li, X., Wang, L., Lei, M., et al. (2021). A critical review on the application of biochar in environmental pollution remediation: Role of persistent free radicals (PFRs). *Journal of Environmental Sciences (China), 108*, 201–216. https://doi.org/10.1016/j.jes.2021.02.021

Luo, L., Xu, C., Chen, Z., & Zhang, S. (2015). Properties of biomass-derived biochars: Combined effects of operating conditions and biomass types. *Bioresource Technology, 192*, 83–89. https://doi.org/10.1016/j.biortech.2015.05.054

Mahmod, S. S., Arisht, S. N., Jahim, J. M., Takriff, M. S., Tan, J. P., Luthfi, A. A. I., et al. (2021). Enhancement of biohydrogen production from palm oil mill effluent (POME): A review. *International Journal of Hydrogen Energy, 47,* 40637–40655. https://doi.org/10.1016/j.ijhydene.2021.07.225

Mohanty, P., Nanda, S., Pant, K. K., Naik, S., Kozinski, J. A., & Dalai, A. K. (2013). Evaluation of the physiochemical development of biochars obtained from pyrolysis of wheat straw, timothy grass and pinewood: Effects of heating rate. *Journal of Analytical and Applied Pyrolysis, 104*, 485–493. https://doi.org/10.1016/j.jaap.2013.05.022

Mumme, J., Srocke, F., & Heeg, K., & Werner, M. (2014). Use of biochars in anaerobic digestion. *Bioresource Technology, 164*, 189–197. https://doi.org/10.1016/j.biortech.2014.05.008

Noor, N. M., Shariff, A., & Abdullah, N. (2012). Slow pyrolysis of cassava wastes for biochar production and characterization. *Iranian Journal of Energy and Environment.*

Nzediegwu, C., Arshad, M., Ulah, A., Naeth, M. A., & Chang, S. X. (2021). Fuel, thermal and surface properties of microwave-pyrolyzed biochars depend on feedstock type and pyrolysis temperature. *Bioresource Technology, 320*, 124282. https://doi.org/10.1016/j.biortech.2020.124282

Oelofse, S. H. H., Polasi, T., Haywood, L., & Musvoto, C. (2021). *Food waste roadmap technical report.* CSIR Report. https://wasteroadmap.co.za/wp-content/uploads/2021/06/17-CSIR-Final_Technical-report_Food-waste.pdf

Oliveira, F. R., Patel, A. K., Jaisi, D. P., Adhikari, S., Lu, H., & Khanal, S. K. (2017). Environmental application of biochar: Current status and perspectives. *Bioresource Technology, 246*, 110–122. https://doi.org/10.1016/j.biortech.2017.08.122

Pan, J., Ma, J., Zhai, L., Luo, T., Mei, Z., & Liu, H. (2019). Achievements of biochar application for enhanced anaerobic digestion: A review. *Bioresource Technology, 292*, 122058. https://doi.org/10.1016/j.biortech.2019.122058

Qin, Y., Wang, H., Li, X., Cheng, J. J., & Wu, W. (2017). Improving methane yield from organic fraction of municipal solid waste (OFMSW) with magnetic rice-straw biochar. *Bioresource Technology, 245*, 1058–1066. https://doi.org/10.1016/j.biortech.2017.09.047

Ren, S., Usman, M., Tsang, D. C. W., O-Thong, S., Angelidaki, I., Zhu, X., & et al. (2020). Hydrochar-facilitated anaerobic digestion: Evidence for direct interspecies electron transfer mediated through surface oxygen-containing functional groups. *Bioresource Technology, 54*, 5755–5766.

Rezaeitavabe, F., Saadat, S., Talebbeydokhti, N., Sartaj, M., & Tabatabaei, M. (2020). Enhancing bio-hydrogen production from food waste in single-stage hybrid dark-photo fermentation by addition of two waste materials (exhausted resin and biochar). *Biomass and Bioenergy, 143*, 105846. https://doi.org/10.1016/j.biombioe.2020.105846

Roberts, D. A., Paul, N. A., Cole, A. J., & de Nys, R. (2015). From waste water treatment to land management: Conversion of aquatic biomass to biochar for soil amelioration and the fortification of crops with essential trace elements. *Journal of Environmental Management, 157*, 60–68. https://doi.org/10.1016/j.jenvman.2015.04.016

Sarkhot, D. V., Ghezzehei, T. A., & Berhe, A. A. (2013). Effectiveness of biochar for sorption of ammonium and phosphate from dairy effluent. *Journal of Environmental Quality, 42*, 1545–1554. https://doi.org/10.2134/jeq2012.0482

Sbizzaro, M., César Sampaio, S., Rinaldo dos Reis, R., de Assis Beraldi, F., Medina Rosa, D., Maria Branco de Freitas Maia, C., et al. (2021). Effect of production temperature in biochar properties from bamboo culm and its influences on atrazine adsorption from aqueous systems. *Journal of Molecular Liquids, 343*, 117667. https://doi.org/10.1016/j.molliq.2021.117667

Sekoai, P. T., Ezeokoli, O., Mufhandu, H., & Habimana, O. (2026). Achieving a circular bioeconomy through fungal biorefineries for single-cell protein production: Advances, breakthroughs, and future directions. *Biofuels, Bioproducts and Biorefining, 20*, 1157–1179. https://doi.org/10.1002/bbb.70121

Sekoai, P. T., Chunilall, V., Msele, K., Buthelezi, L., Johakimu, J., Andrew, J., Zungu, M., Moloantoa, K., Maningi, N., Habimana, O., & Swartbooi, A. (2023). Biowaste biorefineries in South Africa: Current status, opportunities, and research and development needs. *Renewable and Sustainable Energy Reviews, 188*, 113870. https://doi.org/10.1016/j.rser.2023.113870

Sekoai, P. T., Daramola, M. O., Mogwase, B., Engelbrecht, N., Yoro, K. O., du Preez, S. P., et al. (2020). Revising the dark fermentative H_2 research and development scenario – An overview of the recent advances and emerging technological approaches. *Biomass and Bioenergy, 140*, 105673. https://doi.org/10.1016/j.biombioe.2020.105673

Sekoai, P. T., Ghimire, A., Ezeokoli, O. T., Rao, S., Ngan, W. Y., Habimana, O., et al. (2021). Valorization of volatile fatty acids from the dark fermentation waste streams—A promising pathway for a biorefinery concept. *Renewable and Sustainable Energy Reviews, 143*, 110971. https://doi.org/10.1016/j.rser.2021.110971

Sekoai, P. T., Ouma, C. N. M., du Preez, S. P., Modisha, P., Engelbrecht, N., Bessarabov, D. G., et al. (2019). Application of nanoparticles in biofuels: An overview. *Fuel, 237*, 380–397. https://doi.org/10.1016/j.fuel.2018.10.030

Sekoai, P. T., Yoro, K. O., Bodunrin, M. O., Ayeni, A. O., & Daramola, M. O. (2018). Integrated system approach to dark fermentative biohydrogen production for enhanced yield, energy efficiency and substrate recovery. *Reviews in Environmental Science and Bio/Technology, 17*, 501–529. https://doi.org/10.1007/s11157-018-9474-1

Sharma, P., & Melkania, U. (2017). Biochar-enhanced hydrogen production from organic fraction of municipal solid waste using co-culture of *Enterobacter aerogenes* and *E. coli*. *International Journal of Hydrogen Energy, 42*, 18865–18874. https://doi.org/10.1016/j.ijhydene.2017.06.171

Shen, X., Meng, H., Shen, Y., Ding, J., Zhou, H., Cong, H., et al. (2021). A comprehensive assessment on bioavailability, leaching characteristics and potential risk of polycyclic aromatic hydrocarbons in biochars produced by a continuous pyrolysis system. *Chemosphere, 287*, 132116. https://doi.org/10.1016/j.chemosphere.2021.132116

Sugiarto, Y., Sunyoto, N. M. S., Zhu, M., Jones, I., & Zhang, D. (2021). Effect of biochar in enhancing hydrogen production by mesophilic anaerobic digestion of food wastes: The role of minerals. *International Journal of Hydrogen Energy, 46*, 3695–3703. https://doi.org/10.1016/j.ijhydene.2020.10.256

Sun, Y., Xiong, X., He, M., Xu, Z., Hou, D., Zhang, W., & et al. (2021). Roles of biochar-derived dissolved organic matter in soil amendment and environmental remediation: A critical review. *Chemical Engineering Journal, 424*, 130387. https://doi.org/10.1016/j.cej.2021.130387

Sunyoto, N. M. S., Zhu, M., Zhang, Z., & Zhang, D. (2016). Effect of biochar addition on hydrogen and methane production in two-phase anaerobic digestion of aqueous carbohydrates food waste. *Bioresource Technology, 219*, 29–36. https://doi.org/10.1016/j.biortech.2016.07.089

Sunyoto, N. M. S., Zhu, M., Zhang, Z., & Zhang, D. (2017). Effect of biochar addition and initial pH on hydrogen production from the first phase of two-phase anaerobic digestion of carbohydrates food waste. *Energy Procedia, 105*, 379–384. https://doi.org/10.1016/j.egypro.2017.03.329

Tomczyk, A., Sokołowska, Z., & Boguta, P. (2020). Biochar physicochemical properties: Pyrolysis temperature and feedstock kind effects. *Reviews in Environmental Science and Biotechnology, 19*, 191–215. https://doi.org/10.1007/s11157-020-09523-3

van Coller, T. H., & Marx, S. (2022). Techno-economic baseline study for a waste-based biorefinery in South Africa. *Bioresource Technology Reports, 20*, 101280. https://doi.org/10.1016/j.biteb.2022.101280

Wan, Z., Sun, Y., Tsang, D. C. W., Hou, D., Cao, X., Zhang, S., et al. (2020). Sustainable remediation with an electroactive biochar system: Mechanisms and perspectives. *Green Chemistry, 22*, 2688–2711. https://doi.org/10.1039/d0gc00717j

Wang, D., Ai, J., Shen, F., Yang, G., Zhang, Y., Deng, S., et al. (2017). Improving anaerobic digestion of easy-acidification substrates by promoting buffering capacity using biochar derived from vermicompost. *Bioresource Technology, 227*, 286–296. https://doi.org/10.1016/j.biortech.2016.12.060

Wang, G., Li, Q., Dzakpasu, M., Gao, X., Yuwen, C., & Wang, X. C. (2018). Impacts of different biochar types on hydrogen production promotion during fermentative co-digestion of food wastes and dewatered sewage sludge. *Waste Management, 80*, 73–80. https://doi.org/10.1016/j.wasman.2018.08.042

Wang, H., Xiao, K., Yang, J., Yu, Z., Yu, W., Xu, Q., et al. (2020). Phosphorus recovery from the liquid phase of anaerobic digestate using biochar derived from iron-rich sludge: A potential phosphorus fertilizer. *Water Research, 174*, 115629. https://doi.org/10.1016/j.watres.2020.115629

Wang, Z., Liu, K., Xie, L., Zhu, H., Ji, S., & Shu, X. (2019). Effects of residence time on characteristics of biochars prepared via co-pyrolysis of sewage sludge and cotton stalks. *Journal of Analytical and Applied Pyrolysis, 142*, 104659. https://doi.org/10.1016/j.jaap.2019.104659

Weber, K., & Quicker, P. (2018). Properties of biochar. *Fuel, 217*, 240–261. https://doi.org/10.1016/j.fuel.2017.12.054

Wu, B., Yang, Q., Yao, F., Chen, S., He, L., Hou, K., et al. (2019). Evaluating the effect of biochar on mesophilic anaerobic digestion of waste activated sludge and microbial diversity. *Bioresource Technology, 294*, 122235. https://doi.org/10.1016/j.biortech.2019.122235

Xie, R., Zhu, Y., Zhang, H., Zhang, P., & Han, L. (2021). Effects and mechanism of pyrolysis temperature on physicochemical properties of corn stalk pellet biochar based on combined characterization approach of microcomputed tomography and chemical analysis. *Bioresource Technology, 329*, 124907. https://doi.org/10.1016/j.biortech.2021.124907

Yaashikaa, P. R., Kumar, P. S., Varjani, S., & Saravanan, A. (2020). A critical review on the biochar production techniques, characterization, stability and applications for circular bioeconomy. *Biotechnology Reports, 28*, e00570. https://doi.org/10.1016/j.btre.2020.e00570

Yang, G., & Wang, J. (2019). Synergistic enhancement of biohydrogen production from grass fermentation using biochar combined with zero-valent iron nanoparticles. *Fuel, 251*, 420–427. https://doi.org/10.1016/j.fuel.2019.04.059

Yang, H., Huan, B., Chen, Y., Gao, Y., Li, J., & Chen, H. (2016). Biomass-based pyrolytic polygeneration system for bamboo industry waste: Evolution of the char structure and the pyrolysis mechanism. *Energy & Fuels, 30*, 6430–6439. https://doi.org/10.1021/acs.energyfuels.6b00732

Yin, C., Shen, Y., Yuan, R., Zhu, N., Yuan, H., & Lou, Z. (2019). Sludge-based biochar-assisted thermophilic anaerobic digestion of waste-activated sludge in microbial electrolysis cell for methane production. *Bioresource Technology, 284*, 315–324. https://doi.org/10.1016/j.biortech.2019.03.146

You, S., Ok, Y. S., Chen, S. S., Tsang, D. C. W., Kwon, E. E., Lee, J., et al. (2017). A critical review on sustainable biochar system through gasification: Energy and environmental applications. *Bioresource Technology, 246*, 242–253. https://doi.org/10.1016/j.biortech.2017.06.177

Yuan, J. H., Xu, R. K., & Zhang, H. (2011). The forms of alkalis in the biochar produced from crop residues at different temperatures. *Bioresource Technology, 102*, 3488–3497. https://doi.org/10.1016/j.biortech.2010.11.018

Zeng, K., Minh, D. P., Gauthier, D., Weiss-Hortala, E., Nzihou, A., & Flamant, G. (2015). The effect of temperature and heating rate on char properties obtained from solar pyrolysis of beech wood. *Bioresource Technology, 182*, 114–119. https://doi.org/10.1016/j.biortech.2015.01.112

Zhai, S., Li, M., Xiong, Y., Wang, D., & Fu, S. (2020). Dual resource utilization for tannery sludge: Effects of sludge biochars (BCs) on volatile fatty acids (VFAs) production from sludge anaerobic digestion. *Bioresource Technology, 316*, 123903. https://doi.org/10.1016/j.biortech.2020.123903

Zhang, J., Fan, C., & Zang, L. (2017). Improvement of hydrogen production from glucose by ferrous iron and biochar. *Bioresource Technology, 245*, 98–105. https://doi.org/10.1016/j.biortech.2017.08.198

Zhang, J., Lü, F., Luo, C., Shao, L., & He, P. (2014). Humification characterization of biochar and its potential as a composting amendment. *Journal of Environmental Sciences, 26*, 390–397. https://doi.org/10.1016/S1001-0742(13)60421-0

Zhang, J., Yang, M., Zhao, W., Zhang, J., & Zang, L. (2021). Biohydrogen production amended with nitrogen-doped biochar. *Energy & Fuels, 35*, 1476–1487. https://doi.org/10.1021/acs.energyfuels.0c03405

Zhang, L., Lim, E. Y., Loh, K. C., Ok, Y. S., Lee, J. T. E., Shen, Y., et al. (2020). Biochar enhanced thermophilic anaerobic digestion of food waste: Focusing on biochar particle size, microbial community analysis and pilot-scale application. *Energy Conversion and Management, 209*, 112654. https://doi.org/10.1016/j.enconman.2020.112654

Zhao, B., O'Connor, D., Zhang, J., Peng, T., Shen, Z., Tsang, D. C. W., et al. (2018). Effect of pyrolysis temperature, heating rate, and residence time on rapeseed stem derived biochar. *Journal of Cleaner Production, 174*, 977–987. https://doi.org/10.1016/j.jclepro.2017.11.013

Zhao, L., Wu, K. K., Chen, C., Ren, H. Y., Wang, Z. H., Nan, J., et al. (2021). Role of residue cornstalk derived biochar for the enhanced bio-hydrogen production via simultaneous saccharification and fermentation of cornstalk. *Bioresource Technology, 330*, 125006. https://doi.org/10.1016/j.biortech.2021.125006

Zhao, L., Zhang, J., Zhao, W., & Zang, L. (2019). Improved fermentative hydrogen production with the addition of calcium-lignosulfonate-derived biochar. *Energy & Fuels, 33*, 7406–7414. https://doi.org/10.1021/acs.energyfuels.9b01380

Zhou, H., Brown, R. C., & Wen, Z. (2020). Biochar as an additive in anaerobic digestion of municipal sludge: Biochar properties and their effects on the digestion performance. *ACS Sustainable Chemistry & Engineering, 8*, 6391–6401. https://doi.org/10.1021/acssuschemeng.0c00571

Zhu, S., Yang, X., Zhang, Z., Zhang, H., Li, Y., Zhang, Y., et al. (2021). Tolerance of photo-fermentative biohydrogen production system amended with biochar and nanoscale zero-valent iron to acidic environment. *Bioresource Technology, 338*, 125512. https://doi.org/10.1016/j.biortech.2021.125512

3 Microbial Biorefinery-Based Strategies Used in the Valorization of Grape Pomace

Patrick Sekoai, Sizwe Ndlovu, and Olivier Habimana

3.1 INTRODUCTION

Grape (*Vitis vinifera* L.) is one of the most cultivated fruits in the world and is used in the production of wine and other food products (Abreu et al., 2023). It plays a major role in the agricultural market as the global grape production was estimated to be around 73 million tons, with countries such as France, China, Italy, Spain, the USA, and South Africa listed among the top wine-producing nations in the world (Li et al., 2023). During winemaking, the grapes are subjected to various steps such as cleaning, sorting, mashing, and maceration to recover the liquid extracts that are used in the fermentation process, followed by the downstream process that involves the filtration and bottling of wine (Chowdhary et al., 2021). Therefore, tons of wine by-products (grape pomace) are generated each year from wine industries, and these are either discarded in landfills or used in animal feed. The grape pomace comes in the form of seeds, skins, stalks, etc., and represents 10%–30% (w/w) of the processed grape depending on the cultivar, viticulture practices, and the wine making conditions (Favre et al., 2019).

In terms of its chemical constituents, grape pomace has valuable compounds such as fibers, polysaccharides, phenolics, fermentable sugars, and enzymes. Recent scientific efforts have shown that it holds enormous potential as it can be converted/valorized into a wide array of industrially relevant products such as biochemicals, biofuels, and single-cell proteins (SCPs) (Castro et al., 2024). This discourse aligns with the circular bioeconomy framework, which is an innovative strategy that entails the exploitation of our bio-resources in a more sustainable manner, leading to near-zero or zero waste generation (Castro et al., 2024). From a socioeconomic standpoint, circular biobased opportunities could spur many opportunities for agro-food industries such as grape pomace can be used as valuable feedstocks for producing an array

DOI: 10.1201/9781003745150-3

of value-added products, as mentioned above. In addition, this waste valorization strategy is not only beneficial in alleviating environmental pollution but also helps agri-food industries avert the heavy financial levies they incur while trying to manage their waste streams (Kokkinomagoulos and Kandylis, 2023; Marks et al., 2020).

The biorefinery concept is central to the circular bioeconomy agenda – it is analogous to the petroleum refinery but uses biomass as a feedstock instead of crude oil in the case of petroleum refineries (Venkata Mohan et al., 2016). Circular bio-based opportunities are achieved through biorefinery-based technologies, which are used to produce various products such as biomaterials, biofuels, biochemicals, and food, as mentioned above. In this approach, scientific researchers use what is known as the 'closed-loop' and/or 'sequential' biorefinery technologies to synthesize these products in a closed or sequential manner and these technological processes are conducted using chemicals and/or microbial cell factories (Nhleko and Sekoai, 2025; Gaogane et al., 2025).

This chapter shows how microbes could be exploited to synthesize a broad spectrum of market-based products, such as biochemicals, biofuels, proteins, novel foods, and enzymes, through biotechnological strategies. It is hoped that this work will provide deeper insights into the economic opportunities that could be unlocked through the valorization of grape pomace, as there is a paucity of literature that covers this topic.

3.2 THE MAIN BIOACTIVE COMPOUNDS IN GRAPE POMACE

As mentioned earlier, grape pomace consists of various bioactive constituents that are useful in biorefinery-based platforms targeting several high-value-added products. These include phenolics, fermentable sugars, dietary fiber, and bio-oils. These bioactives are elaborated in the following sections.

3.2.1 Phenolic Compounds

Phenolics continue to attract R&D attention due to their remarkable chemical properties that induce antimicrobial, antioxidant, antithrombotic, anticancer, cardioprotective, and vasodilatory effects (Beres et al., 2017). Grape pomace is reported to be one of the richest sources of phenolics as 30%–40% of phenols are only utilized in the winemaking process and the rest remains in the pomace (Natolino and Da Porto, 2020). They are further classified into different groups such as simple phenolics (ferulic acid, caffeic acid, sinapic acid, gallic acid, *p*-coumaric acid, syringic acid, protocatechuic acid, etc.) and polyphenols. The latter is further categorized into flavonoids (myricetin, quercetin, kaempferol, etc.), stilbenes (resveratrol), and tannins (ellagitannins, proanthocyanidins, gallotannins) (Nemzer et al., 2022). The concentration of these compounds varies due to several key factors such as grape cultivars, climatic conditions, and wine-processing conditions (Nemzer et al., 2022). Over the past decade, scientists have conducted characterization studies using various chemical extraction procedures to acquire an in-depth understanding of the biochemical profile of grape pomace and to determine how these bioactive compounds could be useful in various fields. For example, Wang et al. (2010) conducted a chemical characterization study of muscadine grape pomace extract using spectrophotometric techniques. The extracted phenolics were gallic acid, ellagic acid, (–)-epigallocatechin, (–)-epicatechin, myricetin, catechin, and quercetin (Wang et al., 2010). In addition, other bioactives such as cyanidin, delphinidin, petunidin, and malvidin were identified (Wang et al., 2010). Studies have also focused on the characterization of grape pomace derived from white and red wines. A detailed characterization conducted on white wine (*Vitis vinifera* L.) revealed that it primarily consists of phenolics, tannins, monomeric, and oligomeric flavan-3-ol components (González-Centeno et al., 2013). Besides, it was discovered that the whole fraction of grape pomace (skin, stem, and seeds) could be used as a vital source of bioactive compounds as it consisted of all the desirable extractives as opposed to the separation of skins, stems, and seeds, which is a normal procedure that is applied in the literature (González-Centeno et al., 2013). Elsewhere, it was shown that other cultivars of red wine consisted of other valuable bioactive compounds, including monomeric sugars such as glucose, arabinose, xylose, mannose, and galactose, as opposed to the commonly extracted phenolic compounds (Dong et al., 2021). Therefore, these results reiterate the complex chemical constituents present in this plant-based feedstock and how these bioactive compounds can serve as valuable precursors in diverse biotechnological applications. Table 3.1 summarizes the various scientific endeavors that have been undertaken in the extraction of bioactives such as phenols, dietary fibers, polyunsaturated fatty acids, and flavonoids.

TABLE 3.1
The different bioactives compounds extracted from grape pomace

Bioactive compounds extracted	Grape cultivar	Method of extraction	References
Phenolic compounds	*Vitis vinifera*	Microwave-assisted extraction	da Rocha and Noreña (2021)
Anthocyanins, flavonoids, and resveratrol	Croatina grape	Ohmic heating-assisted hydroethanolic extraction	Ferreira-Santos et al. (2024)
Phenolic acids (gallic acid, (+)-catechin, Trolox, *p*-cumaric acid)	BRS Violet grape	Ultrasound-assisted extraction, followed by alginate-Ca^{2+} encapsulation	Bruno Romanini et al. (2021)
Polyphenols, anthocyanins, anthocyanidins, proanthocyanidins, catechins	*Vitis vinifera cv Tannat*	Ultrasound-assisted extraction	González et al, (2020)
Phenolics, triterpenoids, anthocyanin glucosides	Šukac and Agner	Solvents extraction	Pintać et al. (2018)
Polyunsaturated fatty acids, linoleic acid, oleic acid, phenols (polyphenols, proanthocyanidins, catechins)	Nero d'Avola grape	Solvent extraction	Messina et al. (2021)

(Continued)

TABLE 3.1
(Continued)

Bioactive compounds extracted	Grape cultivar	Method of extraction	References
Anthocyanins	BRS Vitória	Pressurized hot water extraction and sequential subcritical water hydrolysis	Ferreira-Santos et al. (2024)
Anthocyanins	*Vitis vinifera* L.	Ultrasonic-assisted extraction and microwave-assisted extraction	Marianne et al. (2024)
Polysaccharides proanthocyanidins	Alicante bouschet grape	Enzyme-assisted extraction	Teles et al. (2019)
Phenolic acids (p-coumaric, gallic, hydroxybenzoic, and ferulic acids), trans-resveratrol, flavan-3-ols [(+)-catechin, (−)-epicatechin and procyanidin dimer B1 and flavonols (myricetin and quercetin)	*Vitis vinifera cv. Graševina*	Supercritical CO_2 extraction, pulsed electric fields-assisted supercritical CO_2 extraction	Ćurko et al. (2023)
Phenolic compounds and dietary fiber	*Cabernet sauvignon*	Steam explosion	Cui et al. (2023)
p-OH-benzoic acid, vanillic acid, gallic acid, epicatechin	*Vitis vinifera*	Supercritical CO_2 extraction coupled with co-solvent	Oliveira et al. (2013)
Resveratrol	*Vitis vinifera*	Supercritical CO_2 extraction	Pascual-Martí et al. (2001)
Rutin and myricetin	Xiahei	Ultrasound-assisted aqueous two-phase extraction	Xiao et al. (2022)

3.2.2 Dietary Fiber

Dietary fibers are carbohydrate compounds that are present in plant cells including grape pomace. Upon further fractionation, they consist of different monomeric units of C_5–C_6 sugars such as glucose, arabinose, fructose, sucrose, mannose, and galactose that have diverse industrial applications (Zhu et al., 2015). Similar to the cereal industry, which is the major source of dietary fiber, fruit processing residues such as grape pomace are also enriched with an array of these beneficial compounds (Zhu et al., 2015). In addition, they have a high composition of soluble dietary content, low calorific content, and desirable functional properties in comparison to dietary fibers that are found in cereal residues (Orlando et al., 2024). It has been reported that grape pomace has a total dietary fiber content of ~70% together with an insoluble dietary fiber content of 20%–78% (Bassani et al., 2020; Nagarajaiah and Prakash, 2016; Nayak et al., 2018).

A detailed characterization study conducted on grape pomace showed that it primarily consisted of 30% polysaccharides (mannan, arabinan, xyloglucan, xylan, cellulose, and galactan), 15% insoluble proanthocyanidins, lignin, proteins, and phenols (Chamorro et al., 2012). These scientific results correlate with other published reports, in which, phenols, polysaccharides, and proteins were also characterized in the grape pomace of white wines (Curiel-Fernández et al., 2023). On the other hand, water-soluble dietary fibers such as pectins, gums, and β-glucans constitute 9%–11% of grape pomace (Chakka and Babu, 2022). Dietary fibers have been shown to confer medicinal benefits such as the reduction in cardiovascular illnesses, anticarcinogenic properties, prevention of diabetes, and improvement in the digestive system (Li and Petrov, 2024).

3.2.3 Grape Seed Oil

In addition to the stalks and skins, the winemaking process generates millions of tons of grape seeds which are usually discarded as waste. Grape seeds contain 3%–20% oil, based on the extracted grape cultivar (Wen et al., 2016). They are also rich in unsaturated fatty acids such as oleic acid and linoleic acid, which have health benefits (dos Santos Freitas et al., 2008). Other compounds that are found in grape seed oil include vitamin E (tocopherols and tocotrienols) and sterols (Rodríguez-Blázquez et al., 2024). The seeds possess numerous health benefits such as anti-cardiovascular diseases, anticarcinogenic activities, reduction in cholesterol, and prevention of sexual impotence (Wen et al., 2016). An earlier study by Matthäus (2008) showed that grape seeds comprise 35% crude fiber, 29% recoverable compounds, 7% moisture, 3% minerals, 7%–20% oils, and 11% protein. Sabir et al. (2012) extracted and examined the bioactive compounds found in 21 grape cultivars and obtained an oil content that ranged from 7% to 22%, with linoleic acid as the main bioactive (53%–69%), followed by oleic acid (16%–31%), palmitic acid (6%–12%), and stearic acid (1%–4%). Hence, grape seeds will play a key role in the advancement of circular opportunities as they have been shown to contain several bioactive substances that are applicable in foods, medicine, and cosmetics (Gitea et al., 2023; Yang et al., 2021).

3.3 SOUTH AFRICA'S WINE INDUSTRY

The South African wine industry is more than three centuries old and is the eighth-largest wine industry in the world in terms of wine production capacity (Moore et al., 2008). It has been predicted that the industry generates around 300,000 direct and indirect employment opportunities (Ferreira and Hunter, 2017). Furthermore, the industry accounts for approximately R36 billion of the country's gross domestic product (GDP), of which ~R19 billion contributes to the economy of the Western Cape Province (Ferreira and Hunter, 2017). The country's 355 wine distillers produce 800–900 million liters of wine each year (Bruwer, 2003). Owing to its major contribution to the country's economy, the local wine industry has established research institutes (e.g., the Institute for Wine Biotechnology, Stellenbosch University) that focus on wine biotechnology to develop innovative technologies that can ensure that the local wine industry remains competitive in the global markets and can withstand harsh climatic conditions. Therefore, the growth of the local wine industry will have severe implications for the environment as the grape pomace is continuously being dumped on landfill sites if circular biobased valorization strategies are not adopted by these industries.

3.4 MICROBIAL BIOREFINERY-BASED STRATEGIES USED TO VALORIZE THE GRAPE POMACE

The valorization of agri-food wastes into high-value-added products (see Figure 3.1) will pave the way for the transition to a circular model (Filippi et al., 2021). Winery wastes consist of phytochemicals/extracts that can serve as valuable biomass-derived precursors for the synthesis of a wide spectrum of industrially relevant products such as biochemicals, biofuels, enzymes, polyhydroxyalkanoates (PHAs), and SCPs. The biosynthesis of these high-value-added products from grape pomace will be covered in this section.

3.4.1 Biofuels

3.4.1.1 Biogas

The grape pomace consists of fermentable sugars that are used as precursors for biogas production via the anaerobic digestion process. A study by El Achkar et al. (2016)

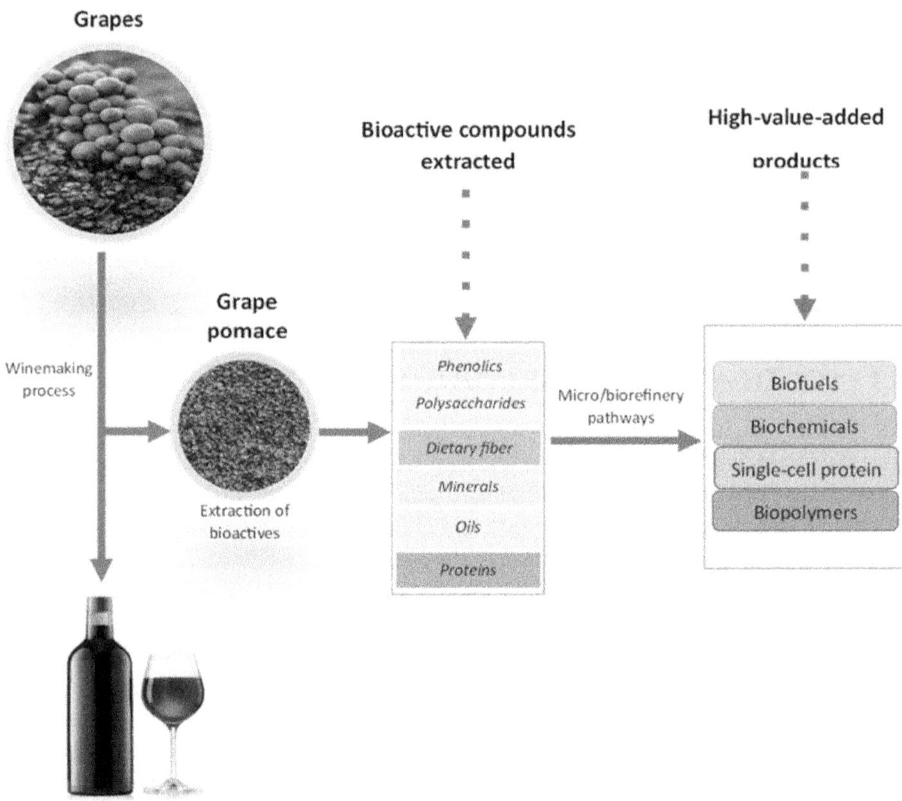

FIGURE 3.1 Depicts the valorization of grape pomace into high-value-added products.

showed that grape pomace (pulp and seeds) was a desirable feedstock for batch fermentations as it resulted in cumulative biogas production of 0.052–0.165 Nm^3/kg chemical oxygen demand. The authors then evaluated the applicability of this substrate for biogas production under continuous conditions where it was revealed that grinding it increased its anaerobic biodegradability resulting in a methane content of 62% (El Achkar et al., 2016). In another study, the grape pomace was first subjected to vigorous chemical extraction to obtain valuable bioactive compounds such as anthocyanins, a class of red-blue pigments (Allison and Simmons, 2018). Thereafter, the grape pomace was applied as a second-generation feedstock for biogas studies (Allison and Simmons, 2018). The methane production yield varied from 190.5 to 213.6 mL/g volatile solids (VS) and these values were found to be comparable with other studies that have used this biomass for biogas in the range of 110–360 mL/g VS (Dinuccio et al., 2010; El Achkar et al., 2016; Failla and Restuccia, 2014).

Recent research is now geared toward the implementation of what is known as the biorefinery concept – this is a new scientific concept that entails the production of high-value-added products through a network of integrated technological pathways using various biomass streams that are acquired from the agricultural sector, municipal sector, food processing sector, and local households. This field is expected to create a myriad of economic opportunities in the 'waste-to-value' innovations as techno-economic studies have shown that millions of dollars could be generated through the establishment of circular-based technologies, which could replace the linear model in the future. Thus, more studies are now drawn toward this concept to improve the process economics involved in both the extraction and upcycling of grape pomace. A multi-purpose biorefinery platform was implemented in the literature to produce polyphenols, volatile fatty acids (VFAs), PHA, and biogas (Martinez et al., 2016). This cascade biorefinery-based model led to the recovery of 2.7 g of total phenols, VFAs yield of 20 g/L, PHAs content of 63%, and methane yield of 113 mL/g VS (Martinez et al., 2016). The extraction of bioactives was achieved through the supercritical CO_2 extraction method, while the PHA production process was obtained using *Cupriavidus necator*, and methane was produced using mixed acidogenic cultures (Martinez et al., 2016).

3.4.1.2 Bioethanol

The monomeric sugars present in grape pomace also make it an ideal carbon source, and scientists are also exploring its use in other biofuels such as bioethanol production. A bioethanol study involving the use of *Debaryomyces nepalensis* as the inoculum source was reported in the literature (Egüés et al., 2013). The grape pomace was first pretreated using a sequential process (autohydrolysis and acid hydrolysis); it was then applied in bioethanol production, where a maximum bioethanol yield of 20.84 g/L was obtained during the fermentation process (Egüés et al., 2013). It was also demonstrated that the grape pomace can be used to achieve a full circular bioeconomy process where the extracted polyphenols served as biofertilizers while the fermentable sugars served as precursors for bioethanol studies (Corbin et al., 2015). Interestingly, bioethanol has also been produced using the solid-state fermentation approach instead of the liquid medium that is commonly used in the literature (Egüés et al., 2013). Although the bioethanol production yields were not comparable to those obtained from the liquid-state fermentation. Nevertheless, these R&D

efforts could provide new insights into bioethanol fermentation strategies. A characterization study that was conducted on various grape cultivars (*Cabernet Sauvignon* and *Sauvignon Blanc*) revealed that 31%–54% w/w of grape pomace consisted of carbohydrates, making it a desirable feedstock for biofuel production technologies such as bioethanol (Corbin et al., 2015). These researchers reported that soluble carbohydrates can be biotechnologically transformed into bioethanol, producing up to 270 L/tonne (Corbin et al., 2015). In addition, this integrated grape pomace upcycling strategy will not only lead to a multi-product value chain but also offer economic, social, and environmental advantages (Corbin et al., 2015).

3.4.1.3 Biohydrogen

Biohydrogen production is the process that yields the cleanest form of hydrogen as its production is mediated by an array of biocatalysts, including strict anaerobes and facultative anaerobes under acidogenic fermentation conditions (Sekoai and Yoro, 2016). Unlike the traditional H_2 production technologies, biohydrogen emits very few pollutants, requires low energy inputs, uses cheap feedstocks in the form of biowastes, and fits well within the circular bioeconomy framework, as it can be used to create socioeconomic opportunities through biowaste upcycling strategies (Sekoai et al., 2018). Over the past decade, many feedstocks have been investigated in biohydrogen process development, and these include agri-food waste, municipal waste, and industrial waste (Sekoai et al., 2021). Grape pomace is one of the biowastes that has caught the attention of many researchers throughout the world, particularly in countries with large winemaking industries. It was shown that the endogenous microbial consortia present in the grape pomace can produce biohydrogen under nutrient-free conditions (François et al., 2021). The heat-treated inoculum found in the grape pomace resulted in a biohydrogen yield of 1.6 mol H_2/mol hexose (François et al., 2021). Moreover, microbial characterization studies showed that bacteria belonging to the genus *Clostridium* were prevalent in the grape pomace and these experimental results could open new scientific innovations in the valorization of winery by-products through biohydrogen production technologies (François et al., 2021). These outcomes were also corroborated in a recent study that demonstrated the possibility of producing biohydrogen using a liquid-gas hollow fiber membrane reactor (Renaudie et al., 2024). This innovative system was implemented to enrich the biohydrogen-producing biofilms and prevent cell washout (Renaudie et al., 2024). As expected, bacterial community studies showed the abundance of *Clostridium* sp., the main biohydrogen-producing bacterial communities that thrive during acidogenic fermentations (Renaudie et al., 2024). Recent studies are also exploring the impact of co-digesting the grape pomace with other feedstocks to enhance the nutritional content, improve the microbial activities during acidogenesis, withstand the formation of microbial inhibitors, etc. (Constantin et al., 2024; Hungría et al., 2021).

3.4.2 Biobased Chemicals

3.4.2.1 Lactic Acid

Grape pomace also consists of soluble carbohydrates that can be used as a carbon source for lactic acid production. A study that investigated the effect of grape pomace as a carbon source using a mutant strain of *Lactiplantibacillus plantarum* showed remarkable results as 92.26% lactic acid yield was obtained from the fermentable sugars obtained from the hydrolysate (Shen et al., 2023). Elsewhere, an acid-hydrolyzed grape pomace was used for lactic acid fermentation (Rivera et al., 2007). It was found that the xylose-rich hydrolysate was a favorable carbon source for lactic acid fermentation, resulting in a lactic acid yield of 5.5 g lactic acid/L (Rivera et al., 2007). Lactic acid is also synthesized to benefit from its antioxidant activities (Rivera et al., 2007). Herein, the grape pomace was used as a substrate for the cultivation of lactic acid-producing inocula (*Lactobacillus paracasei*, *Lactobacillus plantarum*, and *Bifidobacterium breve*) leading to the production of various phenolic compounds (Campanella et al., 2017). Due to its unique chemical properties, lactic acid is not only being exploited as a valuable precursor but also used as a natural preservative in winemaking – it is indigenously present in the grape pomace. Raw grape pomace is usually dried and applied as a probiotic in various applications as opposed to pretreating it to extract the monomeric sugars and later using it for lactic acid production (Anghel et al., 2023).

3.4.2.2 Citric Acid

Citric acid has great economic importance due to its widespread applications (cosmetics, foods, pharmaceuticals, agriculture, etc.). Different agro-food wastes, including grape pomace, produce citric acid using submerged or solid-state fermentation (Reena et al., 2022). A citric acid production study was undertaken in the literature, and it was shown that grape pomace has several essential bioactive compounds that are required for citric acid fermentation using six potent strains of *Aspergillus niger* (Başeğmez, 2021). It was observed that the wild-type AN2 strain was the highest-producing inoculum with 14.32 mg/mL of citric acid under submerged fermentation conditions (Başeğmez, 2021). A solid-state fermentation approach has also been used to produce citric acid using grape pomace as a carbon source (Dhillon et al., 2011).

In this work, 187.96 g/kg dry substrate was produced after 120 hours of fermentation (Dhillon et al., 2011). Its nutrient versatility also enables grape pomace to be supplemented with other agro-food wastes for citric fermentations. It was shown that the mixing of grape pomace and green olive processing waste gave satisfactory results (citric acid yield of 85 g/L) under submerged fermentation (Papadaki and Mantzouridou, 2019). The process was also conducted at scale-up (5 L) using the optimal setpoint conditions that were acquired at flask scale (250 mL), and the results obtained herein were comparable with the literature (Papadaki and Mantzouridou, 2019). Remarkably, the co-digestion of grape pomace and green olive processing waste resulted in 52 g/L of glucose and 58 g/L of fructose, which stimulated the production of *Aspergillus niger* (Papadaki and Mantzouridou, 2019). These results coincide with the literature as it has been shown that the co-digestion of grape pomace with other agro-food wastes provides enhanced nutritional balance and high enzymatic activities compared to mono-digestion processes (Hungría et al., 2021; Kassongo et al., 2020).

3.4.2.3 Succinic Acid

Succinic acid is a C_4 dicarboxylic acid with many industrial applications including cosmetic, food, chemical, pharmaceutical, and agricultural industries (Filippi et al., 2021). Various biomass residues have been explored as potential biobased feedstocks for succinic acid production via fermentation; these include cornstalks (Wang et al., 2011), molasses (Chan et al., 2012), cheese whey (Uysal and Hamamcı, 2021), wheat straw (Lin et al., 2012), brewery waste (Escanciano et al., 2024), sugarbeet (Alexandri et al., 2019), rice husk (Bevilaqua et al., 2015), and bread waste (Leung et al., 2012). Among these raw substrates, grape pomace is also being studied as a potential substrate owing to its rich nutritional composition, as mentioned earlier. In one study, the grape pomace was subjected to vigorous acid and enzyme hydrolysis steps to produce a hydrolysate consisting of free sugars and phenolic extracts that were used to produce succinic acid by *Actinobacillus succinogenes* under fed-batch fermentation conditions (Filippi et al., 2021). Herein, an optimal succinic acid yield of 40.2 g/L was obtained during the fed-batch cultivation conditions (Filippi et al., 2021). Furthermore, this innovative microbiorefinery-based pathway also led to the recovery of other high-value-added products (polyphenols), which have other industrial importance (Filippi et al., 2021). The overall objective of implementing the circular bioeconomy strategy is to propel the industrialization of 'greener' products, thereby contributing to the betterment of humankind and the environment. To this end, recent R&D activities are now focused on investigating the financial consequences of producing succinic acid from 2G feedstocks such as winery waste streams. For example, a techno-economic and life cycle evaluation study revealed that it was economically viable to produce succinic acid from winery wastes. In addition, biorefinery scenarios showed that the valorization of grape pomace was well-suited for the alleviation of environmental pollution (Ioannidou et al., 2022). In summary, the upcycling of winery waste streams could lead to the establishment of novel bioeconomy value chains with new industrial opportunities and effective waste management.

3.5 MICROBIAL-BASED POLYMERS

The continual use of petroleum-derived plastics has contributed to a surge in environmental issues as these are discarded or incinerated. In the last few decades, research has been geared toward finding renewable polymers that can be used as substitutes for fossil-based plastics. This has led to the development of biopolymeric compounds whereby different microbial cell factories are exploited to synthesize these biobased materials (Sekoai et al., 2023). This innovative technological route has led to the synthesis of various biopolymers, including PHA, poly(β-hydroxybutyrate) (PHB), and poly(3-hydroxybutyrate-co-3-hydroxyvalerate) (PHBV) (Biswas et al., 2022). Grape pomace is among the feedstocks that are studied in the production of microbial-based polymers. A microbial-based biorefinery concept was adopted in the use of PHA by Follonier et al. (2014). The authors used *Pseudomonas resinovorans* to achieve a maximum PHA production yield of 21.3 g/L using Solaris grape pomace as the primary feedstock (Follonier et al., 2014). This PHA production technology was further demonstrated at a semi-pilot scale (100 L) using *Pseudomonas putida* KT2400 as the potent inoculum under fed-batch cultivation conditions (Follonier et al., 2015). Other novel biotechnological approaches have been used in this research field. For example, Martinez et al. (2016) adopted a sequential biorefinery-based strategy aimed at extracting volatile fatty acids that were later used as a carbon source for PHA production using the pure culture of a *Cupriavidus necator.* Other scientific studies are tapping into nanotechnology whereby the grape pomace is used to produce succinic acid and PHB-derived nanoparticles (Andler et al., 2024). Many microbial strains such as *Azotobacter beijerinckii*, *Bacillus* sp., *Enterobacter cloacae*, and *Enterobacter cancerogenus* are also used in the production of polymers (Sekoai et al., 2022).

3.6 SINGLE-CELL PROTEIN

Rapid population growth coupled with climate disparities has contributed to a rise in global food shortages. The high reliance on animal- and plant-based proteins has its drawbacks as these nutrient sources require huge hectares of land, use a lot of water, contribute to a carbon footprint, and are expensive (Sekoai et al., 2024). To respond to these bottlenecks, scientists are now leveraging the use of microorganisms to produce protein-rich biomass in the form of SCP which has been shown to have remarkable nutritional properties. Moreover, these nutritional sources have a protein content (30%–70%) that is comparable to other sources of protein, use little land and water, have a short production time, and have a low-fat content (Sekoai et al., 2024). Despite its merits, this biotechnology is plagued by the high operational costs that are incurred during the upstream process as it mainly relies on the use of commercial sugars such as glucose and fructose. In recent years, various attempts have been made to lower the costs by using lignocellulosic-rich biomass, but these technological pursuits have not led to any tangible outcomes/contributions in this scientific domain as costly and energy-intensive biomass fractionation steps are applied to extract the monomeric sugars before the fermentation process targeting the protein-rich biomass (Kumar et al., 2024). With the advent of the circular bioeconomy era, scientific researchers are continually exploring other desirable organic waste streams that could serve as sustainable feedstocks for the production of SCP. Ideally, the biowaste should be highly accessible to cater to pilot-scale studies, should be rich in carbohydrate content to ensure that there is a high availability of sugars for microbial growth, and should require inexpensive pretreatment regimes. Winery waste streams fit well within these criteria as they are highly abundant and rich in numerous fermentable sugars that are desirable for the upstream process. Studies are being undertaken to evaluate the feasibility of producing SCP from winery wastes through microbial biorefinery-based strategies (Chowdhary et al., 2021). Nonetheless, these scientific studies are preliminary in nature, but it is hoped that these R&D findings will provide new perspectives on SCP technologies. Therefore, it is expected that these microbial biorefinery-based activities will provide new opportunities in 'waste-to-value' innovations while accelerating global protein production and tackling the food crisis.

3.7 RECOMMENDATIONS FOR FUTURE STUDIES

The valorization of grape pomace via microbial-based biorefinery strategies is primarily conducted at lower technology readiness levels (TRLs), meaning that these R&D endeavors are preliminary in nature and are not truly representative of the proposed scientific processes. For this reason, more microbial-based biorefineries targeting the abovementioned high-value-added products from grape pomace should be undertaken at pilot-scale studies, as this will pave the way for the commercialization of these microbial-driven technologies. Researchers can also take advantage of advanced molecular biology tools to engineer potent microbial strains to enable the optimization and scalability of these bioprocesses, while inhibiting unwanted metabolic pathways. Other key strategies that must not be overlooked include the use of optimal bioreactor designs, selection of the suitable pretreatment regimes, the use of immobilized inocula, the use of co-cultures, and effective downstream processes. Furthermore, scientists must conduct techno-economic studies to examine the technical and economic implications of pursuing the proposed microbial-based strategies. It is also imperative for scientists to establish a strong collaborative framework to overcome the prevailing technological bottlenecks in this domain. Integrated biorefinery-based strategies involving expertise in chemistry, chemical engineering, and molecular biology must be adopted to unravel other value chains that could be established from winery waste streams. Knowledge in chemistry will be instrumental in the identification of robust analytical tools that can be used for the characterization of the various bioactive compounds present in the grape pomace except for the known carbohydrates that are already used in microbial-based biorefineries. Expertise in chemical engineering can be used to tackle technological barriers while identifying scientific processes with realistic economic outcomes.

3.8 CONCLUSION

Grape pomace has a lot of untapped socioeconomic opportunities that could be explored through the implementation of circular bioeconomy strategies, as evidenced in this chapter. Microbial-based biorefineries will therefore play a pivotal role in the attainment and/or advancement of circular biobased endeavors as these microbial-driven processes can produce many high-value-added products such as biofuels, biochemicals, SCPs, and biopolymers, using grape pomace as the carbon source. Furthermore, circular biobased opportunities will allow agro-food industries including wine producers to substitute the traditional waste management practices with greener biobased technologies in order to alleviate environmental pollution and the high costs of treating these waste streams. This discourse is reiterated in the United Nations' Sustainable Development Goals and also coincides with our South African 'Greener Economy' initiatives.

REFERENCES

Abreu, T., Jasmins, G., Bettencourt, C., Teixeira, J., Câmara, J.S., & Perestrelo, R. (2023). Tracing the volatilomic fingerprint of grape pomace as a powerful approach for its valorization. *Current Research in Food Science*, *7, 100608*. https://doi.org/10.1016/j.crfs.2023.100608

Alexandri, M., Schneider, R., Papapostolou, H., Ladakis, D., Koutinas, A., & Venus, J. (2019). Restructuring the conventional sugar beet industry into a novel biorefinery: Fractionation and bioconversion of sugar beet pulp into succinic acid and value-added coproducts. *ACS Sustainable Chemistry & Engineering*, *7*(7), 6569–6579. https://doi.org/10.1021/acssuschemeng.8b04874

Allison, B.J., & Simmons, C.W. (2018). Obtaining multiple coproducts from red grape pomace via anthocyanin extraction and biogas production. *Journal of Agricultural and Food Chemistry*, *66*(30), 8045–8053. https://doi.org/10.1021/acs.jafc.8b02250

Andler, R., González-Arancibia, F., Vilos, C., Sepulveda-Verdugo, R., Castro, R., Mamani, M., Valdés, C., Arto-Paz, F., Díaz-Barrera, A., & Martínez, I. (2024). Production of poly-3-hydroxybutyrate (PHB) nanoparticles using grape residues as the sole carbon source. *International Journal of Biological Macromolecules*, *261,129649*. https://doi.org/10.1016/j.ijbiomac.2024.129649

Anghel, L., Milea, A. Ştefania, Constantin, O.E., Barbu, V., Chiţescu, C., Enachi, E., Râpeanu, G., Mocanu, G.-D., & Stănciuc, N. (2023). Dried grape pomace with lactic acid bacteria as a potential source for probiotic and antidiabetic value-added powders. *Food Chemistry: X*, *19, 100777*. https://doi.org/10.1016/j.fochx.2023.100777

Başeğmez, H.İ.O. (2021). Effect of waste grape pomace on citric acid bioproduction by submerged fermentation. *Emirates Journal of Food and Agriculture*, *33*(12), 992–999. https://doi.org/10.9755/ejfa.2021.v33.i12.2793

Bassani, A., Alberici, N., Fiorentini, C., Giuberti, G., Dordoni, R., & Spigno, G. (2020). Hydrothermal treatment of grape skins for sugars, antioxidants and soluble fibers production. *Chemical Engineering Transactions*, *79*, 127–132. https://doi.org/10.3303/CET2079022

Beres, C., Costa, G.N.S., Cabezudo, I., da Silva-James, N.K., Teles, A.S.C., Cruz, A.P.G., Mellinger-Silva, C., Tonon, R.V., Cabral, L.M.C., & Freitas, S.P. (2017). Towards integral utilization of grape pomace from winemaking process: A review. *Waste Management*, *68*, 581–594. https://doi.org/10.1016/j.wasman.2017.07.017

Bevilaqua, D.B., Montipó, S., Pedroso, G.B., & Martins, A.F. (2015). Sustainable succinic acid production from rice husks. *Sustainable Chemistry and Pharmacy*, *1*, 9–13. https://doi.org/10.1016/j.scp.2015.09.001

Biswas, M.C., Jony, B., Nandy, P.K., Chowdhury, R.A., Halder, S., Kumar, D., Ramakrishna, S., Hassan, M., Ahsan, M.A., Hoque, M.E., & Imam, M.A. (2022). Recent advancement of biopolymers and their potential biomedical applications. *Journal of Polymers and the Environment*, *30*(1), 51–74. https://doi.org/10.1007/s10924-021-02199-y

Bruno Romanini, E., Misturini Rodrigues, L., Finger, A., Perez Cantuaria Chierrito, T., Regina da Silva Scapim, M., & Scaramal Madrona, G. (2021). Ultrasound assisted extraction of bioactive compounds from BRS Violet grape pomace followed by alginate-Ca^{2+} encapsulation. *Food Chemistry*, *338, 128101*. https://doi.org/10.1016/j.foodchem.2020.128101

Bruwer, J. (2003). South African wine routes: Some perspectives on the wine tourism industry's structural dimensions and wine tourism product. *Tourism Management*, *24*(4), 423–435. https://doi.org/10.1016/S0261-5177(02)00105-X

Campanella, D., Rizzello, C.G., Fasciano, C., Gambacorta, G., Pinto, D., Marzani, B., Scarano, N., De Angelis, M., & Gobbetti, M. (2017). Exploitation of grape marc as functional substrate for lactic acid bacteria and bifidobacteria growth and enhanced antioxidant activity. *Food Microbiology*, *65*, 25–35. https://doi.org/10.1016/j.fm.2017.01.019

Castro, L.E.N., Barroso, T.L.C.T., Sganzerla, W.G., Costa, J.M., Saia, F.T., Colpini, L.M.S., & Forster-Carneiro, T. (2024). Subcritical water hydrolysis of grape pomace as a sustainable pretreatment for anaerobic digestion in a biorefinery concept. *Fuel*, *363*, 130899. https://doi.org/10.1016/j.fuel.2024.130899

Chakka, A.K., & Babu, A.S. (2022). Bioactive compounds of winery by-products: Extraction techniques and their potential health benefits. *Applied Food Research*, *2*(1), 100058. https://doi.org/10.1016/j.afres.2022.100058

Chamorro, S., Viveros, A., Alvarez, I., Vega, E., & Brenes, A. (2012). Changes in polyphenol and polysaccharide content of grape seed extract and grape pomace after enzymatic treatment. *Food Chemistry*, *133*(2), 308–314. https://doi.org/10.1016/j.foodchem.2012.01.031

Chan, S., Kanchanatawee, S., & Jantama, K. (2012). Production of succinic acid from sucrose and sugarcane molasses by metabolically engineered Escherichia coli. *Bioresource Technology*, *103*(1), 329–336. https://doi.org/10.1016/j.biortech.2011.09.096

Chowdhary, P., Gupta, A., Gnansounou, E., Pandey, A., & Chaturvedi, P. (2021). Current trends and possibilities for exploitation of grape pomace as a potential source for value addition. *Environmental Pollution*, *278*, 116796. https://doi.org/10.1016/j.envpol.2021.116796

Constantin, O.E., Stoica, F., Raţu, R.N., Stănciuc, N., Bahrim, G.E., & Râpeanu, G. (2024). Bioactive components, applications, extractions, and health benefits of winery by-products from a circular bioeconomy perspective: A review. *Antioxidants*, *13*(1), 100. https://doi.org/10.3390/antiox13010100

Corbin, K.R., Hsieh, Y.S.Y., Betts, N.S., Byrt, C.S., Henderson, M., Stork, J., DeBolt, S., Fincher, G.B., & Burton, R.A. (2015). Grape marc as a source of carbohydrates for bioethanol: Chemical composition, pre-treatment and saccharification. *Bioresource Technology*, *193*, 76–83. https://doi.org/10.1016/j.biortech.2015.06.030

Cui, W., Wang, Y., Sun, Z., Cui, C., Li, H., Luo, K., & Cheng, A. (2023). Effects of steam explosion on phenolic compounds and dietary fiber of grape pomace. *LWT*, *173,114350.* https://doi.org/10.1016/j.lwt.2022.114350

Curiel-Fernández, M., Bueno-Herrera, M., Guadalupe, Z., Ayestarán, B., & Pérez-Magariño, S. (2023). Chemical characterization of polysaccharide extracts obtained from pomace by-products of different white grape varieties. *Molecules*, *28*(19), 6770. https://doi.org/10.3390/molecules28196770

Ćurko, N., Lukić, K., Tušek, A.J., Balbino, S., Vukušić Pavičić, T., Tomašević, M., Redovniković, I.R., & Ganić, K.K. (2023). Effect of cold pressing and supercritical CO_2 extraction assisted with pulsed electric fields pretreatment on grape seed oil yield, composition and antioxidant characteristics. *LWT*, *184*, 114974. https://doi.org/10.1016/j.lwt.2023.114974

da Rocha, C.B., & Noreña, C.P.Z. (2021). Microencapsulation and controlled release of bioactive compounds from grape pomace. *Drying Technology*, *39*(8), 1018–1032. https://doi.org/10.1080/07373937.2020.1741004

Dhillon, G.S., Brar, S.K., Verma, M., & Tyagi, R.D. (2011). Enhanced solid-state citric acid bio-production using apple pomace waste through surface response methodology. *Journal of Applied Microbiology*, *110*(4), 1045–1055. https://doi.org/10.1111/j.1365-2672.2011.04962.x

Dinuccio, E., Balsari, P., Gioelli, F., & Menardo, S. (2010). Evaluation of the biogas productivity potential of some Italian agro-industrial biomasses. *Bioresource Technology*, *101*(10), 3780–3783. https://doi.org/10.1016/j.biortech.2009.12.113

Dong, X., Zhu, C.P., Huang, G.Q., & Xiao, J.X. (2021). Fractionation and structural characterization of polysaccharides derived from red grape pomace. *Process Biochemistry*, *109*, 37–45. https://doi.org/10.1016/j.procbio.2021.06.022

dos Santos Freitas, L., Jacques, R.A., Richter, M.F., Silva, A.L.d., & Caramão, E.B. (2008). Pressurized liquid extraction of vitamin E from Brazilian grape seed oil. *Journal of Chromatography A*, *1200*(1), 80–83. https://doi.org/10.1016/j.chroma.2008.02.067

Egüés, I., Serrano, L., Amendola, D., De Faveri, D.M., Spigno, G., & Labidi, J. (2013). Fermentable sugars recovery from grape stalks for bioethanol production. *Renewable Energy*, *60*, 553–558. https://doi.org/10.1016/j.renene.2013.06.006

El Achkar, J.H., Lendormi, T., Hobaika, Z., Salameh, D., Louka, N., Maroun, R.G., & Lanoisellé, J.L. (2016). Anaerobic digestion of grape pomace: Biochemical characterization of the fractions and methane production in batch and continuous digesters. *Waste Management*, *50*, 275–282. https://doi.org/10.1016/j.wasman.2016.02.028

Escanciano, I.A., Ladero, M., Blanco, Á., & Santos, V.E. (2024). Succinic acid production by Actinobacillus succinogenes using acid and enzymatic hydrolysates of potato and beer wastes and repeated batch operation. *Biomass and Bioenergy*, *181*, 107034. https://doi.org/10.1016/j.biombioe.2023.107034

Failla, S., & Restuccia, A. (2014). Methane potentials from grape marc by a laboratory scale plant. *Applied Mathematical Sciences*, *8*(129–132), 6665–6678. https://doi.org/10.12988/ams.2014.474551

Favre, G., Hermosín-Gutiérrez, I., Piccardo, D., Gómez-Alonso, S., & González-Neves, G. (2019). Selectivity of pigments extraction from grapes and their partial retention in the pomace during red-winemaking. *Food Chemistry*, *277*, 391–397. https://doi.org/10.1016/j.foodchem.2018.10.085

Ferreira-Santos, P., Nobre, C., Rodrigues, R.M., Genisheva, Z., Botelho, C., & Teixeira, J.A. (2024). Extraction of phenolic compounds from grape pomace using ohmic heating: Chemical composition, bioactivity and bioaccessibility. *Food Chemistry*, *436, 137780.* https://doi.org/10.1016/j.foodchem.2023.137780

Ferreira, S.L.A., & Hunter, C.A. (2017). Wine tourism development in South Africa: A geographical analysis. *Tourism Geographies*, *19*(5), 676–698. https://doi.org/10.1080/14616688.2017.1298152

Filippi, K., Georgaka, N., Alexandri, M., Papapostolou, H., & Koutinas, A. (2021). Valorisation of grape stalks and pomace for the production of bio-based succinic acid by Actinobacillus succinogenes. *Industrial Crops and Products*, *168*, 113578. https://doi.org/10.1016/j.indcrop.2021.113578

Follonier, S., Goyder, M.S., Silvestri, A.C., Crelier, S., Kalman, F., Riesen, R., & Zinn, M. (2014). Fruit pomace and waste frying oil as sustainable resources for the bioproduction of medium-chain-length polyhydroxyalkanoates. *International Journal of Biological Macromolecules*, *71*, 42–52. https://doi.org/10.1016/j.ijbiomac.2014.05.061

Follonier, S., Riesen, R., & Zinn, M. (2015). Pilot-scale production of functionalized mcl-PHA from grape pomace supplemented with fatty acids. *Chemical and Biochemical Engineering Quarterly*, *29*(2), 113–121. https://doi.org/10.15255/CABEQ.2014.2251

François, E., Dumas, C., Gougeon, R.D., Alexandre, H., Vuilleumier, S., & Ernst, B. (2021). Unexpected high production of biohydrogen from the endogenous fermentation of grape must deposits. *Bioresource Technology*, *320*, 124334. https://doi.org/10.1016/j.biortech.2020.124334

Gaogane, G.J., Sekoai, P., & Trois, C. (2025). Sustainable hydrogen production via dark fermentation of protein- and lipid-rich municipal organic waste: Digestate reuse and ammonia mitigation strategies. *Fermentation*, *11*(11), 623. https://doi.org/10.3390/fermentation11110623

Gitea, M.A., Bungau, S.G., Gitea, D., Pasca, B.M., Purza, A.L., & Radu, A.F. (2023). Evaluation of the phytochemistry–therapeutic activity relationship for grape seeds oil. *Life*, *13*(1), 178. https://doi.org/10.3390/life13010178

González-Centeno, M.R., Jourdes, M., Femenia, A., Simal, S., Rosselló, C., & Teissedre, P.L. (2013). Characterization of polyphenols and antioxidant potential of white grape pomace byproducts (Vitis vinifera L.). *Journal of Agricultural and Food Chemistry*, *61*(47), 11579–11587. https://doi.org/10.1021/jf403168k

González, M., Barrios, S., Budelli, E., Pérez, N., Lema, P., & Heinzen, H. (2020). Ultrasound assisted extraction of bioactive compounds in fresh and freeze-dried Vitis vinifera cv Tannat grape pomace. *Food and Bioproducts Processing*, *124*, 378–386. https://doi.org/10.1016/j.fbp.2020.09.012

Hungría, J., Siles, J.A., Chica, A.F., Gil, A., & Martín, M.A. (2021). Anaerobic co-digestion of winery waste: Comparative assessment of grape marc waste and lees derived from organic crops. *Environmental Technology*, *42*(23), 3618–3626. https://doi.org/10.1080/09593330.2020.1737735

Ioannidou, S.M., Filippi, K., Kookos, I.K., Koutinas, A., & Ladakis, D. (2022). Techno-economic evaluation and life cycle assessment of a biorefinery using winery waste streams for the production of succinic acid and value-added co-products. *Bioresource Technology*, *348*, 126295. https://doi.org/10.1016/j.biortech.2021.126295

Kassongo, J., Shahsavari, E., & Ball, A.S. (2020). Co-digestion of grape marc and cheese whey at high total solids holds potential for sustained bioenergy generation. *Molecules*, *25*(23), 5754. https://doi.org/10.3390/molecules25235754

Kokkinomagoulos, E., & Kandylis, P. (2023). Grape pomace, an undervalued by-product: industrial reutilization within a circular economy vision. *Reviews in Environmental Science and Biotechnology*, 22(3), 739–773. https://doi.org/10.1007/s11157-023-09665-0

Kumar, R., Raj, T., Næss, G., Sørensen, M., & Dhawan, V. (2024). Opportunities and challenges in single-cell protein production using lignocellulosic material. *Biofuels, Bioproducts and Biorefining*, *18*(1), 310–321. https://doi.org/10.1002/bbb.2563

Leung, C.C.J., Cheung, A.S.Y., Zhang, A.Y.Z., Lam, K.F., & Lin, C.S.K. (2012). Utilisation of waste bread for fermentative succinic acid production. *Biochemical Engineering Journal*, *65*, 10–15. https://doi.org/10.1016/j.bej.2012.03.010

Li, C.F., Wang, X.C., Chu, Z.Y., & Peng, P. (2023). A low-cost and high precision determination method of 87Sr/86Sr ratios for red wine using thermal ionization mass spectrometry without column separation. *Journal of Analytical Atomic Spectrometry*, *39*(1), 235–243. https://doi.org/10.1039/d3ja00322a

Li, X., & Petrov, M.S. (2024). Dietary fibre for the prevention of post-pancreatitis diabetes mellitus: A review of the literature and future research directions. *Nutrients*, *16*(3), 435. https://doi.org/10.3390/nu16030435

Lin, C.S.K., Luque, R., Clark, J.H., Webb, C., & Du, C. (2012). Wheat-based biorefining strategy for fermentative production and chemical transformations of succinic acid. *Biofuels, Bioproducts and Biorefining*, *6*(1), 88–104. https://doi.org/10.1002/bbb.328

Marianne, L.C., Lucía, A.G., de Jesús, M.S.M., Eric Leonardo, H.M., & Mendoza-Sánchez, M. (2024). Optimization of the green extraction process of antioxidants derived from grape pomace. *Sustainable Chemistry and Pharmacy*, *37*, 101396. https://doi.org/10.1016/j.scp.2023.101396

Marks, J., Kirkel, J., Sekoai, P., Enweremadu, C., & Daramola, M. (2020). Effect of combining different substrates and inoculum sources on bioelectricity generation and COD removal in a two-chambered microbial fuel cell: A preliminary investigation. *Environmental and Climate Technologies*, *24*(2), 67–78. https://doi.org/10.2478/rtuect-2020-0055

Martinez, G.A., Rebecchi, S., Decorti, D., Domingos, J.M.B., Natolino, A., Del Rio, D., Bertin, L., Da Porto, C., & Fava, F. (2016). Towards multi-purpose biorefinery platforms for the valorisation of red grape pomace: production of polyphenols, volatile fatty acids, polyhydroxyalkanoates and biogas. *Green Chemistry*, *18*(1), 261–270. https://doi.org/10.1039/c5gc01558h

Matthäus, B. (2008). Virgin grape seed oil: Is it really a nutritional highlight? *European Journal of Lipid Science and Technology*, *110*(7), 645–650. https://doi.org/10.1002/ejlt.200700276

Messina, C.M., Manuguerra, S., Catalano, G., Arena, R., Cocchi, M., Morghese, M., Montenegro, L., & Santulli, A. (2021). Green biotechnology for valorisation of residual biomasses in nutraceutic sector: Characterization and extraction of bioactive compounds from grape pomace and evaluation of the protective effects in vitro. *Natural Product Research*, *35*(2), 331–336. https://doi.org/10.1080/14786419.2019.1619727

Moore, J.P., Divol, B., Young, P.R., Nieuwoudt, H.H., Ramburan, V., du Toit, M., Bauer, F.F., & Vivier, M.A. (2008). Wine biotechnology in South Africa: Towards a systems approach to wine science. *Biotechnology Journal*, *3*(11), 1355–1367. https://doi.org/10.1002/biot.200800145

Nagarajaiah, S.B., & Prakash, J. (2016). Chemical composition and bioactivity of pomace from selected fruits. *International Journal of Fruit Science*, *16*(4), 423–443. https://doi.org/10.1080/15538362.2016.1143433

Natolino, A., & Da Porto, C. (2020). Kinetic models for conventional and ultrasound assistant extraction of polyphenols from defatted fresh and distilled grape marc and its main components skins and seeds. *Chemical Engineering Research and Design*, *156*, 1–12. https://doi.org/10.1016/j.cherd.2020.01.009

Nayak, A., Bhushan, B., Rosales, A., Turienzo, L.R., & Cortina, J.L. (2018). Valorisation potential of cabernet grape pomace for the recovery of polyphenols: Process intensification, optimisation and study of kinetics. *Food and Bioproducts Processing*, *109*, 74–85. https://doi.org/10.1016/j.fbp.2018.03.004

Nemzer, B., Kalita, D., Yashin, A.Y., & Yashin, Y.I. (2022). Chemical composition and polyphenolic compounds of red wines: Their antioxidant activities and effects on human health—A review. *Beverages*, *8*(1), 1. https://doi.org/10.3390/beverages8010001

Nhleko, L., & Sekoai, P.T. (2025). Sugarcane bagasse: A sustainable feedstock for biorefinery portfolios in South Africa. *Fermentation*, *11*(9), 489. https://doi.org/10.3390/fermentation11090489

Oliveira, D.A., Salvador, A.A., Smânia, A., Smânia, E.F.A., Maraschin, M., & Ferreira, S.R.S. (2013). Antimicrobial activity and composition profile of grape (Vitis vinifera) pomace extracts obtained by supercritical fluids. *Journal of Biotechnology*, *164*(3), 423–432. https://doi.org/10.1016/j.jbiotec.2012.09.014

Orlando, G., Almanza-Oliveros, A., Bautista-Hernández, I., Castro-López, C., Aguilar-Zárate, P., Meza-Carranco, Z., Rojas, R., Michel, M.R., & Cristian Martínez-Ávila, G.G. (2024). Grape pomace-advances in its bioactivity, health benefits, and food applications. *Foods*, *13*(4), 580. https://doi.org/10.3390/foods13040580

Papadaki, E., & Mantzouridou, F.T. (2019). Citric acid production from the integration of Spanish-style green olive processing wastewaters with white grape pomace by Aspergillus niger. *Bioresource Technology*, *280*, 59–69. https://doi.org/10.1016/j.biortech.2019.01.139

Pascual-Martí, M.C., Salvador, A., Chafer, A., & Berna, A. (2001). Supercritical fluid extraction of resveratrol from grape skin of Vitis 6inifera and determination by HPLC. *Talanta*, *5*, 735–740. https://doi.org/10.1016/S0039-9140(01)00319-8

Pintać, D., Majkić, T., Torović, L., Orčić, D., Beara, I., Simin, N., Mimica–Dukić, N., & Lesjak, M. (2018). Solvent selection for efficient extraction of bioactive compounds from grape pomace. *Industrial Crops and Products*, *111*, 379–390. https://doi.org/10.1016/j.indcrop.2017.10.038

Reena, R., Sindhu, R., Athiyaman Balakumaran, P., Pandey, A., Awasthi, M.K., & Binod, P. (2022). Insight into citric acid: A versatile organic acid. *Fuel*, *327*, 125181. https://doi.org/10.1016/j.fuel.2022.125181

Renaudie, M., Dumas, C., Vuilleumier, S., & Ernst, B. (2024). Spontaneous dark fermentation in a pre-seeded liquid-gas membrane bioreactor: Impact of wine and coffee biowaste microflora on continuous biohydrogen production. *Journal of Cleaner Production*, *437*, 140759. https://doi.org/10.1016/j.jclepro.2024.140759

Rivera, O.M.P., Moldes, A.B., Torrado, A.M., & Domínguez, J.M. (2007). Lactic acid and biosurfactants production from hydrolyzed distilled grape marc. *Process Biochemistry*, *42*(6), 1010–1020. https://doi.org/10.1016/j.procbio.2007.03.011

Rodríguez-Blázquez, S., Pedrera-Cajas, L., Gómez-Mejía, E., Vicente-Zurdo, D., Rosales-Conrado, N., León-González, M.E., Rodríguez-Bencomo, J.J., & Miranda, R. (2024). The potential of plum seed residue: Unraveling the effect of processing on phytochemical composition and bioactive properties. *International Journal of Molecular Sciences*, *25*(2), 1236. https://doi.org/10.3390/ijms25021236

Sabir, A., Unver, A., & Kara, Z. (2012). The fatty acid and tocopherol constituents of the seed oil extracted from 21 grape varieties (Vitis spp.). *Journal of the Science of Food and Agriculture*, *92*(9), 1982–1987. https://doi.org/10.1002/jsfa.5571

Sekoai, P., Ezeokoli, O., Yoro, K., Eterigho-Ikelegbe, O., Habimana, O., Iwarere, S., Daramola, M., & Ojumu, T. (2022). The production of polyhydroxyalkanoates using volatile fatty acids derived from the acidogenic biohydrogen effluents: An overview. *Bioresource Technology Reports*, *18*, 101111. https://doi.org/10.1016/j.biteb.2022.101111

Sekoai, P.T., Awosusi, A.A., Yoro, K.O., Singo, M., Oloye, O., Ayeni, A.O., Bodunrin, M., & Daramola, M.O. (2018). Microbial cell immobilization in biohydrogen production: a short overview. *Critical Reviews in Biotechnology*, *38*(2), 157–171. https://doi.org/10.1080/07388551.2017.1312274

Sekoai, P.T., Chunilall, V., Msele, K., Buthelezi, L., Johakimu, J., Andrew, J., Zungu, M., Moloantoa, K., Maningi, N., Habimana, O., & Swartbooi, A. (2023). Biowaste biorefineries in South Africa: Current status, opportunities, and research and development needs. *Renewable and Sustainable Energy Reviews*, *188*, 113870. https://doi.org/10.1016/j.rser.2023.113870

Sekoai, P.T., Ghimire, A., Ezeokoli, O.T., Rao, S., Ngan, W.Y., Habimana, O., Yao, Y., Yang, P., Yiu Fung, A.H., Yoro, K.O., Daramola, M.O., & Hung, C.H. (2021). Valorization of volatile fatty acids from the dark fermentation waste streams-A promising pathway for a biorefinery concept. *Renewable and Sustainable Energy Reviews*, *143*, 110971. https://doi.org/10.1016/j.rser.2021.110971

Sekoai, P.T., Roets-Dlamini, Y., O'Brien, F., Ramchuran, S., & Chunilall, V. (2024). Valorization of food waste into single-cell protein: An innovative technological strategy for sustainable protein production. *Microorganisms*, *12*(1), 166. https://doi.org/10.3390/microorganisms12010166

Sekoai, P.T., & Yoro, K.O. (2016). Biofuel development initiatives in Sub-Saharan Africa: Opportunities and challenges. *Climate*, *4*(2), 33. https://doi.org/10.3390/cli4020033

Shen, Y., Kang, B., Lu, Y., Du, X., Qin, C., Li, J., Zhao, Z., Yu, R., Shi, S., & Han, L. (2023). Production of optical pure L-lactic acid from Cabernet Sauvignon grape pomace by engineered Lactiplantibacillus plantarum. *Frontiers in Energy Research*, *11*, 1228827. https://doi.org/10.3389/fenrg.2023.1228827

Teles, A.S.C., Chávez, D.W.H., Oliveira, R.A., Bon, E.P.S., Terzi, S.C., Souza, E.F., Gottschalk, L.M.F., & Tonon, R.V. (2019). Use of grape pomace for the production of hydrolytic enzymes by solid-state fermentation and recovery of its bioactive compounds. *Food Research International*, *120*, 441–448. https://doi.org/10.1016/j.foodres.2018.10.083

Uysal, U., & Hamamcı, H. (2021). Succinic acid production from cheese whey via fermentation by using alginate immobilized Actinobacillus succinogenes. *Bioresource Technology Reports*, *16*, 100829. https://doi.org/10.1016/j.biteb.2021.100829

Venkata Mohan, S., Nikhil, G.N., Chiranjeevi, P., Nagendranatha Reddy, C., Rohit, M.V., Kumar, A.N., & Sarkar, O. (2016). Waste biorefinery models towards sustainable circular bioeconomy: Critical review and future perspectives. *Bioresource Technology*, *215*, 2–12. https://doi.org/10.1016/j.biortech.2016.03.130

Wang, D., Li, Q., Yang, M., Zhang, Y., Su, Z., & Xing, J. (2011). Efficient production of succinic acid from corn stalk hydrolysates by a recombinant Escherichia coli with ptsG mutation. *Process Biochemistry*, *46*(1), 365–371. https://doi.org/10.1016/j.procbio.2010.09.012

Wang, X., Tong, H., Chen, F., & Gangemi, J.D. (2010). Chemical characterization and antioxidant evaluation of muscadine grape pomace extract. *Food Chemistry*, *123*(4), 1156–1162. https://doi.org/10.1016/j.foodchem.2010.05.080

Wen, X., Zhu, M., Hu, R., Zhao, J., Chen, Z., Li, J., & Ni, Y. (2016). Characterisation of seed oils from different grape cultivars grown in China. *Journal of Food Science and Technology*, *53*(7), 3129–3136. https://doi.org/10.1007/s13197-016-2286-9

Xiao, Y., Sun, L., Loke Show, P., Hu, H., Han, Y., & Song, C. (2022). Simultaneous extraction and preliminary purification of polyphenols from grape pomace using an aqueous two-phase system exposed to ultrasound irradiation: Process characterization and simulation. *Frontiers in Nutrition*, *9*, 993475. https://doi.org/https://doi.org/10.3389/fnut.2022.993475

Yang, C., Shang, K., Lin, C., Wang, C., Shi, X., Wang, H., & Li, H. (2021). Processing technologies, phytochemical constituents, and biological activities of grape seed oil (GSO): A review. *Trends in Food Science and Technology*, *116*, 1074–1083. https://doi.org/10.1016/j.tifs.2021.09.011

Zhu, F., Du, B., Zheng, L., & Li, J. (2015). Advance on the bioactivity and potential applications of dietary fibre from grape pomace. *Food Chemistry*, *186*, 207–212. https://doi.org/10.1016/j.foodchem.2014.07.057

4 *Bacillus* Species Bioprocessing for Circular Bioeconomy

Exploring Fats, Oils, and Grease Waste Valorization

Thando Ndlovu, Adebowale Aladejana, Ayodeji Adedoyin, Lebang Kgosietsile, Pako Motshwarakgole, Tumelo Bosena, Pako Manyala, Tidimalo Coetzee, and Sizwe Ndlovu

4.1 INTRODUCTION

Sustainable consumption and production aim to balance economic growth with environmental responsibility. By using resources efficiently, we can improve living standards without harming the planet. A key strategy towards achieving this is the circular bioeconomy (CBE), which transforms renewable biological materials into valuable products such as food, biofuels, and other goods (European Commission, 2012). This approach emphasises a strong quality of life for people while minimising economic, environmental, and societal costs. However, as the bioeconomy expands, environmental pressures and land-use challenges become a concern. Efficiently designed processes that ensure optimal resource utilisation and minimal environmental output are essential for a sustainable bioeconomy. While research on biomass valorisation is growing, the connection between organic waste management and the circular economy needs further exploration. The CBE approach emphasises the need for a more holistic and integrated focus on the synergies between the bioeconomy and the circular economy to achieve sustainable development (Gaogane et al., 2025).

Cooking oils and fats, both plant- and animal-based, are the primary contributors to FOG accumulation in wastewater systems (Chauhan et al., 2025). They pose a significant challenge to urban infrastructure, causing severe damage to sewer systems. This is costly and disruptive for municipalities, businesses, and the environment. The problem is exacerbated by the increase in urbanisation, changes in dietary habits, and the growth of the food service industry. While eliminating FOG from cooking is impractical, its

 DOI: 10.1201/9781003745150-4

disposal in sewers is unacceptable and requires action to rethink a sustainable way of managing this waste (Mosholi and Cloete, 2018). A prevalent 'out of sight, out of mind' attitude towards waste disposal exacerbates the issue. The frying process, combined with metal ions from food and cleaning products, accelerates FOG formation through chemical reactions such as aggregation, crystallisation, and saponification (Chauhan et al., 2025). FOG, when improperly disposed of in sanitary sewers, can accumulate with other organic matter, forming deposits that clog pipes. The growing number of food service establishments (FSEs) and rising consumer demand for fatty foods have intensified this problem. These deposits reduce sewer capacity, leading to blockages and overflows that pose significant health and environmental risks.

Bacillus species have emerged as promising workhorses for transforming FOG waste into valuable products (Tzirita et al., 2018; Roets-Dlamini et al., 2022). These bacteria excel at breaking down a wide range of compounds through the production of extracellular enzymes (Danilova and Sharipova, 2020). These enzymes initiate the degradation process by converting complex molecules into simpler ones, which the bacteria can then consume. The rich nutrient content of food waste, including FOG, makes it an ideal substrate for microbial processing (Andler and Goddard, 2018). In the case of FOG, bacterial enzymes convert fats into fatty acids, serving as a nutrient source (Tzirita et al., 2018). Additionally, many *Bacillus* species produce biosurfactants that could enhance the breakdown of FOG by increasing substrate accessibility (Amodu et al., 2014; Roets-Dlamini et al., 2022).

By leveraging the versatile capabilities of *Bacillus* species, researchers and industries can explore innovative strategies for the valorisation of FOG waste, transforming a challenging waste stream into a valuable resource. The integration of *Bacillus*-based bioprocessing approaches into waste management systems can lead to the development of more sustainable and circular solutions for the utilisation of FOG waste. This chapter explores the potential of *Bacillus* species to transform FOG waste into valuable products. By examining the production of bioactive compounds by these bacteria, we aim to identify opportunities for the discovery and development of antimicrobial agents, bioremediation agents, and technologies for recovering hydrocarbons and essential elements from the environment.

4.2 COMPOSITION OF FOG

Fats, oils, and greases (FOGs) are primarily composed of triglycerides, which are molecules made up of glycerol and three fatty acids. These fatty acids can be saturated, monounsaturated, or polyunsaturated. Saturated fatty acids (SFAs), such as palmitic and stearic acids, have linear and tightly packed structures, resulting in solid fats at room temperature. Monounsaturated fatty acids (MUFAs), like oleic acid, have one double bond, creating a kink that prevents tight packing and results in liquid oils. Polyunsaturated fatty acids (PUFAs), including linoleic and alpha-linolenic acids, have multiple double bonds, further disrupting the packing and influencing the properties of the oils (Gunstone, 2004). The composition of FOGs varies based on their source. Animal fats predominantly contain saturated and monounsaturated fatty acids, while plant oils are rich in unsaturated fatty acids, especially oleic and linoleic acids. Marine oils often include long-chain PUFAs like eicosapentaenoic acid (EPA) and docosahexaenoic acid (DHA). Greases exhibit a mixed composition of saturated and unsaturated fatty acids, depending on their origin (Gunstone, 2004).

4.3 PHYSIOLOGICAL AND BIOCHEMICAL PROPERTIES OF *BACILLUS*

Bacillus is a genus of rod-shaped, Gram-positive bacteria renowned for their ubiquity in diverse environments, primarily soil and water. As of 2020, approximately over 270 species have been identified (Amaresan et al., 2020). A defining characteristic of members of the genus *Bacillus* is their ability to form endospores, highly resistant structures enabling survival under extreme conditions such as desiccation, heat, and radiation (Celandroni et al., 2019). This resilience allows *Bacillus* to thrive in various niches, including those with high temperatures (55°C–60°C, e.g., *Bacillus thermophilus*) or high salinity levels (2%–5% NaCl, e.g., *Bacillus halodurans*), indicating their adaptability compared to other bacterial genera that may have limited environmental tolerances (Burgess et al., 2017; Donio et al., 2013). Members of the genus *Bacillus* can either be obligate aerobes, which require oxygen, or facultative anaerobes, which can survive in both the presence and absence of oxygen. This diverse range of metabolic processes distinguishes them from other bacterial genera. Although most *Bacillus* species are non-pathogenic, some species, such as *Bacillus anthracis* and *Bacillus cereus*, are known to cause diseases in humans and animals (Ehling-Schulz et al., 2019).

Bacillus species exhibit significant diversity based on ecological, genetic, and phenotypic features. The genus's wide distribution and environmental adaptability contribute to this significant diversity. The genus comprises over 200 named species with a wide variety of biochemical and physiological characteristics (Amaresan et al., 2020). Members of this genus have been found to exhibit extensive genetic variety even within a single species, such as *Bacillus cereus* (Mohamadou et al., 2013). Strains that have been isolated from disparate settings display unique physiological characteristics and genetic relatedness. In contrast to more genetically conserved riverbank strains, desert strains typically exhibit higher levels of genetic variation and adaptation, indicating that environmental factors are important in determining intra-species diversity (Polonca, 2020). Furthermore, *Bacillus* species can be differentiated based on their motility, size, shape, colony colour, and biochemical characteristics such as catalase production and carbon source utilisation. Their ecological versatility is demonstrated by their widespread presence in diverse environments (Dame et al., 2021). Beyond their ecological significance, *Bacillus* species hold substantial biotechnological value. Many species are employed in industrial applications, including enzyme synthesis, bioremediation, and agricultural probiotics (Balaji et al., 2020; Dame et al., 2021). Their ability to produce various enzymes also makes them valuable in waste management and food processing (Zalma and El-Sharoud, 2021).

4.4 BIODEGRADATION AND BIOTRANSFORMATION OF FOG BY *BACILLUS* SPECIES

Bacillus species are versatile microorganisms with applications ranging from pharmaceuticals, medicine, and agriculture, driven by their diverse physiology and extensive

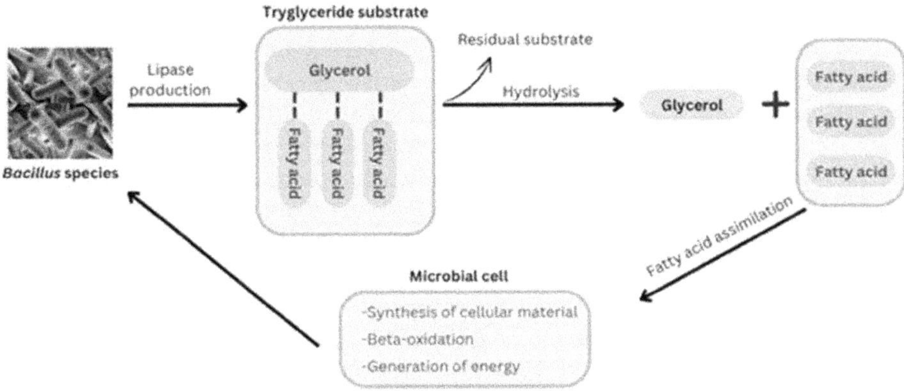

FIGURE 4.1 An illustration of *Bacillus* species role in the lipase-mediated biodegradation of triglycerides [adapted from Tzirita et al. (2019) and Tzirita et al. (2021)].

metabolite production, including antimicrobial agents and enzymes (Celandroni et al., 2019). They have shown great potential in the bioremediation of FOGs in a variety of waste streams, particularly in effluents from food processing (Tzirita et al., 2018; Roets-Dlamini et al., 2022). These bacteria metabolise a broad spectrum of organic matter, including carbohydrates, proteins, and FOGs. They effectively degrade FOG components such as butter, olive oil, and waste cooking fats through enzyme production (Tzirita et al., 2019). For example, *B. licheniformis* and *B. subtilis* have been reported to produce lipases to hydrolyse triglycerides into fatty acids and glycerol, as shown in Figure 4.1, a process influenced by factors such as temperature, pH, and nutrient availability (Gurd et al., 2020; Court et al., 2021). Furthermore, *Bacillus* species can use FOGs as sources of carbon for growth and energy production (Roets-Dlamini et al., 2022). They can produce acetyl-CoA for the citric acid cycle and energy production by metabolising the fatty acids produced from triglyceride hydrolysis through β-oxidation (Gurd et al., 2020). Additionally, most of the species under this genus are known to form biofilms that aid their adherence to and degradation of FOGs. The biofilm matrix protects them and enhances their ability to utilise FOGs as a source of carbon (Court et al., 2021).

4.4.1 *Bacillus* Strains Effective in FOG Waste Treatment

Bacillus subtilis is an extensively used bacterium because of its robust lipase production and capacity to degrade various types of fats in wastewater. In a study by Lauprasert et al. (2017), *B. subtilis* strain TISTR1248 was found to be effective (64.51%) in fat degradation, and this was attributed to its lipase production activity. Additionally, recent studies have demonstrated that *B. subtilis* can effectively lower biochemical oxygen demand (BOD) and FOG concentrations in wastewater treatment (Lauprasert et al., 2017; Roets-Dlamini et al., 2022). *Bacillus licheniformis* NCIMB9375 is also known to possess a high lipid degradation capacity (Tzirita et al., 2018). This strain utilised FOGs as a carbon source and could easily adjust to varying wastewater compositions. Moreover, *B. thermoleovorans* has been established as a thermophilic bacterium that produces enzymes at higher temperatures, making it ideal for the degradation of FOGs in warmer wastewater environments. *Bacillus cereus* variants have also been demonstrated to have good bioremediation potential for utilisation in the degradation of FOGs (Karim et al., 2019; Roets-Dlamini et al., 2022). In a study by Hassan et al. (2018), *B. cereus* HSS strain was found to have the highest lipolytic hydrolysis among 20 isolates with a removal efficiency of oil and grease at 94.7%. Likewise, *B. cereus* EM8 strain was able to biodegrade up to 40% of slop oil as reported by Marchut-Mikolajczyk et al. (2020).

Synergistic interactions between *Bacillus* strains and other microorganisms or additives have been investigated for enhanced FOG degradation. Co-culturing *Bacillus* species with *Pseudomonas putida* CP1 resulted in enhanced FOG degradation (Tzirita et al., 2018). The *Pseudomonas* strain preferentially utilises unsaturated fatty acids, while *Bacillus* species hydrolyse fats, leading to a synergistic effect that promotes rapid degradation of both saturated and unsaturated fats (Tzirita et al., 2019). Similarly, Loperena et al. (2009) found that a mixture of *Acinetobacter*, *Bacillus* strains, and *Pseudomonas putida* strain VII exhibited higher removal efficiencies for protein (93%) and fat (72%) in dairy wastewaters compared to a commercial inoculum (54% and 93%). Additives such as surfactants can improve the bioavailability of fats and oils, making them more vulnerable to microbial degradation by *Bacillus* species. These surfactants work by emulsifying fats, which enhances their accessibility for microbial action.

Lipopeptides are a type of biosurfactant that plays a crucial role in the biodegradation of FOGs by lowering surface tension and enhancing the solubility of hydrophobic compounds. This process makes FOGs more accessible for microbial degradation (Wong et al., 2025). Key lipopeptides involved in this degradation include surfactin, iturin, and fengycin. By reducing the surface tension of water, lipopeptides facilitate the spread and penetration of water into hydrophobic compounds, thereby increasing the surface area available for enzymatic action (Abedi et al., 2026). Additionally, lipopeptides enhance FOG solubility by forming micelles around hydrophobic molecules, making them more accessible to microbial enzymes. This integration of lipopeptides into lipid membranes helps break down complex fat structures into smaller, more manageable molecules.

4.4.2 *Bacillus* Species Are Promising Bioprocessing for FOG Waste Valorisation

Bacillus sp. is among the currently known promising bioprocessing agents used in the valorisation of FOGs (Deaver and Popat, 2022). FOG are waste products from industrial effluents, agricultural wastes, food processing industries, domestic wastes, and biotechnological establishments (Wallace et al., 2017; Balaji et al., 2020). FOG wastes are difficult and challenging waste streams confronting the biotechnology, pharmaceutical, effluent and environmental organisations. The FOG wastes generated by these companies, if not adequately managed, have the potential to cause infrastructural challenges, sewage blockage, and pollution (Murtaza et al., 2024). The use of a circular economy is necessary to tackle FOG waste problems. According to Macarthur and Heading (2019), a circular economy is an economy that is restorative and regenerative by design, with the aims of keeping yields, components, and materials at their highest utility and values permanently, thereby differentiating between biological and technological cycles. The use of a circular economy is necessary as it brings about a reduction in greenhouse gas emissions, reuse and recycling alongside conservation, waste management via biological systems, and pollution reduction (Sekoai and Habimana, 2025). In addition to various organisms such as *Saccharomyces cerevisiae*, *Rhodotorula gracilis*, *Rhodotorula mucilaginosa*, *Alcaligenes faecalis*, *Yarrowia lipolytica*, *Cryptococcus curvatus*, and *Escherichia coli* used in the valorisation of FOG wastes, *Bacillus* species, including *Bacillus cereus*, *Bacillus thuringiensis*, *Bacillus subtilis*, along with *Lactobacillus plantarum*, and *Actinobacillus succinogenes*, possess the metabolic potential required to convert waste into valuable products (Lopes et al., 2020; Roets-Dlamini et al., 2022; Utlang et al., 2024). A sustainable contribution to the CBE is made by utilising lipid wastes (such as fats and oils) produced worldwide as low-cost raw materials (substrates) in different industries to produce these value-added products. The following sections provide a detailed discussion of several products, including biosurfactants, various enzymes, and biofuels.

4.4.3 Biosynthesis of Biosurfactants from FOG Substrates by *Bacillus* Species

Biosurfactants, which are naturally derived surfactants mainly produced by bacteria (Tripathi et al., 2020), have emerged as promising alternatives to synthetic surfactants (Kavyarathna et al., 2023). Recent studies have harnessed the potential of *Bacillus* species to convert FOG waste into valuable biosurfactants (see Table 4.1). These microorganisms efficiently utilise fatty acids from FOG as a carbon source, transforming them into glycerol and fatty acids through biodegradation. Subsequently, glycerol is converted into rhamnose, which combines with fatty acids to form the amphipathic structure of biosurfactants (Pardhi et al., 2022). Renowned for their ability to reduce surface tension and enhance the solubility of hydrophobic compounds, biosurfactants possess diverse applications across various industries, as shown in Figure 4.2, including environmental remediation, cosmetics, and pharmaceuticals (Sundaram et al., 2024). Unlike synthetic surfactants, which often rely on petroleum-based feedstocks and contribute to environmental concerns such as toxicity and poor biodegradability (Kavyarathna et al., 2023), biosurfactants offer a sustainable and eco-friendly solution. As a result, there is growing interest in developing innovative methods to optimise biosurfactant production from FOG waste and minimise its environmental impact, as illustrated in Figure 4.2.

Bacillus species, particularly known for producing cyclic lipopeptides such as surfactins (see Table 4.1) (Narisetty et al., 2022), demonstrate significant potential in the synthesis of biosurfactants. For example, *Bacillus* sp. HIP3 has been shown to effectively utilise cooking oil to produce surfactin, a compound useful in heavy metal removal (Hisham et al., 2019). Similarly, *B. amyloliquefaciens* HM618 produced iturin from food waste (rice, vegetables, and small amounts of soup), demonstrating broad-spectrum antifungal activity and potential for plant disease control (Zhao et al., 2019; Wang et al., 2020; Miao et al., 2022). Currently, there is a high demand

TABLE 4.1
Biosurfactant-producing *Bacillus* species utilizing waste oil substrates [adapted from Gautam et al. (2023)]

Source of FOG waste	Concentration of oil waste used	*Bacillus* species	Biosurfactant type	Biosurfactant yield (g/L)	References
Waste frying oil	2%	*Bacillus* sp.	Surfactin	9.5	Hisham et al. (2019)
Waste frying oil	5%	*Bacillus pumilus*	Surfactin	5.7	Oliveira and Garcia-Cruz (2013)
Food waste	6%	*Bacillus amyloliquefaciens* HM618	Iturin	0.006	Miao et al. (2022)
Waste frying rice bran oil	4%	*Bacillus subtilis* MTCC 2423	Surfactin	4.67	Vedaraman and Venkatesh (2011)
Crude oil	10%	*Bacillus velezensis*	Rhamnolipid	3.12	Sultana et al. (2024)
Sunflower seed cake	5%	*Bacillus subtilis*	Surfactin	1.19	Ciurko et al. (2022)
Olive oil mill waste	10%	*Bacillus subtilis*	Surfactin	0.001	Ramírez et al. (2015)
Waste sunflower oil	5%	*Bacillus subtilis* MTCC 2423	Surfactin	0.7	Vedaraman and Venkatesh (2011)
Sunflower oil cake	Not specified	*Bacillus amyloliquefaciens*	Iturin	0.85	Kumar et al. (2017)
Waste cooking oil	2%	*Bacillus subtilis*	Fengycin	0.3	Zhao et al. (2024)

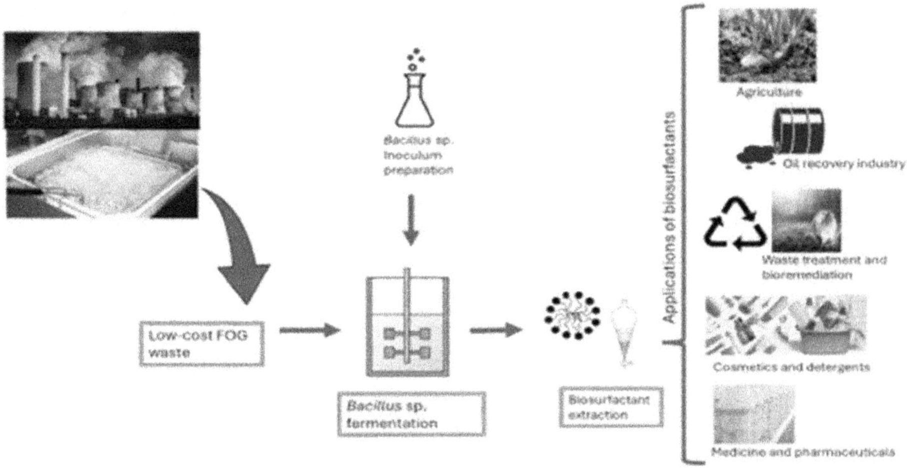

FIGURE 4.2 Utilisation of FOG waste by *Bacillus* sp. for biosurfactant production [adapted from Liepins et al. (2021)].

for the exploitation of FOG wastes. *Bacillus* species have been shown to drastically reduce FOG wastes via the production of enzymes called lipases, with the capacity to convert fats into fatty acids and glycerol depending on temperature and pH conditions (Kavyarathna et al., 2023; Zhao et al., 2024). For example, *B. subtilis* is among the *Bacillus* species that shows high lipase potential and is widely used to break down FOG restaurants waste, thereby serving as a substrate for the biotechnological production of various types of biosurfactants including diverse analogues of surfactins, as shown in Table 4.1 (Abdel-Mawgoud et al., 2008; Vedaraman and Venkatesh, 2011 Zhao et al., 2024). Durval et al. (2021) reported that *Bacillus* strains, such as *B. cereus*, *B. licheniformis*, *B. pumilus*, and *B. megaterium*, support the hydrolysis of triglycerides in FOG wastes with the capability of producing polyhydroxyalkanoates (PHAs) or their by-products. Additionally, *Bacillus* strains also aid in the conversion of waste into bioplastic, biofuel, and biodiesel (Khattab et al., 2022).

The biosurfactant produced is effective in emulsifying oils, which can enhance industrial cleaning processes and aid in wastewater bioremediation (Nayak et al., 2020; Malkapuram et al., 2021). Biosurfactants, acting as natural cleaning agents, can transport and dissolve soil contaminants in water, thereby enhancing environmental bioremediation processes. This not only enhances soil quality but also presents a potential alternative to harmful surfactants commonly found in pesticides (Oliveira et al., 2026). Ciurko et al. (2022) showed that *Bacillus* strains can effectively produce surfactin using sunflower seed cake as a substrate (Table 4.1), indicating significant potential for oil recovery applications. Surfactin aids in oil extraction from reservoirs by emulsifying the oil, facilitating its mobilisation and release into water or other extraction fluids (Ciurko et al., 2022). Furthermore, Cho et al. (2022) reported that biosurfactants, specifically sophorolipids with antibacterial properties, are being used as additives in herbicide formulations to control phytopathogens. Sophorolipids exhibit antifungal activity, helping to prevent root diseases caused by phytopathogenic fungi. Beyond their industrial and agricultural uses, biosurfactants have demonstrated significant potential in medicine and cosmetics due to their biocompatibility and targeted effects. In medical applications, biosurfactants offer promising therapeutic benefits with their antibacterial, antifungal, anticancer, and antiadhesive properties (Vatsa et al., 2010). These properties make them suitable for use in oral care, wound healing, and dermatological treatments. In the cosmetic industry, biosurfactants are valued for their surface-active properties, which allow them to function effectively as wetting agents, cleansers, detergents, and foaming agents (see Figure 4.2; Adu et al., 2020). Consumers increasingly seek non-irritating products with low cytotoxicity, and biosurfactants are particularly significant in this regard due to their non-ionic nature (Gautam et al., 2023).

4.4.4 Biosynthesis of Biofuels by Bacterial Strains

Adverse growth conditions such as extreme temperatures, high pH, high salinity, and nutrient starvation enhance endospore formation in *B. licheniformis* and *B. pumilus*, thereby leading to high yield in biofuel or biodiesel production (Malode et al., 2021). Biodiesel production can be achieved using esterification and transesterification methods via fatty acids such as oleic, palmitic, and stearic acids. However, to attain optimum biodiesel production, triacyl glycerides (TAGs) remain a major requirement (Abomohra et al., 2020; Arya et al., 2022). *Bacillus subtilis* has been genetically engineered for biofuel production by acetone-butanol-ethanol fermentation using different substrates, coupled with suitable metabolic pathways aside from lactic and glutamic acids (Jiang et al., 2019; Sandeep et al., 2024). Kim et al. (2020) recorded an overall yield of 1.1 g/L of biofuel with 80% isobutanol production in a genetically engineered mixed culture (4:1) of *E. coli* and *B. subtilis* carrying the genes that included *alsS* (acetolactate synthase), *kivD* (2-ketoisovalerate decarboxylase), *ilvC* (ketol-acid reducto-isomerase), *ilvD* (dihydroxy acid dehydratase), and *yqhD* (alcohol dehydrogenase). The *B. subtilis* expressing *leuDH* (leucine dehydrogenase), *kivD*, and *yqhD* genes promptly processes branched-chain amino acids (BCAAs) into the biofuels isobutanol, 2-methylbutanol, and 3-methylbutanol distinctly. However, Najjar et al. (2021) recorded catalytic lipase activity of 13.14 ± 0.03 U/mL with *B. circulans* and with a 98.4% yield of biodiesel production.

4.4.5 Biosynthesis of PHAs Using *Bacillus*

The escalating production and use of petroleum-based synthetic polymers for packaging and other applications pose significant global challenges. In response, a worldwide shift towards more circular and sustainable materials is underway (Moshood et al., 2022). One promising strategy for transforming waste into value is the microbial production of biopolymers from FOG waste using *Bacillus* species. PHAs are a class of bio-polyesters that can be produced by various microorganisms, including bacteria, fungi, and green microalgae. While multiple organisms can synthesise these biopolymers, bacteria remain the primary producers, with polyhydroxybutyrate (PHB) being the predominant biosynthetic pathway (Raza et al., 2018). The production of PHAs is influenced by several factors, such as the bacterial strain, carbon substrate, accumulation levels, and the costs associated with downstream processing (Lee et al., 2000; Silva et al., 2021). Specifically, *Bacillus* species, including *B. cereus* and *B. megaterium*, have been shown to synthesise PHAs by converting fatty acids and glycerol derived from FOG waste into these value-added biopolymers (Dalton et al., 2022). PHAs are biodegradable, mechanically strong, and thermoplastic, making them ideal for diverse applications, especially in medicine. Their biodegradability offers a sustainable alternative to traditional plastics, reducing environmental pollution. By decomposing into organic matter, PHAs can be returned to the soil, contributing to an eco-friendlier solution to plastic waste (Behera et al., 2022; Cueva-Almendras et al., 2022). They are used in hospitals, clinics, and pharmaceutical settings for wound treatment, anticancer materials, orthopaedic and vascular implants, and drug delivery equipment. The PHAs also have applications in food packaging (Dalton et al., 2022; Ngwasiri et al., 2022).

Recent research has focused on optimising PHA production by selecting *Bacillus* strains that can efficiently degrade FOG waste under controlled conditions to maximise PHA output (Shah and Kumar, 2021; Cueva-Almendras et al., 2022; Dalton et al., 2022). *Bacillus* species convert carbon-rich FOG waste into PHAs through three primary metabolic pathways: acetyl-CoA to 3-hydroxybutyryl-CoA, fatty acid beta-oxidation, and fatty acid biosynthesis (Saratale et al., 2021). In a study conducted by Povolo et al. (2012), *B. altitudinis* and *B. pumilus* were cultured in a medium characterised by high carbon content and low nitrogen levels, using corn oil (2.0% v/v) as the sole carbon source. The researchers observed PHA accumulation of 0.1% in *B. altitudinis* and 0.5% in *B. pumilus* (w/w cell dry mass). Similarly, Gómez Cardozo et al. (2016) investigated the production of PHB, a type of PHA, using native strains of *Bacillus* sp. and *Bacillus megaterium*. They observed that *B. megaterium*, when grown in media supplemented with glycerol, exhibited the highest biomass production (1.23 ± 0.05 g/L) and PHA yield, reaching 3.06 g/L under conditions of 25°C, 15 g/L glycerol, and a pH of 7.0.

Although PHAs hold great promise for various industries, their widespread adoption has been hindered by the high production costs resulting from low yields and inefficient substrate conversion (Chen et al., 2020; Dalton et al., 2022). To address these challenges, recent research has focused on optimising PHA production by bioprospecting for microbial strains capable of efficiently degrading FOG waste under controlled conditions, to maximise PHA output (Shah and Kumar, 2021; Cueva-Almendras et al., 2022). Sohail et al. (2020) demonstrated the ability of *B. tequilensis*, isolated from oil fields, to produce substantial amounts of PHA when grown on glycerol (12.4 g/L) and animal fat (14.2 g/L) as carbon sources. This suggests that the selection of microbes with the capacity to effectively utilise low-cost waste feedstocks, such as FOG, could be a promising approach to improving the economic viability of PHA production.

4.5 ENZYMATIC BREAKDOWN OF FOP BY *BACILLUS* SP.

Bacillus species produce a variety of enzymes that play a significant role in the biodegradation of FOGs. The primary enzymes that are involved in the degradation processes of FOGs include lipases, proteases, amylases, and esterases. The coordinated activity of these enzymes ensures the efficient biodegradation of FOG by breaking down complex organic materials into simpler molecules that can be further metabolised by the bacterial cells.

i. **Lipases** - The primary role of lipase enzymes produced by *Bacillus* species is to break down triglycerides (fats and oils). These lipases catalyse the hydrolysis of triglycerides, cleaving them into glycerol and free fatty acids (Figure 4.1) (Sharma et al., 2001). The lipase enzyme recognises and binds to the triglyceride substrate at the lipid-water interface. The enzyme's active site, typically containing a catalytic triad of serine, histidine, and aspartate (Harwood and Kikuchi, 2022; Vojnovic et al., 2024), facilitates the nucleophilic attack on the ester bonds in the triglyceride. Water molecules are then added to the ester bonds, breaking them and releasing one molecule of glycerol and three molecules of fatty acids (Sharma et al., 2001). The released glycerol can be taken up by the cell, phosphorylated to glycerol-3-phosphate, and then enter glycolysis as dihydroxyacetone phosphate (DHAP) (Deutscher, 2008). The free fatty acids are transported into the cell, activated to acetyl-CoA derivatives, and then undergo beta-oxidation to produce acetyl-CoA for the citric acid cycle (Voet and Voet, 2010).

   Lipases from *Bacillus* species play a crucial role in making fatty acids accessible to bacterial cells, as oil-degrading microorganisms commonly produce these enzymes extracellularly. Utilising FOG waste as a substrate for lipase production offers several advantages, including reduced production costs compared to using expensive oils or carbon and nitrogen sources (Mhlongo et al., 2021). Substituting expensive oils or carbon and nitrogen sources with waste oils can significantly lower the production expenses, as it has been shown that high contents of waste frying oil in the growth medium enhance lipase production by *B. cereus* (Awad et al., 2015).

   Lipases from various *Bacillus* species have been reported to exhibit antimicrobial activity against significant phytopathogens. For example, lipase from *Bacillus* strain 739, along with other hydrolytic enzymes such as chitinases, has shown mycolytic activity against *Bipolaris sorokiniana* (Aktuganov et al., 2007), while *B. subtilis* A101 (Saha et al., 2012) demonstrated antifungal properties against *Fusarium solani*, the causative agent of wilt disease in plants. These findings highlight the potential application of *Bacillus* species lipases as biocontrol agents against various plant pathogens. Additionally, lipases purified from *B. methylotrophicus* PS3 (Sharma et al., 2017) were identified as promising candidates for the laundry detergent industry, with tests showing complete removal of grease, butter, and vegetable oil stains compared to washing with water or conventional soap. In the food processing industry, there is an increasing demand for environmentally friendly and economically viable technologies, particularly for the treatment of fatty wastewater. A study by Balaji et al. (2020) evaluated the effectiveness of a lipase enzyme from the halotolerant *Bacillus* sp. VITL8 in reducing the oil and grease content in effluents from industries such as dairy, bakery, and poultry.

ii. **Proteases:** *Bacillus* is a significant source of proteases, producing high yields of neutral and alkaline enzymes with notable stability. These enzymes are resistant to extreme temperatures, pH, organic solvents, detergents, and oxidising compounds (Contesini et al., 2018). The primary role of proteases produced by *Bacillus* species is to break down the protein components which are associated with the FOG. They do this by hydrolysing the peptide bonds in proteins, breaking them down into smaller peptides and amino acids (Gupta et al., 2002). The protease enzyme would recognise and bind to specific sequences in proteins. The active site, which often contains serine or cysteine residues, would attack the peptide bond, forming a tetrahedral intermediate (Harwood and Kikuchi, 2022; Vojnovic et al., 2024). The water molecules would break the peptide bond thus releasing the smaller peptides and free amino acids. The smaller peptides are then further hydrolysed by intracellular peptidases into free amino acids.

iii. **Amylases:** The primary role of amylases produced by *Bacillus* species is to break down the polysaccharides present in FOGs by hydrolysing starches and other polysaccharides into simpler sugars such as maltose and glucose (Pandey et al., 2000). Amylase binds to the glycosidic bonds in polysaccharides. The active site of amylase would facilitate the addition of water molecules to the glycosidic bonds, breaking them and releasing oligosaccharides, disaccharides, and monosaccharides (Gupta et al., 2002). Glucose is taken up by the cell and metabolised through glycolysis, producing pyruvate, which is subsequently converted into acetyl-CoA. The acetyl-CoA then enters the citric acid cycle (Voet and Voet, 2010).

iv. **Esterases:** The primary role of esterases produced by *Bacillus* species is the hydrolysis of ester bonds in complex lipids. They do this by breaking down the esters into their corresponding alcohols and acids. Esterases would bind to ester bonds in complex lipids. The active site of esterase would catalyse the addition of water molecules to the ester bonds, breaking them and releasing alcohols and fatty acids in the process (Jaeger et al., 1999). The alcohols can be further oxidised to aldehydes and then to acids, which can then enter various metabolic pathways. The fatty acids produced during the esterase action would be taken up by the cell and would undergo beta-oxidation, producing acetyl-CoA. This would begin the citric acid cycle (Voet and Voet, 2010).

4.6 OPPORTUNITIES ASSOCIATED WITH *BACILLUS*-BASED FOG WASTE VALORISATION

i. **Environmental benefits:** The widespread adoption of *Bacillus*-based FOG waste valorisation offers many possibilities, promoting the principles of the CBE. An important benefit is the transformation of waste into valuable goods (Tzirita et al., 2018; Roets-Dlamini et al., 2022), which leads to a decrease in environmental pollution and lessens the strain on current waste management systems and landfills. *Bacillus* species can effectively convert FOG waste, which is commonly considered a troublesome waste stream due to its tendency to block sewage systems and its challenging disposal process, into valuable and biologically active substances (Roets-Dlamini et al., 2022). These products include biofuels (Tzirita, 2012), bioplastics (Sekoai et al., 2022), and compounds such as enzymes (e.g., lipases and amylases) (Roets-Dlamini et al., 2022) and proteases (Tzirita et al., 2021), which have significant industrial applications in sectors such as energy (Tzirita, 2012), food, and detergents (Tzirita, 2012; Roets-Dlamini et al., 2022).

Furthermore, *Bacillus* species are known for their ability to produce biosurfactants, which have applications in the bioremediation of sites contaminated with oil (Pan et al., 2026). *Bacillus*-based valorisation's conversion of waste into useful products, not only reduces waste but also improves resource efficiency and sustainability. Consequently, the implementation of a large-scale approach could not only effectively decrease pollution by managing FOG waste, but also offer effective remedies for addressing and resolving current contaminations. This dual strategy for environmental protection offers a comprehensive solution to pollution problems caused by FOGs. In addition, the use of biological treatment for FOG waste has a smaller environmental impact in terms of carbon emissions compared to conventional waste management techniques such as incineration or landfilling (Faisal et al., 2025). Hence, the widespread adoption of this bioprocess has the potential to effectively decrease the total amount of greenhouse gas emissions linked to waste management.

ii. **Economic benefits:** Another significant opportunity lies in the economic benefits associated with this bioprocess. The biotechnological use of *Bacillus* species for FOG waste valorisation has the potential to generate new revenue streams as well as business and job opportunities. The bioactive compounds and metabolites generated by *Bacillus* species during FOG degradation possess potential applications in diverse industries. Biodiesel, for instance, is commonly utilised for power generation (Rosas-Vega et al., 2026), bioplastics are popular in the plastic industry (Khattab et al., 2022), and lipases and other enzymes can be extracted and commercially utilised in detergents, food processing, and pharmaceutical industries (Guncheva and Zhiryakova, 2011; Balaji et al., 2020). Contesini et al. (2018) stated in their review that the genus *Bacillus* is the most significant bacterial provider of proteases. These proteases can produce copious amounts of neutral and alkaline proteolytic enzymes that possess exceptional characteristics, including high resistance to extreme temperatures, pH levels, organic solvents, detergents, and oxidising compounds. This transformation promotes economic growth by decreasing dependence on fossil resources and minimising expenses associated with waste management. Furthermore, the scalability of *Bacillus*-based processes is enhanced by the strong and adaptable characteristics of these bacterial species. These species are recognised for their exceptional resilience and capacity to flourish in various environmental conditions (Aqel et al., 2024). This characteristic makes them well-suited for use in large-scale industrial applications, as it guarantees a consistent and uninterrupted provision of bioactive compounds. The valorisation process can produce valuable bioproducts, such as biosurfactants, that can be used in environmental remediation, enhanced oil recovery, and as more environmentally friendly alternatives to chemical surfactants in consumer products (Pandit et al., 2026). These valuable products have substantial economic potential. By converting waste into valuable resources, this approach advocates for the conversion, thereby promoting resource efficiency and diminishing dependence on virgin raw materials. This can result in the development of a more sustainable industrial ecosystem, in which waste generated from one process is used as input for another, thereby closing material loops, conserving natural resources, fostering a circular economy, and potentially lowering production costs.

iii. **Technological advancements:** Implementing *Bacillus* FOG valorisation on a large scale would stimulate innovation in bioprocessing and biotechnology. It would stimulate advancements in bioreactor design (Ushani et al., 2020), process optimisation (Formenti et al., 2014; Ousaadi et al., 2021), and downstream processing techniques (Pérez-Rivero et al., 2019; Haque et al., 2022). These advancements could potentially have a ripple effect, providing advantages to other sectors within the field of industrial biotechnology and bioengineering. Moreover, the scalability of *Bacillus*-based processes may enable the development of modular and decentralised waste treatment systems, leading to more effective and localised waste management solutions. This can revolutionise the way we manage waste, especially in urban areas and industrial regions.

iv. **Agricultural applications:** Some *Bacillus* species produce metabolites that possess bio-pesticidal properties (El-Bendary, 2006; Vatsa et al., 2010). The large-scale valorisation of FOG waste has the potential to produce eco-friendly substitutes for chemical pesticides (Aktuganov et al., 2007; Jagadeesan et al., 2022), thereby promoting the shift towards more sustainable agricultural methods. In addition, specific by-products resulting from the FOG valorisation process can be utilised as biofertilisers. Luo et al. (2022) found that *B. amyloliquefaciens* enhances plant growth through the production of hormones, volatile organic compounds (VOCs), and siderophores, which leads to increased availability of nutrients in the soil and the suppression of soil pathogens. Another research study that analysed the complete genome of *B. amyloliquefaciens* discovered gene clusters that are responsible for the production of secondary metabolites through non-ribosomal synthesis. Additionally, it identified genes associated with the production of plant growth promotion hormones like indole-3-acetic acid (IAA) and acetoin secretion (Qin et al., 2015). This ability to create biofertiliser from *Bacillus* FOG valorisation results in a circular system in which urban waste contributes to food production. This completes the cycle between urban waste and agricultural input, promoting sustainability in both sectors.

v. **Broader impacts:** Other waste valorisation methods can be designed based on the successful large-scale implementation of the *Bacillus*-based FOG valorisation technology. It has the potential to demonstrate the feasibility and advantages of biotechnological waste management solutions, which could lead to the adoption of similar methods for other types of organic waste and various industrial applications (Guncheva and Zhiryakova, 2011). Furthermore, this approach can induce a broader and more profound revolution in waste management paradigms, by redirecting the focus from simply disposing of waste to recovering and using it as a resource. This shift has the potential to have significant and wide-ranging effects on how societies perceive and handle waste in the future.

vi. **Alignment with global sustainability goals**: *Bacillus*-based valorisation of waste contributes to the achievement of global sustainability goals by simultaneously tackling various environmental issues. Examples of positive environmental impacts include the reduction of greenhouse gas emissions through waste-to-resource conversion (Sharma et al., 2020; Surendra et al., 2022), the improvement of soil and water quality through bioremediation (Zhao et al., 2009; Lauprasert et al., 2017), and the promotion of sustainable agricultural practices through biopesticides (Aktuganov et al., 2007; Jagadeesan et al., 2022) and biofertilisers (Qin et al., 2015). The incorporation of these bioprocesses into current waste management systems can greatly expedite the shift towards a more circular and sustainable bioeconomy.

4.7 CHALLENGES IN INDUSTRIALISING *BACILLUS*-BASED FOG VALORISATION

The CBE aims to optimise the use of biological resources by transforming waste into valuable products. FOG waste holds significant potential for conversion into biosurfactants, various enzymes, or bioenergy sources such as biogas and biodiesel. *Bacillus* species offer an environmentally friendly approach to FOG valorisation by producing lipases that break down fats into easily digestible components for other microorganisms (Tzirita et al., 2019). Additionally, *Bacillus* species enhance FOG biodegradability through emulsification. However, commercial-scale implementation of *Bacillus*-based FOG valorisation faces challenges that require careful consideration to ensure efficiency, smooth operations, and economic viability. Implementing *Bacillus*-based valorisation of FOG waste on a large scale poses several challenges and limitations, which need to be addressed for the technology to be viable and sustainable. The bioprocessing of FOG waste presents a promising solution for treating and reducing the environmental impacts of this waste stream. However, several challenges have slowed the implementation of this approach. The use of *Bacillus* species in the bioremediation of FOG waste has been studied due to their ability to produce lipase enzymes, emulsify FOG, and generate other valuable products. Yet, there are key obstacles to scaling up the use of *Bacillus* for this purpose. FOG waste is generated across various industries, with food processing facilities such as restaurants being major contributors. The composition of FOG waste can vary significantly depending on its source, whether domestic or industrial (Tzirita et al., 2019; Teixeira et al., 2020). This variability can impact the efficiency of *Bacillus* species, as the waste may contain high levels of contaminants such as heavy metals or food scraps that could inhibit bacterial growth (Gurd et al., 2020).

Market acceptance and regulatory adherence are paramount for the commercial success of FOG-derived products. Strict quality and purity standards, such as those outlined in European specifications, must be met to ensure product safety and marketability (Abomohra et al., 2020; Jacob-Lopes et al., 2022). Nevertheless, the heterogeneous nature of FOG waste can compromise product consistency and regulatory compliance, especially for products such as biodiesel, fatty acids (to produce soaps), glycerol, and biopolymers (Mohapatra et al., 2017; Jacob-Lopes et al., 2022). Scaling up the bioprocess presents additional challenges. Optimal laboratory conditions may not be directly transferable to large-scale operations, necessitating careful adjustment of parameters such as temperature, pH, and aeration (Jacob-Lopes et al., 2022). Moreover, the prolonged incubation periods often required by *Bacillus* species for FOG degradation can hinder high-throughput production (Al-Dhabi et al., 2020; Teixeira et al., 2020).

Preserving the purity of *Bacillus* species cultures or well-defined microbial communities is essential for optimal bioprocess performance. Contamination with extraneous microorganisms from FOG waste or process impurities can hinder *Bacillus* activity, compromise product quality, and potentially introduce harmful substances (Tzirita et al., 2019). The specialised nature of FOG waste necessitates custom-designed bioreactors, increasing operational costs. Additional hurdles to large-scale FOG waste bioprocessing include the complexities of waste collection and transportation, bioreactor maintenance, and developing reliable markets for bio-derived products (Iorhemen et al., 2016; Jacob-Lopes et al., 2022).

4.8 STRATEGIES FOR IMPROVING THE EFFICIENCY OF *BACILLUS*-BASED BIOPROCESSES

i. **FOG cost-effective collection systems:** Urban areas with a high concentration of FOG-generating facilities provide an ideal setting for efficient waste management. Establishing centralised collection points can streamline logistics and ensure a consistent supply of FOG for bioprocessing (Jacob-Lopes et al., 2022). Collaborative networks among local businesses can optimise the collection routes and minimise overlaps. Smaller, mobile collection units can effectively gather FOG from smaller sources. Effective pre-treatment is crucial for smooth bioprocessing. Screening and filtration can remove solid impurities, while alkaline hydrolysis can break down lipids into more easily digestible components. Supplementing the process with commercial lipase can further enhance microbial access to FOG (Khatib et al., 2023). Fed-batch fermentation optimises substrate levels, preventing growth inhibition and maximising microbial productivity (Fan, 2014). Adequate aeration and mixing within bioreactors sustain aerobic conditions essential for *Bacillus* growth. Tailored purification methods are crucial to reduce production costs. Distillation and adsorption refine biodiesel, while crystallisation and solvent extraction purify bioplastics (Abomohra et al., 2020; Jacob-Lopes et al., 2022). Integrated approaches, such as *in-situ* product recovery and membrane filtration, enhance efficiency, and product quality (Iorhemen et al., 2016; Abomohra et al., 2020).

ii. **Genetic engineering and enhanced strain development:** *Bacillus* species offer considerable potential as cell factories for transforming FOG waste into profitable value-added products while safeguarding the environment. Genetic engineering can enhance FOG breakdown and product formation by overexpressing genes encoding enzymes such as lipases (Tolibia et al., 2023). Modifying genetic control mechanisms can also create *Bacillus* strains optimised for FOG degradation and bioproduct generation. To deepen our understanding of FOG breakdown processes, genomic and metabolomic analyses can be employed to identify metabolic limitations and refine the process. Moreover, evolving *Bacillus* strains under industrial conditions can lead to the development of robust microorganisms suitable for large-scale FOG bioprocessing, as current research primarily operates on a laboratory scale.

iii. **Process optimisation and integration:** Developing integrated models that incorporate *Bacillus*-based FOG conversion into a broader CBE is crucial. These models should investigate how the process can be linked with other waste streams to optimise resource recovery and minimise environmental harm. To enhance bioprocessing efficiency and lower costs, bioreactors equipped with real-time monitoring and control systems for critical parameters such as pH, temperature, and aeration are essential.

iv. **Bacterial enzyme immobilisation and stability:** Prolonged use can diminish the stability of *Bacillus*-derived enzymes and extracts. To enhance stability and reusability, lipases can be immobilised on solid supports using techniques such as adsorption or covalent bonding (Datta et al., 2013). Additionally, optimising process conditions and employing additives such as surfactants or micelles can prolong enzyme activity, thereby reducing enzyme-production costs and the overall operational expenses (Datta et al., 2013; Minteer, 2017; Khatib et al., 2023).

v. **Strain selection and optimisation of growth conditions:** *Bacillus* species, including *B. subtilis, B. circulans, B. licheniformis*, and *B. amyloliquefaciens*, are well-suited for FOG bioprocessing due to their natural ability to thrive in FOG-rich environments. These *Bacillus* species have been frequently isolated from such conditions, demonstrating their robustness and adaptability to high FOG concentrations (Hendricks, 2015; Tzirita et al., 2018; Al-Dhabi et al., 2020). One of the primary challenges is the selection and optimisation of *Bacillus* strains capable of efficiently degrading and converting FOG waste into valuable products. Not all *Bacillus* species possess the same capabilities, and even within the same species, different strains can exhibit significant variability in their metabolic activities and product yields (Formenti et al., 2014; Aqel et al., 2024). The process of isolating, screening, and genetically optimising suitable strains is labour-intensive and time-consuming. Additionally, maintaining the stability and performance of these strains over long periods in industrial settings can be challenging due to potential genetic drift or contamination (Vilas-Boas et al., 2002; Chilcutt and Tabashnik, 2004). Optimising growth conditions is crucial for maximising *Bacillus* species' performance in FOG bioprocessing. Research indicates that temperatures between 30°C and 37°C and a pH of approximately 6 are optimal for the growth of *Bacillus* and the production of lipases (Fan, 2014; Khatib et al., 2023). These conditions can be maintained in bioreactors to enhance productivity. FOG waste composition varies, necessitating nutrient optimisation. To support *Bacillus* growth and enzyme production, essential nutrients such as carbon, nitrogen, and phosphorus may need to be supplemented (Tzirita et al., 2019). Efficient bioreactors are essential for FOG treatment. Stirred tank bioreactors provide adequate mixing, aeration, and oxygen supply, while removing toxic gases (Fan, 2014). Continuous and fed-batch culture systems can also be employed to optimise cell density and prevent substrate inhibition, enhancing FOG degradation (Lauprasert et al., 2017). To optimise FOG bioprocessing, a consortium of microorganisms may be more effective than single strains. Combining *Bacillus* spp. with other bacteria, such as *Pseudomonas putida*, can enhance fat degradation by utilising the diverse metabolic capabilities of different organisms (Tzirita et al., 2018). Additionally, genetic engineering can be employed to improve the performance of *Bacillus* strains for FOG bioprocessing. Modifying regulatory genes can lead to increased surfactant production and other desirable traits (Tolibia et al., 2023).

vi. **Process scale-up:** To successfully transition laboratory-scale processes to industrial-scale operations, numerous obstacles must be surmounted. When dealing with larger-scale production, we often face challenges related to the inefficiencies of selected species that become more pronounced as the scale of production increases. Additionally, we encounter inefficiencies that arise from the equipment used during the scaling-up process (Dholiya et al., 2023). One of the primary challenges in scaling up a process is the limitations imposed by the organisms involved (Paddon et al., 2013). When the organisms responsible for producing the desired secondary metabolites that are effective in FOG valorisation are subjected to large-scale production environments, they often undergo alterations or fail to function effectively due to stress (Vilas-Boas et al., 2002; Chilcutt and Tabashnik, 2004). Instances have been documented where the insulin- and enzyme-production cell factory, *Escherichia coli,* has shown an upsurge in the production of undesirable products following exposure to stressful conditions such as inadequate mixing, fluctuating temperatures, changes in pH, and imbalances in nutrient and oxygen levels (Caspeta et al., 2009; Tripathi, 2016; Dholiya et al., 2023). These factors diminish the quality of the products and escalate the production expenses because of the supplementary purification procedure. These stress conditions have varying effects on the growth of organisms, causing depletion zones in the media and saturation zones. This leads to disrupted stability of plasmids, uneven formation of products, increased stress, and decreased product quality. Ensuring ideal and uniform reaction conditions for microorganisms, minimising microbial stress, optimising mass and heat transfer, and preventing contamination are intricate tasks (Paddon et al., 2013; Dholiya et al., 2023). The monitoring and adjustment of parameters such as pH, temperature, oxygen levels, and nutrient supply often necessitate the use of advanced bioreactor designs and control systems. These factors contribute to the complexity and operational expenses associated with large-scale bioprocessing. If these parameters are not properly monitored and adjusted, it can lead to decreased efficiency and inconsistent product qualities, which are significant considerations for large-scale industrial applications.

4.9 ECONOMIC FEASIBILITY

To fully assess the sustainability of *Bacillus*-based FOG processing, a comprehensive evaluation of its economic and environmental impacts is crucial (Jacob-Lopes et al., 2022). With few to no studies conducted to assess the market for *Bacillus*-based FOG products, a thorough market analysis to understand product demand, consumer preferences, regulatory compliance, and competitive dynamics is highly required. *Bacillus*-based FOG products may face tough competition from traditionally used chemical alternatives for FOG, which may be deemed more effective (Mohapatra et al., 2017). A detailed cost-benefit analysis should determine the financial feasibility of scaling up the bioprocess, considering both initial and ongoing costs, revenue streams, and return on investment. Integrating renewable energy sources, such as solar power, into bioprocessing facilities can enhance economic efficiency and reduce environmental impact. Adopting a circular economy framework that emphasises the conversion of industrial by-products such as glycerol into value-added products is critical for maximising resource efficiency and minimising environmental impact. Establishing strategic partnerships between waste generators and product end-users can further enhance waste valorisation pathways, streamline management processes, and strengthen the market uptake of bioproducts (Marx and Venter, 2014; Jacob-Lopes et al., 2022). Within this context, *Bacillus* species demonstrate remarkable potential for transforming FOG waste into commercially valuable biomolecules, thereby supporting the development of a resilient CBE. However, realising this potential at scale requires addressing key challenges related to process optimisation, cost-effective waste collection and pre-treatment, reactor design, and efficient downstream product recovery.

i. **Large-scale operations:** The economic feasibility of large-scale *Bacillus*-based FOG waste valorisation is another substantial obstacle. Although the concept is promising from a circular economy perspective, the expenses associated with the establishment and operation of such facilities can be substantial (Abomohra et al., 2020). The economic feasibility is contingent upon the equilibrium between the market value of the end products and operational costs (Tabatabaei et al., 2019; Abomohra et al., 2020). The expenses related to microbial cultivation, waste collection and preprocessing infrastructure, bioreactor systems, downstream processing equipment, and ongoing operational costs for substrate preparation can be significant (Meyer et al., 2017; Wang et al., 2019; Doddapaneni and Kikas, 2021). Moreover, variations in market demand and prices for biofuels, bioactive compounds, enzymes, and other bioproducts can impact the profitability of the process. The economic feasibility of the process is largely contingent upon the value of the final products and the efficiency of the conversion process (Doddapaneni and Kikas, 2021). Often, products made from FOG waste may have to contend with well-established alternatives in the market, presenting a significant challenge. It is crucial to create bioprocesses that are both cost-effective and efficient to be competitive with conventional waste management and product synthesis methods.

ii. **Downstream processing:** Extraction and purification of valuable bioconversion products can be a significant challenge in large-scale operations. Depending on the target products (e.g., enzymes, biosurfactants, biodiesel, or bioplastics), complex and costly downstream processing, which includes microbial cell separation, product extraction, purification, and/or formulation, may be required (Marx and Venter, 2014; Jacob-Lopes et al., 2022). The efficiency and cost-effectiveness of these processes have a significant impact on the overall viability of the valorisation approach (Jacob-Lopes et al., 2022). The products are frequently produced in dilute aqueous solutions containing cell mass, metabolic by-products, and various salts, making product recovery an important part of the processing. The level of product purification varies depending on the type of product and its intended use. Industrial enzymes, for example, are frequently separated from the cell mass and concentrated through ultrafiltration to produce a crude extract (Yang, 2007). On the other hand, antibiotics and therapeutic proteins for human use must be very pure and require extensive downstream processing using common purification steps such as microfiltration and ultrafiltration (to remove cell mass and other debris or retain larger molecular weight proteins), concentration by evaporation, centrifugation, or reverse osmosis, crystallisation by cooling, evaporation, or pH adjustment, distillation, solvent extraction, and/or chromatography (Yang, 2007). Purification costs vary, ranging from 5% to 90% of total production costs, and are typically proportional to product value.

iii. **Regulatory and environmental concerns:** Large-scale bioprocessing facilities must follow strict environmental and safety regulations. This is especially applicable to processes involving genetically modified organisms, which could be created to improve the efficiency of FOG waste valorisation. The utilisation of genetically modified *Bacillus* strains, if employed, may give rise to concerns regarding biosafety and biosecurity, necessitating thorough risk evaluations and regulatory authorisations (Hosseini et al., 2018). Davison (2005) states that there have been a few instances of genuine bioremediation field trials involving living genetically modified bacteria. This is primarily due to the challenges of adhering to environmental regulations in both the United States and Europe. Furthermore, not all naturally occurring soil bacteria are suitable as bioremediation agents from a biosafety perspective. The Environmental Protection Agency (EPA) of the United States, for example, declined the use of *Burkholderia cepacia* as an environmental agent for bioremediation due to its dual nature as a human pathogen associated with cystic fibrosis and its resistance to multiple antibiotics (Holmes et al., 1998; Davison, 2005). Furthermore, it is imperative to effectively manage the disposal of residual biomass and by-products to avoid detrimental environmental consequences. Incorporating environmental sustainability and regulatory compliance into the entire process increases the level of complexity in implementing large-scale projects. Adhering to environmental, health, and safety regulations can be intricate and expensive.

iv. **Competition with existing waste management systems:** Valorisation of FOG waste by *Bacillus* faces competition from both well-established and emerging waste management technologies. Some examples of these include physical-chemical treatment methods (Fox, 1996; Bhargava, 2016), alternative biological processes (Sanghamitra et al., 2021; Jacob-Lopes et al., 2022), and thermal treatments such as incineration with energy recovery (Penner et al., 1988; Grosso et al., 2010; Fruergaard et al., 2010). Each of these options possesses distinct benefits and may be favoured in specific circumstances. It is essential to showcase the exceptional efficiency, cost-effectiveness, and environmental advantages of *Bacillus*-based processes in various situations to encourage their widespread use. Additionally, there may be challenges in integrating *Bacillus*-based FOG waste valorisation into the existing waste management infrastructure, whereas its integration with existing systems used in the collection and processing of industrial and municipal waste is essential. The integration must take logistical considerations, such as the handling, storage, and transportation of bioproducts and waste, into account. To get FOG waste ready for microbial processing, it is essential to set up a reliable and consistent supply chain and do the necessary pre-treatment (Hidalgo et al., 2012; Ravindran and Jaiswal, 2016).

v. **Technological and knowledge gaps:** There are still technological and knowledge gaps in determining the full potential and limitations of *Bacillus* species for FOG waste valorisation. Continuous research and development are required to discover new strains, optimise metabolic pathways, and create innovative bioprocessing technologies. Collaboration among academia, industry, and government agencies is critical for progressing the field and closing gaps through interdisciplinary research and knowledge sharing. Furthermore, despite the adaptability of *Bacillus* species, their technical limitations constrain their application in FOG waste valorisation. For example, whereas many *Bacillus* strains generate lipases capable of breaking down fats and oils (Sangeetha et al., 2011; Al Mohaini et al., 2022), the efficiency of this process may not always be ideal for all types of FOG waste. According to Baena et al. (2022), lipases, despite their advantages, are typically challenging to retrieve and are susceptible to various reaction conditions such as temperature, pH, ionic strength, and impurities. Consequently, these factors restrict their practical use in industries. Additionally, certain persistent substances in waste streams may resist degradation by lipase, leading to incomplete valorisation (Baena et al., 2022). Furthermore, the production of some high-value compounds from FOG waste may necessitate complex metabolic engineering of *Bacillus* strains, which can be technically challenging and subject to regulatory oversight. Addressing these challenges will necessitate a collaborative effort among researchers, industry stakeholders, and policymakers to create sustainable, efficient, and economically viable *Bacillus*-based bioprocessing technologies for FOG waste valorisation.

4.10 CONCLUSION

Bacillus species present a promising biotechnological route for the valorisation of FOG waste into high-value bioproducts, thereby advancing the principles of a CBE. Their robust metabolic capacity to degrade and utilise lipid-rich substrates align well with sustainable waste management strategies. However, realising their full potential requires overcoming several technical and economic challenges, including low product yields, process scale-up constraints, and cost inefficiencies. Addressing these limitations through targeted strain engineering, process intensification, and techno-economic optimisation could significantly enhance process feasibility. The widespread adoption of *Bacillus*-based FOG waste valorisation represents both opportunities and challenges. On one hand, it promises substantial environmental benefits through waste reduction and resource recovery, economic gains via bioproduct diversification and new value chains, and technological advancements in microbial bioprocessing. On the other hand, the complexity of integrating such systems at industrial-scale demands coordinated efforts among academia, industry stakeholders, and regulatory bodies. Ultimately, as research and innovation continue to refine *Bacillus*-mediated bioprocesses, these systems could play a pivotal role in reducing pollution, fostering sustainable production, and accelerating the global transition towards a resilient CBE.

REFERENCES

Abdel-Mawgoud, A.M., Aboulwafa, M.M., & Hassouna, N.A.-H. (2008). Characterization of surfactin produced by *Bacillus subtilis* isolate BS5. *Applied Biochemistry and Biotechnology*, 150, 289–303.

Abedi, E., & Wietstock, P.C. (2026). A comparative analysis of specific surface area and molecular structure in CO_2-microbubble-assisted starch hydrolysis for the effective binding of α-amylase to starch. *Innovative Food Science & Emerging Technologies*, 109, 104461.

Abomohra, A.E.F., Elsayed, M., Esakkimuthu, S., El-Sheekh, M., & Hanelt, D. (2020). Potential of fat, oil and grease (FOG) for biodiesel production: A critical review on the recent progress and future perspectives. *Progress in Energy and Combustion Science*, 81, 100868.

Adu, S.A., Naughton, P.J., Marchant, R., & Banat, I.M. (2020). Microbial biosurfactants in cosmetic and personal skincare pharmaceutical formulations. *Pharmaceutics*, 12(11), 1099.

Aktuganov, G.E., Galimzyanova, N.F., Melent'Ev, A.I., & Kuz'mina, L.Y. (2007). Extracellular hydrolases of strain *Bacillus* sp. 739 and their involvement in the lysis of micromycete cell walls. *Microbiology*, 76, 413–420.

Al Mohaini, M., Farid, A., Muzammal, M., Ghazanfar, S., Dadarsnia, A., Abdulkhaliq, J.A., … & Salmah, I. (2022). Enhancing lipase production of *Bacillus salmalaya* strain 139SI using different carbon sources and surfactants. *Applied Microbiology*, 2, 237–247.

Al-Dhabi, N.A., Esmail, G.A., & Arasu, M.V. (2020). Enhanced production of biosurfactant from *Bacillus subtilis* strain Al-Dhabi-130 under solid-state fermentation using date molasses from Saudi Arabia for bioremediation of crude-oil-contaminated soils. *International Journal of Environmental Research and Public Health*, 17, 8446.

Amaresan, N., Kumar, M.S., Annapurna, K., Kumar, K., & Sankaranarayanan, A. (Eds.). (2020). *Beneficial microbes in agro-ecology: Bacteria and fungi*. Academic Press.

Amodu, O.S., Ntwampe, S.K., & Ojumu, T.V. (2014). Emulsification of hydrocarbons by biosurfactant: Exclusive use of agrowaste. *Bioresources*, 9, 3508–3525.

Andler, S.M., & Goddard, J.M. (2018). Transforming food waste: how immobilized enzymes can valorize waste streams into revenue streams. *NPJ Science of Food*, *2*(1), 19.

Aqel, H., Farah, H., & AL-Hunaiti, A. (2024). Ecological versatility and biotechnological promise: Comprehensive characterization of the isolated thermophilic *Bacillus* strains. *PLoS ONE*, 19, e0297217.

Arya, P.S., Yagnik, S.M., Rajput, K.N., Panchal, R.R., & Raval, V.H. (2022). Valorization of agro-food wastes: Ease of concomitant-enzymes production with application in food and biofuel industries. *Bioresource Technology*, 361, 127738.

Awad, G.E.A., Mostafa, H., Danial, E.N., Abdelwahed, N.A.M., & Awad, H.M. (2015). Enhanced production of thermostable lipase from *Bacillus cereus* ASSCRC-P1 in waste frying oil-based medium using statistical experimental design. *Journal of Applied Pharmaceutical Science*, 5, 7–15.

Baena, A., Orjuela, A., Rakshit, S.K., & Clark, J.H. (2022). Enzymatic hydrolysis of waste fats, oils and greases (FOGs): Status, prospective, and process intensification alternatives. *Chemical Engineering and Processing-Process Intensification*, 175, 108930.

Balaji, L., Chittoor, J.T., & Jayaraman, G. (2020). Optimization of extracellular lipase production by halotolerant *Bacillus* sp. VITL8 using factorial design and applicability of enzyme in pretreatment of food industry effluents. *Preparative Biochemistry and Biotechnology*, 50, 708–716.

Behera, S., Priyadarshanee, M., & Das, S. (2022). Polyhydroxyalkanoates, the bioplastics of microbial origin: Properties, biochemical synthesis, and their applications. *Chemosphere*, 294, 133723.

Bhargava, A. (2016). Physico-chemical wastewater treatment technologies: An overview. *International Journal of Research in Education and Science*, 4, 5308–5319.

Burgess, S.A., Flint, S.H., Lindsay, D., Cox, M.P., & Biggs, P.J. (2017). Insights into the *Geobacillus stearothermophilus* species based on phylogenomic principles. *BMC Microbiology*, 17, 1–12.

Caspeta, L., Flores, N., Pérez, N.O., Bolívar, F., & Ramírez, O.T. (2009). The effect of heating rate on *Escherichia coli* metabolism, physiological stress, transcriptional response, and production of temperature-induced recombinant protein: A scale-down study. *Biotechnology and Bioengineering*, 102, 468–482.

Celandroni, F., Vecchione, A., Cara, A., Mazzantini, D., Lupetti, A., & Ghelardi, E. (2019). Identification of *Bacillus* species: Implication on the quality of probiotic formulations. *PLoS ONE*, 14, e0217021.

Chauhan, S., Mohanty, A. & Meena, S.S. (2025). Unlocking the potential of rhamnolipids: production via agro-industrial waste valorization, market insights, recent advances, and applications. *Biomass Conversion and Biorefinery*, 15, 28847–28876.

Chen, G.Q., Chen, X.Y., Wu, F.Q., & Chen, J.C. (2020). Polyhydroxyalkanoates (PHA) toward cost competitiveness and functionality. *Advanced Industrial and Engineering Polymer Research*, 3, 1–7.

Chilcutt, C.F., & Tabashnik, B.E. (2004). Contamination of refuges by *Bacillus thuringiensis* toxin genes from transgenic maize. *Proceedings of the National Academy of Sciences*, 101, 7526–7529.

Cho, W.Y., Ng, J.F., Yap, W.H., & Goh, B.H. (2022). Sophorolipids—Bio-based antimicrobial formulating agents for applications in food and health. *Molecules*, 27(17), 5556.

Ciurko, D., Czyżnikowska, Ż., Kancelista, A., Łaba, W., & Janek, T. (2022). Sustainable production of biosurfactant from agro-industrial oil wastes by *Bacillus subtilis* and its potential application as antioxidant and ACE inhibitor. *International Journal of Molecular Sciences*, 23, 10824.

Contesini, F.J., Melo, R.R.D., & Sato, H.H. (2018). An overview of *Bacillus* proteases: From production to application. *Critical Reviews in Biotechnology*, 38, 321–334.

Court, E.K., Chaudhuri, R.R., Kapoore, R.V., Villa, R.X., Pandhal, J., Biggs, C.A., & Stafford, G.P. (2021). Looking through the fog: Microbiome characterization and lipolytic bacteria isolation from a fatberg site. *Microbiology*, 167, 001117.

Cueva-Almendras, L.C., Alvarado, J.C.A., Fuentes-Olivera, A.J., Llontop-Bernabé, K.S., Cerna, C.E.Q., Rodriguez-Soto, J.C., … & Alvarez, M.A.Q. (2022). Production of polyhydroxyalkanoate by *Bacillus thuringiensis* isolated from agricultural soils of Cascas-Peru. *Brazilian Archives of Biology and Technology*, 65, e22220107.

Dalton, B., Bhagabati, P., De Micco, J., Padamati, R.B., & O'Connor, K. (2022). A review on biological synthesis of the biodegradable polymers polyhydroxyalkanoates and the development of multiple applications. *Catalysts*, 12, 319.

Dame, Z.T., Rahman, M., & Islam, T. (2021). *Bacilli* as sources of agrobiotechnology: Recent advances and future directions. *Green Chemistry Letters and Reviews*, 14, 1905080.

Danilova, I., & Sharipova, M. (2020). The practical potential of bacilli and their enzymes for industrial production. *Frontiers in Microbiology*, 11, 1782.

Datta, S., Christena, L.R., & Rajaram, Y.R. (2013). Enzyme immobilization: an overview on techniques and support materials. *Biotech*, 3, 1–9.

Davison, J. (2005). Risk mitigation of genetically modified bacteria and plants designed for bioremediation. *Journal of Industrial Microbiology and Biotechnology*, 32, 639–650.

Deaver, J.A., & Popat, S.C. (2022). Fats, oils, and grease (FOG): Opportunities, challenges, and economic approaches. In M.C. Depra, E. Jacob-Lopes, & L.Q. Zepka (Eds.), *Handbook of waste biorefinery: Circular economy of renewable energy* (1st ed., pp. 285–308). Springer International Publishing.

Deutscher, M.P. (2008). Degradation of RNA in bacteria: Comparison of mRNA and stable RNA. *Nucleic Acids Research*, 36, 659–666.

Dholiya, K., Parmar, A., & Bashir, H.S. (2023). Scale-up: Lab to commercial scale. In A.K. Bhatt, R.K. Bhatia, & T.C. Bhalla (Eds.), *Basic biotechniques for bioprocess and bioentrepreneurship* (1st ed., pp. 341–353). Academic Press.

Doddapaneni, T.R.K.C., & Kikas, T. (2021). Thermochemical and biochemical treatment strategies for resource recovery from agri-food industry wastes. In R. Bhat (Ed.), *Valorization of agri-food wastes and by-products* (1st ed., pp. 787–807). Academic Press.

Donio, M.B.S., Ronica, S.F.A., Viji, V.T., Velmurugan, S., Jenifer, J.A., Michaelbabu, M., & Citarasu, T. (2013). Isolation and characterization of halophilic *Bacillus* sp. BS3 able to produce pharmacologically important biosurfactants. *Asian Pacific Journal of Tropical Medicine*, 6, 876–883.

Durval, I.J., Ribeiro, B.G., Aguiar, J.S., Rufino, R.D., Converti, A., & Sarubbo, L.A. (2021). Application of a biosurfactant produced by *Bacillus cereus* UCP 1615 from waste frying oil as an emulsifier in a cookie formulation. *Fermentation*, 7, 189.

Ehling-Schulz, M., Lereclus, D., & Koehler, T.M. (2019). The *Bacillus cereus* group: *Bacillus* species with pathogenic potential. *Microbiology spectrum*, 7, 10–1128.

El-Bendary, M.A. (2006). *Bacillus thuringiensis* and *Bacillus sphaericus* biopesticides production. *Journal of Basic Microbiology*, 46, 158–170.

European Commission: Directorate-General for Research and Innovation. (2012). *Innovating for sustainable growth: A bioeconomy for Europe*. Publications Office.

Faisal, S., Almutairi, A.W., Saif, I., Ting, L., Wang, Q., Mustafa, A., & Ebaid, R. (2025). Seaweed valorization as anaerobic co-substrate with fat, oil, and grease: Biomethane potential and microbial dynamics. *Bioresource Technology*, 418, 132158.

Fan, G. (2014). Biodegradation of fat, oil, & grease (FOG) in wet wells. [Master of Science dissertation, University of Alberta] University of Alberta education and research archives: https://era.library.ualberta.ca/items/56a43a9f-9d91-4261-9bc5-869b42506964

Formenti, L.R., Nørregaard, A., Bolic, A., Hernandez, D.Q., Hagemann, T., Heins, A.L., & Gernaey, K.V. (2014). Challenges in industrial fermentation technology research. *Biotechnology Journal*, 9(6), 727–738.

Fox, R.D. (1996). Physical/chemical treatment of organically contaminated soils and sediments. *Journal of the Air & Waste Management Association*, 46, 391–413.

Fruergaard, T., Christensen, T.H., & Astrup, T. (2010). Energy recovery from waste incineration: Assessing the importance of district heating networks. *Waste Management,* 30, 1264–1272.

Gaogane, G.J., Sekoai, P., & Trois, C. (2025). Sustainable hydrogen production via dark fermentation of protein- and lipid-rich municipal organic waste: Digestate reuse and ammonia mitigation strategies. *Fermentation*, 11(11), 623.

Gautam, K., Sharma, P., Gaur, V.K., Gupta, P., Pandey, U., Varjani, S., … & Chang, J.S. (2023). Oily waste to biosurfactant: A path towards carbon neutrality and environmental sustainability. *Environmental Technology & Innovation*, 30, 103095

Gómez Cardozo, J.R., Mora Martínez, A.L., Yepes Pérez, M., & Correa Londoño, G.A. (2016). Production and characterization of polyhydroxyalkanoates and native microorganisms synthesized from fatty waste. *International Journal of Polymer Science*, 2016, 1–12.

Grosso, M., Motta, A., & Rigamonti, L. (2010). Efficiency of energy recovery from waste incineration, in the light of the new Waste Framework Directive. Waste Management, 30, 1238–1243.

Guncheva, M., & Zhiryakova, D. (2011). Catalytic properties and potential applications of Bacillus lipases. *Journal of Molecular Catalysis B: Enzymatic*, 68(1), 1–21.

Gunstone, F.D. (2004). *The chemistry of oils and fats: Sources, composition, properties, and uses.* Blackwell Publishing.

Gupta, R., Beg, Q.K., & Lorenz, P. (2002). Bacterial alkaline proteases: Molecular approaches and industrial applications. *Applied Microbiology and Biotechnology*, 59, 15–32.

Gurd, C., Villa, R., & Jefferson, B. (2020). Understanding why fat, oil, and grease (FOG) bioremediation can be unsuccessful. *Journal of Environmental Management*, 267, 110647.

Haque, M.A., Priya, A., Hathi, Z.J., Qin, Z.-H., Mettu, S., & Lin, C.S.K. (2022). Advancements and current challenges in the sustainable downstream processing of bacterial polyhydroxyalkanoates. *Current Opinion in Green and Sustainable Chemistry*, 36, 100631.

Harwood, C.R., & Kikuchi, Y. (2022). The ins and outs of *Bacillus* proteases: activities, functions and commercial significance. *FEMS Microbiology Reviews*, 46(1), fuab046.

Hassan, S.W.M., El Latif, H.H.A., & Ali, S.M. (2018). Production of cold-active lipase by free and immobilized marine *Bacillus cereus* HSS: Application in wastewater treatment. *Frontiers in Microbiology*, 9, 2377.

Hendricks, A.A. (2015). Isolation and characterisation of lipolytic bacteria and investigation of their ability to degrade fats, oils, and grease in grain distillery wastewater. [Master of Science dissertation, Stellenbosch University] Stellenbosch University repository: https://hdl.handle.net/10019.1/97059

Hidalgo, D., Sastre, E., Gómez, M., & Nieto, P. (2012). Evaluation of pre-treatment processes for increasing biodegradability of agro-food wastes. *Environmental Technology*, 33, 1497–1503.

Hisham, N.H.M.B., Ibrahim, M.F., Ramli, N., & Abd-Aziz, S. (2019). Production of biosurfactant produced from used cooking oil by *Bacillus* sp. HIP3 for heavy metals removal. *Molecules*, 24, 2617.

Holmes, A., Govan, J., & Goldstein, R. (1998). Agricultural use of *Burkholderia* (*Pseudomonas*) *cepacia*: A threat to human health? *Emerging Infectious Diseases*, 4, 221.

Hosseini, S., Curilovs, A., & Cutting, S.M. (2018). Biological containment of genetically modified *Bacillus subtilis*. *Applied and Environmental Microbiology*, 84, e02334–17.

Iorhemen, O., Hamza, R., & Tay, J. (2016). Membrane Bioreactor (MBR) technology for wastewater treatment and reclamation: membrane fouling. *Membranes*, 6, 33.

Jacob-Lopes, E., Zepka, L.Q., & Deprá, M.C. (2022). *Handbook of waste biorefinery: Circular economy of renewable energy*. Springer International Publishing Springer.

Jaeger, K.E., Dijkstra, B.W., & Reetz, M.T. (1999). Bacterial biocatalysts: Molecular biology, three-dimensional structures, and biotechnological applications of lipases. *Annual Review of Microbiology*, 53, 315–351.

Jagadeesan, Y., Meenakshisundaram, S., Saravanan, V., & Balaiah, A. (2022). Greener and sustainable biovalorization of poultry waste into peptone via Bacto-enzymatic digestion: A breakthrough chemical-free bioeconomy waste management approach. *Waste and Biomass Valorization*, 13, 3197–3219.

Jiang, Y., Wu, R., Zhou, J., He, A., Xu, J., Xin, F., … & Dong, W. (2019). Recent advances of biofuels and biochemicals production from sustainable resources using co-cultivation systems. *Biotechnology for Biofuels*, 12, 1–12.

Karim, A., Islam, M.A., Yousuf, A., Khan, M.M.R., & Faizal, C.K.M. (2019). Microbial lipid accumulation through bioremediation of palm oil mill wastewater by *Bacillus cereus*. *ACS Sustainable Chemistry and Engineering*, 7, 14500–14508.

Kavyarathna, V., Somashekhar, R., Pashumathi, N.S., Revathi, K.B., & George, K.S. (2023). Applications of biosurfactants produced from food-related waste as substrate and their role in renewable resource utilization. In I. Inamuddin, & C.O. Adetunji (Eds.), *Applications of next generation biosurfactants in the food sector* (pp. 403–447). Elsevier.

Khatib, M.F.R.A., Alqedra, F., & Alam, M.Z. (2023). Biodegradation of fats, oil and grease using microorganisms isolated from palm oil mill effluent. *IIUM Engineering Journal*, 24, 1–10.

Khattab, A.E.-N., Nashy, E., Abdel-Razek, A.G., & Abdel-Fattah, A. (2022). Enhancement of lipase production based on improved *Bacillus licheniformis* for catabolizing of edible oil wastes. *Egyptian Journal of Chemistry*, 65, 723–732.

Kim, S.Y., Yang, Y.-H., & Choi, K.-Y. (2020). Bioconversion of plant hydrolysate biomass into biofuels using an engineered *Bacillus subtilis* and *Escherichia coli* mixed-whole cell biotransformation. *Biotechnology and Bioprocess Engineering*, 25, 477–484.

Kumar, P.N., Swapna, T.H., Khan, M.Y., Reddy, G., & Hameeda, B. (2017). Statistical optimization of antifungal iturin A production from *Bacillus amyloliquefaciens* RHNK22 using agro-industrial wastes. *Saudi Journal of Biological Sciences*, 24(7), 1722–1740.

Lauprasert, P., Chansirirattana, J., & Paengjan, J. (2017). Effect of selected bacteria as bioremediation on the degradation of fats oils and greases in wastewater from cafeteria grease traps. *European Journal of Sustainable Development*, 6, 181–186.

Lee, S.Y., Wong, H.H., Choi, J.i., Lee, S.H., Lee, S.C., & Han, C.S. (2000). Production of medium-chain-length polyhydroxyalkanoates by high-cell-density cultivation of Pseudomonas putida under phosphorus limitation. *Biotechnology and Bioengineering*, 68, 466–470.

Liepins, J., Balina, K., Soloha, R., Berzina, I., Lukasa, L.K., & Dace, E. (2021). Glycolipid biosurfactant production from waste cooking oils by yeast: Review of substrates, producers and products. *Fermentation*, 7, 136.

Loperena, L., Ferrari, M.D., Díaz, A.L., Ingold, G., Pérez, L.V., Carvallo, F., … & Lareo, C. (2009). Isolation and selection of native microorganisms for the aerobic treatment of simulated dairy wastewaters. *Bioresource Technology*, 100, 1762–1766.

Lopes, M., Miranda, S.M., Belo, I. (2020). Microbial valorisation of waste cooking oils for valuable compounds production–A review. *Critical Reviews in Environmental Science and Technology*, 50, 2583–2616.

Luo, L., Zhao, C., Wang, E., Raza, A., & Yin, C. (2022). *Bacillus amyloliquefaciens* as an excellent agent for biofertilizer and biocontrol in agriculture: An overview for its mechanisms. *Microbiological Research*, 259, 127016.

Macarthur, E., & Heading, H. (2019). How the circular economy tackles climate change. *Ellen MacArthur Found*, 1, 1–71.
Malkapuram, S.T., Sharma, V., Gumfekar, S.P., Sonawane, S., Sonawane, S., Boczkaj, G., & Seepana, M.M. (2021). A review on recent advances in the application of biosurfactants in wastewater treatment. *Sustainable Energy Technologies and Assessments*, 48, 101576.
Malode, S.J., Prabhu, K.K., Mascarenhas, R.J., Shetti, N.P., & Aminabhavi, T.M. (2021). Recent advances and viability in biofuel production. *Energy Conversion and Management*: X, 10, 100070.
Marchut-Mikolajczyk, O., Drożdżyński, P., & Struszczyk-Świta, K. (2020). Biodegradation of slop oil by endophytic *Bacillus cereus* EN18 coupled with lipase from *Rhizomucor miehei* (Palatase®). *Chemosphere*, 250, 126203.
Marx, S., & Venter, R. (2014). Evaluation of waste process grease as feedstock for biodiesel production. *Waste and Biomass Valorization*, 5, 75–86.
Meyer, H.P., Minas, W., & Schmidhalter, D. (2017). Industrial-scale fermentation. In C. Wittmann, & J.C. Liao (Eds.), *Industrial biotechnology: Products and processes* (pp.1–53). Wiley.
Miao, C.H., Wang, X.F., Qiao, B., Xu, Q.M., Cao, C.Y., & Cheng, J.S. (2022). Artificial consortia of *Bacillus amyloliquefaciens* HM618 and *Bacillus subtilis* for utilizing food waste to synthetize iturin A. *Environmental Science and Pollution Research*, 29(48), 72628–72638.
Mhlongo, S.I., Ezeokoli, O.T., Roopnarain, A., Ndaba, B., Sekoai, P.T., Habimana, O., & Pohl, C.H. (2021). The potential of single-cell oils derived from filamentous fungi as alternative feedstock sources for biodiesel production. *Frontiers in Microbiology*, 12, 637381.
Minteer, S.D. (2017). *Enzyme stabilization and immobilization*. Springer New York.
Mohamadou, B.-A., Mbofung, C.M., & Barbier, G. (2013). Genotypic and phenotypic diversity among *Bacillus* species isolated from Mbuja, a Cameroonian traditional fermented condiment. *African Journal of Biotechnology*, 12(12), 1335–1343.
Mohapatra, S., Maity, S., Dash, H.R., Das, S., Pattnaik, S., Rath, C.C., & Samantaray, D. (2017). *Bacillus* and biopolymer: Prospects and challenges. *Biochemistry and Biophysics Reports*, 12, 206–213.
Mosholi, T., & Cloete, C.E. (2018). Fats, oils and greases in effluent streams from shopping centres. *WIT Transactions on Ecology and the Environment*, 215, 465–476.
Moshood, T.D., Nawanir, G., Mahmud, F., Mohamad, F., Ahmad, M.H., & AbdulGhani, A. (2022). Sustainability of biodegradable plastics: New problem or solution to solve the global plastic pollution? *Current Research in Green and Sustainable Chemistry*, 5, 100273.
Murtaza, B., Rahman, M.S., Xu, C.C., Zhu, T., & Qin, W. (2024). Environmental impact associated with oil and grease and their emerging mitigation strategies. *Waste and Biomass Valorization,* 15, 3913–3928.
Najjar, A., Hassan, E.A., Zabermawi, N., Saber, S.H., Bajrai, L.H., Almuhayawi, M.S., & Harakeh, S. (2021). Optimizing the catalytic activities of methanol and thermotolerant *Kocuria flava* lipases for biodiesel production from cooking oil wastes. *Scientific Reports*, 11(1), 13659.
Narisetty, V., Adlakha, N., Singh, N.K., Dalei, S.K., Prabhu, A.A., Nagarajan, S., … & Kumar, V. (2022). Integrated biorefineries for repurposing of food wastes into value-added products. *Bioresource Technology*, 363, 127856.
Nayak, N.S., Purohit, M.S., Tipre, D.R., & Dave, S.R. (2020). Biosurfactant production and engine oil degradation by marine halotolerant *Bacillus licheniformis* LRK1. *Biocatalysis and Agricultural Biotechnology*, 29, 101808.
Ngwasiri, P.N., Ambindei, W.A., Adanmengwi, V.A., Ngwi, P., Mah, A.T., Ngangmou, N.T., … & Aba, E.R. (2022). A review paper on agro-food waste and food by-product valorization into value added products for application in the food industry: Opportunities and challenges for Cameroon bioeconomy. *Asian Journal of Biotechnology and Bioresource Technology*, 8(3), 32–61.

Oliveira, J.G.D., & Garcia-Cruz, C.H. (2013). Properties of a biosurfactant produced by *Bacillus pumilus* using vinasse and waste frying oil as alternative carbon sources. *Brazilian Archives of Biology and Technology*, 56, 155–160.

Oliveira, K.W.d., Faccioli, Y.E.d. S., de Araújo, G.P., Converti, A., da Silva, R.d.C.F.S., & Sarubbo, L.A. (2026). From the problem of corrosion to green solutions: The role of biosurfactants as anti-corrosion agents. *Materials*, 19(4), 743.

Ousaadi, M.I., Merouane, F., Berkani, M., Almomani, F., Vasseghian, Y., & Kitouni, M. (2021). Valorization and optimization of agro-industrial orange waste for the production of enzyme by halophilic *Streptomyces* sp. *Environmental Research*, 201, 111494.

Paddon, C.J., Westfall, P.J., Pitera, D.J., Benjamin, K., Fisher, K., McPhee, D., & Newman, J.D. (2013). High-level semi-synthetic production of the potent antimalarial artemisinin. *Nature*, 496(7446), 528–532.

Pan, Y., Wang, W., Dai, C., Zhou, M., Zhang, L., Sun, L., Cai, M., & He, R. (2026). Bacillus-derived lipopeptides as green biosurfactants: Structures, functions, production, and safety for sustainable applications. *Comprehensive Reviews in Food Science and Food Safety*, 25(3), e70508.

Pandey, A., Nigam, P., Soccol, C.R., Soccol, V.T., Singh, D., & Mohan, R. (2000). Advances in microbial amylases. *Biotechnology and Applied Biochemistry*, 31(2), 135–152.

Pandit, N.K., & Meena, S.S. (2026). Exploring sustainable biosurfactant production through waste valorization: Emerging research trends and industrial applications. *Waste and Biomass Valorization*, 17, 585–619.

Pardhi, D.S., Panchal, R.R., Raval, V.H., Joshi, R.G., Poczai, P., Almalki, W.H., & Rajput, K.N. (2022). Microbial surfactants: A journey from fundamentals to recent advances. *Frontiers in Microbiology*, 13, 982603.

Penner, S.S., Chang, D.P.Y., Goulard, R., & Lester, T. (1988). Waste incineration and energy recovery. *Energy*, 13(12), 845–851.

Pérez-Rivero, C., López-Gómez, J.P., & Roy, I. (2019). A sustainable approach for the downstream processing of bacterial polyhydroxyalkanoates: State-of-the-art and latest developments. *Biochemical Engineering Journal*, 150, 107283.

Polonca, S. (2020). Environment shapes the intra-species diversity of *Bacillus subtilis* isolates. *Microbial Ecology*, 79(4), 853–864.

Povolo, S., Romanelli, M.G., Fontana, F., Basaglia, M., & Casella, S. (2012). Production of polyhydroxyalkanoates from fatty wastes. *Journal of Polymers and the Environment*, 20, 944–949.

Qin, Y., Han, Y., Shang, Q., & Li, P. (2015). Complete genome sequence of *Bacillus amyloliquefaciens* L-H15, a plant growth promoting rhizobacteria isolated from cucumber seedling substrate. *Journal of Biotechnology*, 200, 59–60.

Ramírez, I.M., Tsaousi, K., Rudden, M., Marchant, R., Alameda, E.J., Román, M.G., & Banat, I. (2015). Rhamnolipid and surfactin production from olive oil mill waste as sole carbon source. *Bioresource Technology*, 198, 231–236.

Ravindran, R., & Jaiswal, A.K. (2016). A comprehensive review on pre-treatment strategy for lignocellulosic food industry waste: Challenges and opportunities. *Bioresource Technology*, 199, 92–102.

Raza, Z.A., Abid, S., & Banat, I.M. (2018). Polyhydroxyalkanoates: Characteristics, production, recent developments and applications. *International Biodeterioration & Biodegradation*, 126, 45–56.

Roets-Dlamini, Y., Moonsamy, G., Lalloo, R., & Ramchuran, S. (2022). Use of *Bacillus* spp in the bioremediation of fats, oils and greases (FOG's), and other waste substrates in food processing effluents. *Biocatalysis and Agricultural Biotechnology*, 42, 102351.

Rosas-Vega, F. E., Terán-Hilares, R., de Moura Campos, S., Santos Costa, G., Cavali, M., Lima Serra, J., Vega Arroyo, F. S., Tanobe, V., Sekoai, P., Pozzan, R., & Martínez-Burgos, W. J. (2026). Integrated enzymatic processes for next-generation biodiesel production. *Next Sustainability, 7*, 100320.

Saha, D., Purkayastha, G.D., Ghosh, A., Isha, M., & Saha, A. (2012). Isolation and characterization of two new *Bacillus subtilis* strains from the rhizosphere of eggplant as potential biocontrol agents. *Journal of Plant Pathology*, 94, 109–118.

Sandeep, B., Baranwal, U., & Mandpe, A. (2024). Biofuel production from microbial sources: Advances and challenges. *Harnessing Microbial Potential for Multifarious Applications*, 323–357.

Sangeetha, R., Arulpandi, I., & Geetha, A. (2011). Bacterial lipases as potential industrial biocatalysts: An overview. *Research Journal of Microbiology*, 6(1), 1.

Sanghamitra, P., Mazumder, D., & Mukherjee, S. (2021). Treatment of wastewater containing oil and grease by biological method-A review. *Journal of Environmental Science and Health, Part A*, 56(4), 394–412.

Saratale, R.G., Cho, S.K., Saratale, G.D., Kadam, A.A., Ghodake, G.S., Kumar, M., & Shin, H.S. (2021). A comprehensive overview and recent advances on polyhydroxyalkanoates (PHA) production using various organic waste streams. *Bioresource Technology*, 325, 124685.

Sekoai, P.T., & Habimana, O. (2026). Biochar-driven circular bioeconomy in South Africa: Waste valorization for climate and socioeconomic resilience. *Scientific African*, 32, e03390.

Sekoai,P.,Ezeokoli,O.,Yoro,K.,Eterigho-Ikelegbe,O.,Habimana,O.,Iwarere,S.,Daramola,M.,& Ojumu, T. (2022). The production of polyhydroxyalkanoates using volatile fatty acids derived from the acidogenic biohydrogen effluents: *An overview. Bioresource Technology Reports*, 18, 101111.

Shah, S., & Kumar, A. (2021). Production and characterization of polyhydroxyalkanoates from industrial waste using soil bacterial isolates. *Brazilian Journal of Microbiology*, 52, 715–726.

Sharma, P., Gaur, V.K., Kim, S.H., & Pandey, A. (2020). Microbial strategies for biotransforming food waste into resources. *Bioresource Technology*, 299, 122580.

Sharma, P., Sharma, N., Pathania, S., & Handa, S. (2017). Purification and characterization of lipase by *Bacillus methylotrophicus* PS3 under submerged fermentation and its application in detergent industry. *Journal of Genetic Engineering and Biotechnology*, 15(2), 369–377.

Sharma, R., Chisti, Y., & Banerjee, U.C. (2001). Production, purification, characterization, and applications of lipases. *Biotechnology Advances*, 19(8), 627–662.

Silva, J.B., Pereira, J.R., Marreiros, B.C., Reis, M.A., & Freitas, F. (2021). Microbial production of medium-chain length polyhydroxyalkanoates. *Process Biochemistry*, 102, 393–407.

Sohail, R., Jamil, N., Ali, I., & Munir, S. (2020). Animal fat and glycerol bioconversion to polyhydroxyalkanoate by produced water bacteria. *e-Polymers*, 20(1), 92–102.

Sultana, S., Sultana, R., Al-Mansur, M.A., Akbor, M.A., Bhuiyan, N.A., Ahmed, S., … & Jamal, A.S.I.M. (2024). An industrially potent rhamnolipid-like biosurfactant produced from a novel oil-degrading bacterium, *Bacillus velezensis* S2. *RSC Advances*, 14,24516–24533.

Sundaram, T., Govindarajan, R.K., Vinayagam, S., Krishnan, V., Nagarajan, S., Gnanasekaran, G.R., & Rajamani Sekar, S.K. (2024). Advancements in biosurfactant production using agro-industrial waste for industrial and environmental applications. *Frontiers in Microbiology*, 15, 1357302.

Surendra, K.C., Angelidaki, I., & Khanal, S.K. (2022). Bioconversion of waste-to-resources (BWR-2021): Valorization of industrial and agro-wastes to fuel, feed, fertilizer, and bio-based products. *Bioresource Technology*, 347, 126739.

Tabatabaei, M., Aghbashlo, M., Dehhaghi, M., Panahi, H.K.S., Mollahosseini, A., Hosseini, M., & Soufiyan, M.M. (2019). Reactor technologies for biodiesel production and processing: A review. *Progress in Energy and Combustion Science*, 74, 239–303.

Teixeira, P.D., Silva, V.S., & Tenreiro, R. (2020). Integrated selection and identification of bacteria from polluted sites for biodegradation of lipids. *International Microbiology*, 23(3), 367–380.

Tolibia, S.E.M., Pacheco, A.D., Balbuena, S.Y.G., Rocha, J., & López y López, V.E. (2023). Engineering of global transcription factors in Bacillus, a genetic tool for increasing product yields: A bioprocess overview. *World Journal of Microbiology and Biotechnology*, 39(1), 12.

Tripathi, N.K. (2016). Production and purification of recombinant proteins from *Escherichia coli. ChemBioEng Reviews*, 3(3), 116–133.

Tripathi, V., Gaur, V.K., Dhiman, N., Gautam, K., & Manickam, N. (2020). Characterization and properties of the biosurfactant produced by PAH-degrading bacteria isolated from contaminated oily sludge environment. *Environmental Science and Pollution Research*, 27, 27268–27278.

Tzirita, M. (2012). A characterisation of bioaugmentation products for the treatment of waste fats, oils and grease *(FOG)* [Doctoral dissertation, Dublin City University].

Tzirita, M., Papanikolaou, S., & Quilty, B. (2019). Enhanced fat degradation following the addition of a *Pseudomonas* species to a bioaugmentation product used in grease traps. *Journal of Environmental Sciences*, 77, 174–188.

Tzirita, M., Papanikolaou, S., & Quilty, B. (2021). A study of the suitability of three commercial bioaugmentation products for use in grease traps. *Biomass Conversion and Biorefinery*, 11, 907–924.

Tzirita, M., Papanikolaou, S., Chatzifragkou, A., & Quilty, B. (2018). Waste fat biodegradation and biomodification by *Yarrowia lipolytica* and a bacterial consortium composed of *Bacillus* spp. and *Pseudomonas putida. Engineering in Life Sciences*, 18(12), 932–942.

Ushani, U., Sumayya, A.R., Archana, G., Banu, J.R., & Dai, J. (2020). Enzymes/biocatalysts and bioreactors for valorization of food wastes. In R.J. Banu, G. Kumar, M. Gunasekaran, & S. Kavitha (Eds), *Food waste to valuable resources* (pp. 211–233). Academic Press.

Utlang, N.M.S., Utlang, S.M.S., Paler, E.M.L., Salvatierra, R.C., Paday, J.C., Mugot, D.A., … & Arazo, R.O. (2024). Closing the loop: Valorization of industrial fats, oils, and grease as a biodiesel source. *Circular Economy and Sustainability*, 4, 2397–2412.

Vatsa, P., Sanchez, L., Clement, C., Baillieul, F., & Dorey, S. (2010). Rhamnolipid Biosurfactants as New Players in Animal and Plant Defense against Microbes. *International Journal of Molecular Sciences*, 11(12), 5095–5108.

Vedaraman, N., & Venkatesh, N. (2011). Production of surfactin by Bacillus subtilis MTCC 2423 from waste frying oils. *Brazilian Journal of Chemical Engineering*, 28, 175–180.

Vilas-Boas, G., Sanchis, V., Lereclus, D., Lemos, M.V.F., & Bourguet, D. (2002). Genetic differentiation between sympatric populations of *Bacillus cereus* and *Bacillus thuringiensis. Applied and Environmental Microbiology*, 68(3), 1414–1424.

Voet, D., & Voet, J.G. (2010). *Biochemistry*. John Wiley & Sons.

Vojnovic, S., Aleksic, I., Ilic-Tomic, T., Stevanovic, M., & Nikodinovic-Runic, J. (2024). *Bacillus* and *Streptomyces* spp. as hosts for production of industrially relevant enzymes. *Applied Microbiology and Biotechnology*, 108, 185.

Wallace, T., Gibbons, D., O'Dwyer, M., & Curran, T.P. (2017). International evolution of fat, oil and grease (FOG) waste management–A review. *Journal of Environmental Management*, 187, 424–435.

Wang, Y., Ling, C., Chen, Y., Jiang, X., & Chen, G.Q. (2019). Microbial engineering for easy downstream processing. *Biotechnology Advances*, 37(6), 107365.

Wang, Y., Zhang, C., Liang, J., Wu, L., Gao, W., & Jiang, J. (2020). Iturin A extracted from *Bacillus subtilis* WL-2 affects *Phytophthora infestans* via cell structure disruption, oxidative stress, and energy supply dysfunction. *Frontiers in Microbiology*, 11, 536083.

Wong, J.Y., Ngieng, N.S., Husaini, A., Saat, R., & Hussain, H. (2025). Influence of pH on the biodegradation efficiency of fats, oils, and grease by biosurfactant-producing bacterial consortia. *Biodegradation*, 36, 50.

Yang, S. (2007). *Bioprocessing for value-added products from renewable resources: New technologies and applications*. Elsevier.

Zalma, S.A., & El-Sharoud, W.M. (2021). Diverse thermophilic *Bacillus* species with multiple biotechnological activities are associated within the Egyptian soil and compost samples. *Science Progress*, 104(4), 00368504211055277.

Zhao, H., Zhao, X., Lei, S., Zhang, Y., Shao, D., Jiang, C., ... & Shi, J. (2019). Effect of cell culture models on the evaluation of anticancer activity and mechanism analysis of the potential bioactive compound, iturin A, produced by *Bacillus subtilis*. *Food and Function*, 10(3), 1478–1489.

Zhao, S., Hu, N., Chen, Z., Zhao, B., & Liang, Y. (2009). Bioremediation of reclaimed wastewater used as landscape water by using the denitrifying bacterium *Bacillus cereus*. *Bulletin of Environmental Contamination and Toxicology*, 83, 337–340.

Zhao, Z.Q., Zhang, T., Shi, T.Q., Barceló, D., & Zheng, H.B. (2024). Biodegradation and high-value utilization of waste cooking oil: Strategies and mechanisms for producing lipopeptide biosurfactants using *Bacillus subtilis* YZQ-2. *Journal of Environmental Chemical Engineering*, 12(6), 114549.

5 Emerging Bioaugmentation Strategies for Biogas-Producing Microbes

Patrick Sekoai, Obinna Ezeokoli, and Freedom Tshabuse

5.1 INTRODUCTION

There is now irrefutable evidence for the detrimental effects associated with a high reliance on the fossil-based linear economy. These effects include greenhouse gas emissions, environmental pollution, resource depletion, water shortages, food insecurity, and rising energy costs (Sekoai et al., 2023a, 2023b). To mitigate these effects, there has been a shift toward renewable fuels.

Biogas is one of the cornerstones of renewable fuels that will spearhead the United Nations Sustainable Development Goals while ameliorating the environmental scourges related to the continual disposal of biowastes in landfills (Emezirinwune et al., 2024). The technology of biogas generation is at an advanced stage in developed nations, with more than 18,000 biogas plants in the European Union (EU) alone; this number is expected to increase in the coming years (Menzel et al., 2020). The growth in biogas generation technology is also supported by legislative frameworks such as the EU's Bioeconomy Strategy, Circular Economy Plan, and EU Waste Policy (Zhu et al., 2024). In addition, EU countries are mandated to achieve 30% renewable resources in the energy sector and 10% renewable resources in the transport sector, before the year 2030 (Hughes and Jones, 2020). Likewise, developing nations such as South Africa follow a similar path, as evidenced by recent policies such as the National Development Plan 2030, the South African Department of Science and Innovation's Decadal Plan, the Bioeconomy Strategy, the Waste Economy Master Plan, and the Forestry Master Plan (Sekoai et al., 2023a).

Biogas technology is well aligned with the burgeoning global circular bioeconomy trends due to its (1) ability to valorize the overabundant organic wastes, including industrial and municipal side streams; (2) contributions to the 'waste-to-value' innovations; (3) contributions to social upliftment and economic development in peri-urban

DOI: 10.1201/9781003745150-5

communities; (4) ability to use mesophilic conditions as opposed to energy-intensive processes such as pyrolysis and electrolysis; (5) ability to be converted into high methane content, which can be used in natural gas pipelines; (6) attainment of large-scale implementation, which implies the potential for industrialization; and (7) provision of a more cost-effective route of producing 'clean' energy (Sekoai et al., 2020a, 2020b).

The biogas production process is a sequential four-stage process of anaerobic digestion that is mediated by diverse microorganisms spanning several taxonomic divisions (phyla), including *Pseudomonadota* (synonym *Proteobacteria*), *Bacillota* (synonym *Firmicutes*), *Bacteroidota* (synonym *Bacteroidetes*), and *Euryarchaeota* (Zhao et al., 2024). During the biogas production process, these taxonomic groups establish complex metabolic networks through symbiotic and/or synergistic biochemical relations with other microorganisms that directly or indirectly contribute to the enrichment of biogas production (Sekoai et al., 2022).

5.2 SIGNIFICANCE AND CONTRIBUTION TO KNOWLEDGE

Despite the R&D strides that have been accomplished in this field, there is a paucity of literature that discusses bioaugmentation strategies for the enrichment of biogas-producing microbial communities. This paucity is partly due to the placement of emphasis on ameliorating the engineering dynamics that contribute to the low biogas conversion efficiency in anaerobic biogas systems while overlooking the key microorganisms that regulate this bioprocess. The composition and metabolic variations of the microbial groups in biogas systems have been shown to provide invaluable insights into process stability and effective bioreactor operations (Demirel and Scherer, 2008). From a practical standpoint and toward long-term stable process performance, obtaining an in-depth understanding of the ecophysiology of microbial assemblages that thrive during biogas accumulation is vital, as it will enable researchers and industries to develop robust bioprocess methods that boost the proliferation (and potentially the activity) of these microbes. Yet, this remains an untapped scientific route for biogas production and has not been thoroughly reviewed in the scholarly literature. Therefore, this chapter seeks to review the innovative biochemical bioaugmentation strategies that apply to the optimization of biogas-producing microbial communities. Robust methods, such as cell encapsulation and the use of nanoparticles, multivariate tools, optimal bioreactor designs, and integrated biorefinery processes for the enrichment of biogas-producing microbes, are presented herein. Furthermore, the knowledge gaps and barriers associated with the proposed bioaugmentation strategies are explored. This chapter will provide profound insights into the bioprocess methods that stimulate the biogas-producing microbial consortia and can help biogas experts address the common biochemical disturbances and variations during anaerobic digestion.

5.3 INSIGHTS INTO THE BIOGAS-PRODUCING COMMUNITIES

5.3.1 Hydrolytic Consortia

Hydrolytic bacterial species are mainly associated with the breakdown of substrates. These species establish complex interconnected biochemical networks to convert substrates into monomeric units using secreted enzymes (Lin, 2022). Herein, polymeric substances, such as carbohydrates, lipids, and proteins, are hydrolyzed into monomeric sugars, volatile fatty acids (VFAs), and amino acids (Lin, 2022). These hydrolytic products are later utilized in the fermentation step, which is the second step of the anaerobic process (Sekoai et al., 2018a, 2018b, 2018c). It has been shown that hydrolytic bacteria in anaerobic biogas systems are phylogenetically diverse, but *Bacillota* and *Bacteroidota* are the two dominant phyla (Lim et al., 2020). In addition, hydrolytic bacteria are tolerant to varying environmental conditions such as temperature and pH, thus making them potent biocatalysts for breaking down an array of feedstocks during the anaerobic process (Sharma et al., 2020). These bacterial species secrete various extracellular hydrolases, such as lipases, xylanases, cellulases, pectinases, and amylases, that are responsible for the hydrolysis of feedstocks (Karray et al., 2018). Numerous studies have characterized the hydrolytic bacterial community during anaerobic digestion. Some of these studies have shown that a plethora of microorganisms have a hydrolytic profile. These include *Clostridium* spp., *Bacillus* spp., *Enterobacter* spp., *Klebsiella* spp., *Pseudomonas* spp., *Serratia* spp., and *Lactobacillus* spp. (Mazzucotelli et al., 2013; Aderonke et al., 2017; Alrumman et al., 2018). Studies have also shown that fungal species exhibit a hydrolytic profile. Notable hydrolytic fungal species include those of the genera *Aspergillus*, *Trichoderma*, and *Penicillium* (Harirchi et al., 2022). The hydrolytic efficiency of these microbial cell factories has led to breakthroughs in this scientific domain, as some major biofuel and agricultural corporations have pioneered the genetic engineering of some hydrolytic species belonging to the genera *Trichoderma* and *Aspergillus* (Park et al., 2017; Frisvad et al., 2018; Kumar et al., 2023).

5.3.2 Acidogenic Bacteria

Acidogenic bacteria (also referred to as acidogens) are taxonomically diverse bacterial populations predominantly from the *Bacillota* phylum and a few species are from the *Bacteroidota* and *Pseudomonadota* phyla (Sekoai et al., 2023). These species participate in the second stage of the anaerobic digestion process, where the hydrolytic products (from the first stage) are primarily converted into short-chain VFAs such as acetate, butyrate, valerate, propionate, isobutyrate, hydrogen gas (H_2), and carbon dioxide (CO_2); these are the major constituents of acidogenic fermentation (Sekoai et al., 2018). The conversion of hydrolytic products into short-chain VFAs involves a complex biochemical pathway mediated by an array of microbes, including hydrolytic and non-hydrolytic microbial communities (Sekoai et al., 2020). The diverse microbial communities provide beneficial symbiotic and/or synergistic interactions with one another during the formation of the anaerobic metabolites (Sekoai et al., 2016). The second stage is also dominated by obligate and facultative anaerobes, particularly *Clostridium* spp. and *Bacillus* spp., which are the main acidogenic bacterial species regulating the whole process of acidogenesis (Sekoai et al., 2021). These species are fast-growing and typically thrive under acidic conditions (pH 3.0–5.0) and at short hydraulic retention times (HRT of 8–48 h) in contrast to methanogenic archaea, which are known to proliferate under alkaline conditions (pH 7.0) and at elevated HRTs (Sekoai and Kana, 2015). A deeper understanding of this anaerobic route has assisted scientists who specialize in renewable energy technologies in using these microbial consortia as potent biocatalysts for the biochemical production of H_2. However, renewable H_2 technology is still plagued by low yields, which hinders its large-scale implementation. The optimal setpoint conditions (e.g., substrate concentration, pH, temperature, HRT, mode of fermentation, reactor design, and inoculum type) that contribute to the proliferation of acidogenic microbial consortia are well covered in the literature (Sekoai et al., 2018; Sekoai et al., 2020; Sekoai et al., 2021).

5.3.3 Acetogenic Bacteria

During the third stage of the anaerobic digestion process (termed acetogenesis), acetogenic bacteria play vital roles in mediating the reductive acetogenesis and syntrophic oxidation of organic acids. The activities of acetogenic bacteria during this stage result in the generation of additional fermentative byproducts such as VFAs, H_2, CO_2, and formate (Westerholm et al., 2016; Timmers et al., 2018). Thus, acetogenic bacterial species serve as a vital link between the hydrolysis-acidogenic microbial communities and the methanogenic archaea (Sekoai et al., 2018). The assessment of acetogens in anaerobic systems has revealed the prevalence of different genera, including *Acetobacterium*, *Actinobacterium*, *Acetivibrio*, *Aciditerrimonas*, *Clostridium*, and *Thermoanaerobacter* (Kushkevych et al., 2019; Singh et al., 2021a, 2021b). Therefore, it is evident that acetogenesis is very important as the intermediates produced herein act as key precursors during methanogenesis, whereby hydrogenotrophic methanogens and/or acetoclastic methanogens metabolize these biological constituents to generate biogas as detailed in Section 5.3.4. It has been shown that the cultivation conditions also influence the proliferation of potent acetogenic-producing bacteria. For example, both mesophilic (30°C–37°C) and thermophilic (40°C–60°C) conditions have been shown to be optimal for acetogenic communities, including syntrophic acetate-oxidizing bacteria, with elevated temperatures favoring the growth of these consortia (Westerholm et al., 2016; Singh et al., 2021). Interestingly, these types of bacterial species can withstand high concentrations of ammonia and VFAs during anaerobic fermentations, making them ideal for anaerobic systems (Westerholm et al., 2016).

5.3.4 Methanogenic Archaea

During the final stage of the anaerobic digestion process (termed methanogenesis), methanogenic archaeal species use the fermented metabolites generated from the acidogenic and acetogenic pathways – the metabolites consist primarily of organic acids, H_2, CO_2, and other trace compounds (Conrad, 2007; Prakash et al., 2024; Ungerfeld and Pitta, 2024). These specialized species produce biogas comprising 50%–70% CH_4, 20%–40% CO_2, 5%–10% H_2, and small concentrations of other fermentable byproducts such as ammonia and sulfur dioxide (Rusanowska et al., 2023). The process is mediated by a specialized group of archaeal species that are classified into three groups based on their metabolic pathways: (1) hydrogenotrophic methanogens, which use H_2 as a precursor for methane formation via the hydrogenotrophic biochemical route; (2) acetotrophic/acetoclastic methanogens, which use

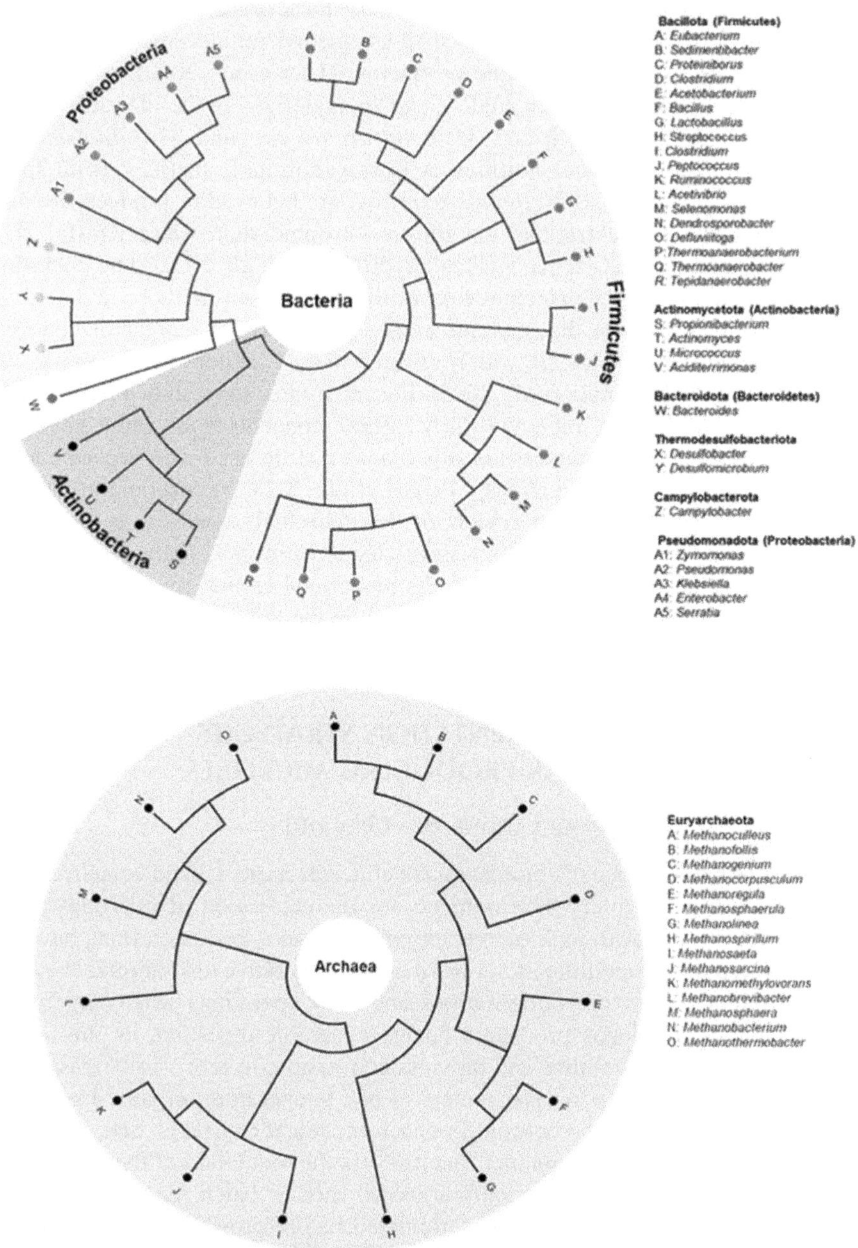

FIGURE 5.1 The phylogenetic relationship of some (A) bacterial and (B) archaeal genera involved in the biogas production process. Phylogenetic relatedness is based on the 16S ribosomal RNA gene. The phylogenetic dendrogram was constructed in Molecular Evolutionary Genetics Analysis (MEGA) version X (Tamura et al., 2021).

acetate as a precursor for methane production via the acetoclastic biochemical route; and (3) methylotrophic methanogens, which use methyl sulfides, methylamines, and methanol as precursors for methane production (Lim et al., 2020). A plethora of archaeal communities have been studied and characterized in the literature. Among these, species affiliated with the *Methanosarcinaceae* and *Methanotrichaceae* (formerly *Methanosaetaceae*) families or hydrogenotrophic methanogenic species (e.g., *Methanospirillum* and *Methanobacterium*) are believed to produce the maximum biomethane concentrations via the acetotrophic route (Kao-Kniffin et al., 2010; Rachbauer et al., 2016; Kougias et al., 2017; Pan et al., 2021). The operational setpoint conditions play a principal role in the enrichment of these archaeal species, which in turn affect the methane content; the optimal bioreactor conditions that contribute to high yields are widely covered in the literature (Glivin et al., 2021; Kasinath et al., 2021; Singh et al., 2023). Recent studies have shown that it is possible to attain high-purity methane (CH_4 > 95%) concentrations using novel scalable biogas upgrading strategies and this biogas can be used as a replacement for natural gas in natural gas pipelines (Gkotsis et al., 2023; Yellezuome et al., 2023; Li et al., 2024a, 2024b). An overview of the microbial consortia, notably bacteria and archaea, involved in the biogas production process is summarized in the phylogenetic dendrogram in Figure 5.1. As mentioned earlier, the bacterial species are predominantly of the *Bacillota* phylum, while the archaeal genera all belong to the *Euryarchaeota* phylum.

5.4 EMERGING BIOAUGMENTATION STRATEGIES TARGETING BIOGAS-PRODUCING MICROBES

5.4.1 Bioreactor Configurations and Operations

In the field of bioprocessing, a bioreactor is defined as a specialized vessel/fermenter that provides suitable micro-environments for the cultivation of microbes (Sekoai et al., 2019a, 2019b). Although bioreactor configurations and operations have been widely covered in the scientific literature, it is still imperative to scrutinize the effects of the different bioreactor configurations and their operations when targeting the bioaugmentation of biogas producers during anaerobic digestion, as this impacts the long-term process stability and biomass conversion, in terms of biogas production. The operational setpoint parameters of pH, temperature, organic loading rate (OLR), and HRT need to be optimal in anaerobic reactor systems, and a bioreactor must be fabricated in such a manner that it favors the regulation of these parametric conditions. Different bioreactor configurations, such as batch reactors (BRs), continuous stirred tank reactors (CSTRs), fluidized bed reactors, anaerobic membrane reactors, up-flow anaerobic sludge blanket reactors (UASBRs), immobilized reactors, expanded granular sludge bed reactors, and hybrid anaerobic reactors have all been explored and the performance of each bioreactor system in terms of biogas concentration is summarized from literature reports in Table 5.1. Among these bioreactors, UASBRs appear to be the most commonly used in biogas studies and this is attributable to their various advantages, including maximum biogas production yields, high chemical oxygen demand removal efficiency, ability to withstand process

TABLE 5.1
Summary studies of biogas production from different feedstocks

Reactor design	Feedstock	Inoculum source	HRT (day)	Temperature (°C)	OLR	pH	CH_4 (%)	References
BR	Pig manure	Anaerobic sludge	–	9–55	–	–	67.9	Tian et al. (2018)
CSTR	Sugar beet	Anaerobic sludge	10	37	3.0	–	–	Moraes et al. (2015)
BR	Coffee waste	Chicken manure and anaerobic sludge	40	35	10.28 g COD/L/day	7.28	79.27	Battista et al. (2016)
UASBR	Coffee wastewater	Seed sludge	0.75	55	1–8 g COD/L/day	–	73	Jung et al. (2012)
UASBR	Coffee effluent	Activated sludge	4.0, 5.2, and 6.2	–	3–5.8 g COD/L/day	7.0	69–89	Bruno and de Oliveira (2013)
UASBR	Vinasse	Granular sludge	26	30	17.05 kg COD/m³/day	7.5	70–74	España-Gamboa et al. (2012)
UASBR	Sugarcane vinasse	Anaerobic sludge	8–24	54–56	45 g COD/L/day	8.64	73 and 83	de Barros et al. (2017)
UAFBR	Coffee wastewater	Anaerobic sludge	1.06	32.9	0.81–4.41 kg/m³/day	4.0–4.8	–	Fia et al. (2012)
BR	OFMSW	Cow dung	30	55	–	7.6	92–97	Sekoai et al. (2020)
UASBR	Food waste	Granular sludge	2.2	35	1.6–13.2 COD/L/day	4.5	67–75	Moon and Song (2011)
UASBR	Brewery spent grain	Seed sludge	0.66	35	–	6.5	55–66	Čater et al. (2015)
BR	Food waste hydrolysate	Seed sludge	–	35	–	6.95	76.42	Pomdaeng et al. (2024)
IR	Glucose	Seed sludge	–	30–70	–	7.0	–	Satar et al. (2017)

Abbreviations: –, not reported; BR, batch reactor; COD, chemical oxygen demand; CSTR, continuous stirred tank reactors; HRT, hydraulic retention time; IR, immobilized reactor; OFMSW, organic fraction of municipal solid waste; OLR, organic loading rate; UAFBR, up-flow anaerobic fixed biofilm reactor; UASBR, up-flow anaerobic sludge blanket reactor.

instabilities, simple operation and low maintenance costs, and fast start-up periods (Daud et al., 2018). It is noteworthy to mention that BRs and CSTRs are also widely applied in biogas studies due to ease of operation and stable long-term process performance, respectively.

5.4.2 Cell Immobilization

Cell immobilization (also known as cell encapsulation) is an approach that entails the entrapment of microbial cells inside a permeable biocarrier support matrix such as natural polymers, carbonaceous materials, inorganic materials, and synthetic polymers (Sekoai et al., 2018). In this approach, the permeable biocarrier matrix is used to encapsulate the biogas-producing microorganisms (Lapponi et al., 2022). In contrast to suspended cells, immobilized cells possess outstanding attributes that make them desirable in anaerobic digestion systems. Immobilized cells are able to withstand growth inhibitors, maintain the pH of the medium, prolong the biogas fermentation process, provide a high-surface-area-to-volume ratio, shorten the lag phase duration, and prevent cell washout (Lapponi et al., 2022). It has been shown that a CH_4-rich (75% v/v) biogas can be achieved using immobilized cells (Senko et al., 2019). In terms of microbial ecology, it has been reported that this anaerobic bioaugmentation strategy is beneficial as it enriches the acidogenic and methanogenic consortia (Jensen et al., 2019). The support materials also promote the proliferation of key biofilm-forming microbial communities such as *Clostridium* spp. and *Methanosarcina* spp., which regulate acidogenesis and methanogenesis (Senko et al., 2019). In a biohythane production process that utilized a reactor system embedded with immobilized cells, high biohythane conversion efficiency, long-term process stability, and resistance to high organic loading were observed (Germec et al., 2020). In addition, the immobilized cells were reusable in subsequent fermentation processes (Germec et al., 2020). In most anaerobic studies involving these biocatalysts, natural polymers, such as collagen, keratin, alginate, agarose, cellulose, and lignin, are usually preferred due to their high permeability, affordability, biodegradability, non-toxic nature, and large surface area (Ivanova et al., 2011). The major barrier associated with these natural polymers is that they suffer from weak mechanical stability, but this issue is addressed by coupling them with strong materials such as ethyl glycol, metals, polyvinyl alcohol, polyvinyl chloride, activated carbons, and mesoporous silica (Yang et al., 1988). Cell immobilization strategies are gaining widespread prominence as they are now being used in other fields such as environmental bioremediation, wastewater treatment, and biofuel technologies.

5.4.3 Applications of Biochar

Biochar is a porous biomass-derived carbonaceous material produced using thermochemical conversion technologies such as pyrolysis, hydrothermal carbonization, and gasification (Wiedner et al., 2013). This material is used as a valuable bio-additive in diverse applications such as agriculture, environmental remediation, wastewater treatment, pharmaceuticals, medicine, and bioprocesses (Monga et al., 2022). There has also been a significant increase in the use of biochar in biogas production studies over the past decade due to its outstanding properties such as large-surface-area-to-volume ratio, tolerance to inhibitors, high utilization of feedstocks, promotion of biofilm-forming communities, provision of a balanced C/N ratio, provision of trace elements, and high ion-exchange capacity and adsorption abilities (Nie et al., 2023). This bio-additive is synthesized from organic feedstocks derived from different sectors such as agriculture, food processing, municipal, and forestry (Ji et al., 2022).

Biochar has numerous chemical compounds such as fixed carbon, volatile compounds, ash, and moisture compounds (Enders et al., 2012). In addition to these attributes, it has remarkable properties such as micro- and macro-elements (calcium, copper, magnesium, potassium, sodium, zinc, and iron) and functional groups (C–O, C=O, COOH, OH, and $-NH_x$), which all contribute to its porosity and reactivity (Leng et al., 2021). Biochar is also characterized by high pore size, pore volume, specific surface area (SSA), pH, electrical conductivity (EC), and cation exchange capacity (CEC) (Leng et al., 2021). Depending on the type of feedstock used, the pore size varies from 17 to 300 nm (Tan et al., 2018). It has also been shown that the SSA and pore volume of biochars typically fall within the range of 8–132 m^2/g and 0.016–0.083 cm^3/g, respectively (Leng et al., 2021). But when suitable feedstocks alongside the optimal parameters are used in pyrolysis, the SSA and pore volume can reach up to 490.8 m^2/g (Chen et al., 2008) and 0.25 cm^3/g (Yang et al., 2016), respectively. In contrast, the pH of biochar tends to be in the alkaline range of 7.0–10.0 as it is affected by the setpoint parameters, such as pyrolysis temperature and feedstock type. The EC and CEC values for biochar range between 2.2 and 5.0 mS/cm and 5 and 50 $cmol^+/kg$, respectively (María De la Rosa et al., 2018; Tarelho et al., 2020).

A study that investigated the role of biochar on the performance of the biogas production process showed that this bio-additive enhanced the acclimatization of microbial communities during the four stages of anaerobic digestion and also maximized the biogas concentration by up to 26% (Awode et al., 2022). The authors reported a maximum biogas fraction of approximately 68% (Awode et al., 2022). Interestingly, the biochar could withstand the formation of bioprocess inhibitors such as H_2S and ammonia (Awode et al., 2022). Elsewhere, it was reported that supplementing anaerobic biogas systems with biochar positively contributed to sustainable agricultural practices as it yielded an NPK-rich biodigestate that can be used as a substitute for non-environmentally friendly commercial fertilizers (Sekoai et al., 2022; Sekoai et al., 2023). In terms of its toxicity, the biochar-treated digestate is considered to be safe as it does not consist of pathogenic microorganisms that pose a threat to both humans and the environment (Maleki Shahraki and Mao, 2022). The success of biochar has resulted in the establishment of pilot-scale biochar-treated reactors targeting maximum biogas yields and a nutrient-rich digestate that serves as a biofertilizer (Li et al., 2018; Li et al., 2020; Borth et al., 2022). Data from these studies imply that biochar has the potential for industrial or real-world applications.

5.4.4 Application of Nanoparticles

The field of nanotechnology has garnered enormous attention from different research fields due to the exquisite attributes possessed by nanoparticles. Such attributes include miniaturized size (1–100 nm), large surface-to-volume ratio, strong catalytic activity, high adsorption ability, reusability, prolonged fermentation process, resistance to biochemical inhibitors, maintenance of the bioprocess pH, high biodegradation of substrates, and maximum biogas yield (Sekoai et al., 2019; Abdelwahab et al., 2020). Studies have revealed that nanoparticles boost direct interspecies electron transfer (DIET) phenomenon caused by redox reactions, which, in turn, stimulate the activity of methanogens (Park et al., 2018). In comparison to bulk materials, nanoparticles can permeate cell membranes and improve several pathways, resulting in high methane production rates, shortened lag phases, and high substrate conversion (Ajay et al., 2020). To date, several nano-based additives have been explored in biogas production systems. These include metallic nanoparticles, carbon-based nanoparticles, inorganic nanoparticles, organic-based nanoparticles, and nanocomposites (Dikshit et al., 2021). Among these, metallic nanoparticles (e.g., silver nanoparticles and iron nanoparticles) and inorganic nanoparticles (e.g., iron oxide and zinc oxide) are commonly used in biogas enrichment methods as they have been proven to improve the exchange of electrons, promote agglomeration, enhance DIET, stimulate hydrolysis, maximize the methanation process, and induce EC (Ajay et al., 2020). A remarkable biogas production increment of between 66% and 129% has been reported with the use of iron oxide nanoparticles in several anaerobic digestion processes (Wang et al., 2016; Abdelsalam et al., 2017; Khalid et al., 2019). Such a remarkable increment underscores the importance of these additives in the enrichment of biofuel processes. Nevertheless, there is some skepticism about the toxicity of nano-based additives to humans and the environment (Khan et al., 2022). Such skepticism has prompted the search for economically viable and environmentally benign alternatives among researchers, as evidenced by the adoption of recent biomass-derived nanoparticles that are being used in several studies (Mohammadzadeh et al., 2022; Osterberg et al., 2023).

5.4.5 Process Optimization via Multivariate Tools

For many years, researchers relied on classical statistical tools such as the well-known OFAT (one-factor-at-a-time) method, which suffers from many drawbacks, such as overlooking the synergistic impact of the setpoint parameters, requirements for large experiments, and increased time and capital expenditure. To circumvent these inadequacies, scientists are tapping into the use of robust multivariate tools such as response surface methodology (RSM) to acquire deeper insights into the individualistic and synergistic effects of the key operational parameters, such as pH, temperature, HRT, OLR, and substrate concentration, on biogas production processes. These multivariate tools employ the design of experiments (DoEs) software packages such as Plackett-Burman Design, Central Composite Design, Box-Behnken Design, Full Factorial Design, and Mixture Design (Sekoai et al., 2020). Additionally, these tools involve a five-step process that includes: (1) experimental design using DoE software consisting of lower and upper limits; (2) the use of the experimental data to build and validate the empirical model; (3) the use of the built model to elucidate the individual and interactive effects of the setpoint parameters using three-dimensional plots; (4) the use of the built model to predict the outcome (dependent variable) based on the input variables (independent variables); and (5) the experimental validation of the built model to confirm the authenticity of the model based on experimental findings (Sekoai et al., 2020). Multivariate tools have played an instrumental role in determining the optimal set of variables/parameters that generate maximum biogas yields (Odejobi et al., 2022; Habchi et al., 2023; Ingabire et al., 2023; Otieno et al., 2023). The detailed understanding of the desirable sets of operational conditions has led to the establishment of various large-scale biogas production facilities using mesophilic and thermophilic conditions (Mao et al., 2015; Kemausuor et al., 2018; Wu et al., 2021). The RSM tools are advantageous as they reduce the number of experimental trials and only focus on the most important sets of variables and their corresponding outputs, making these bioprocesses cost-effective (Djimtoingar et al., 2022). However, RSM tools are linear polynomial models, which implies that they do not consider the complexity, non-linearities, and unsteady nature of bioprocesses, and therefore do not depict the actual behavior/dynamics of biogas systems (Sewsynker-Sukai et al., 2017). To address this limitation, researchers are now leveraging advanced computer modeling and optimization techniques such as artificial neural networks (ANNs) and genetic algorithms (GAs) (Li et al., 2024; Rahimieh et al., 2024). ANNs have been shown to possess exquisite features as they can be used in microbial-based processes devoid of previous data such as kinetic and metabolic fluxes and can emulate the human brain by using interconnected neurons with outstanding learning, analysis, and adaptation mechanisms (Nagata and Chu, 2003). They also consist of integrated layers that assist with the complex networks between the input parameters (independent variables) and the output variables (dependent variables). Recent studies are now coupling linear models (RSM) with non-linear models (ANNs) as an innovative empirical modeling strategy for gauging the complex linearities and non-linearities that are established by microbial fermentation systems such as anaerobic digestion (Elfghi, 2016; Ram Talib et al., 2019). It has also been shown that the RSM-ANN-GA hybrid approach could provide greater accuracy and high predictability (Aung et al., 2022).

5.4.6 Closed-Loop Biorefinery Technologies

Innovations are also geared toward the adoption of closed-loop (one-pot) biorefinery-based strategies to improve the economics of biogas technologies, as it has been discovered that diverse high-value-added products can be produced during the anaerobic digestion process. These R&D trends are in alignment with the '3Rs' (reduce-reuse-recycle) mantras of the circular bioeconomy, and this will ensure that our bio-resources are utilized in a more sustainable manner, leading to near-zero or zero waste accumulation and minimization of environmental pollution. Due to the diverse microbial species that coexist within the biogas fermenter, other metabolic pathways favor the synthesis of biopolymers, single-cell proteins, organic acids, and biofertilizers (Sekoai et al., 2024). From a socioeconomic standpoint, this technology is highly desirable in bioprocessing as it facilitates the hydrolysis, saccharification, and fermentation processes in one fermenter, thus, eliminating the laborious and cost-intensive two-stage sequential steps of hydrolysis and fermentation in separate fermenters. For many years, scientists viewed the anaerobic digestate produced from biogas systems as waste, but recent studies have shown that biogas technology can achieve a full circular bioeconomy route as the NPK-rich digestate can be used as a biofertilizer that can replace non-environmentally friendly commercial fertilizers. In a 2,000 m^3 anaerobic reactor, it was observed that it is economically viable to implement a multi-output biorefinery system (biogas and biopolymers) rather than a single-product recovery route focusing on biogas alone (Valentino et al., 2019). These attempts will propel the development of biogas technologies as this strategy has shown positive economic outcomes in techno-economic analyses (Rajendran and Han, 2022; Wang et al., 2023).

5.4.7 Co-digestion of Feedstocks

Microbial growth is promoted by the presence of nutritional factors, including the carbon and nitrogen sources in the biogas fermenter. Co-digestion is an approach that not only assists in obtaining a balanced C/N ratio but also helps in ensuring that the ammonia level is within the acceptable range (<200 mg/L) (Nzila, 2017). It has been previously demonstrated that this method enhances the formation of precursors used in the final stage of anaerobic digestion, resulting in a biogas increment of 20%–400% over mono-digestion processes (Cavinato et al., 2010; Shah et al., 2015). Microbial ecology studies have observed the proliferation of predominant communities such as *Clostridium* spp. and *Methanosarcina* spp. when the C/N ratio is within the optimal range of 20–30 (Wang et al., 2017; Khayum et al., 2018). Co-digested substrates also contribute to the alkalinity of the fermentation medium and promote synergistic interactions among the diverse biofilm-forming microbial communities during biogas production. This is a promising strategy that is used to enhance biogas yields during the anaerobic digestion of various organic waste streams, and it is economically feasible.

5.4.8 Real-Time Bioprocess Monitoring and Control

With the emergence of advanced computer packages such as artificial intelligence, bioprocess engineers are now tapping into computer modeling tools to obtain invaluable insights into bioprocess dynamics during biogas fermentation systems. These advanced computer packages or software tools are integrated with sensors and actuators, which are specifically designed to regulate the key operational setpoint parameters, such as pH, temperature, dissolved oxygen level, and optical density, during the duration of the biogas fermentation process (Sekoai and Kana, 2013; Sekoai and Daramola, 2018). Real-time monitoring is crucial in the field of bioprocessing as this provides information on the process behavior and how the microbial cell factories are acclimatizing to the conditions of the new system (pH, temperature, organic acids, etc.), which in turn affects the overall process performance. The monitoring devices are coupled with actuators that are programmed to respond to minor changes caused by the various metabolic fluxes, which tend to alter the optimal conditions due to the formation of undesired/unwanted biochemical inhibitors. As such, industrial biogas plants incorporate these bioprocess monitoring and regulating strategies to maintain long-term process stability that leads to maximum biogas outputs (Ward et al., 2011). Such maximum biogas outputs are achieved by stimulating the

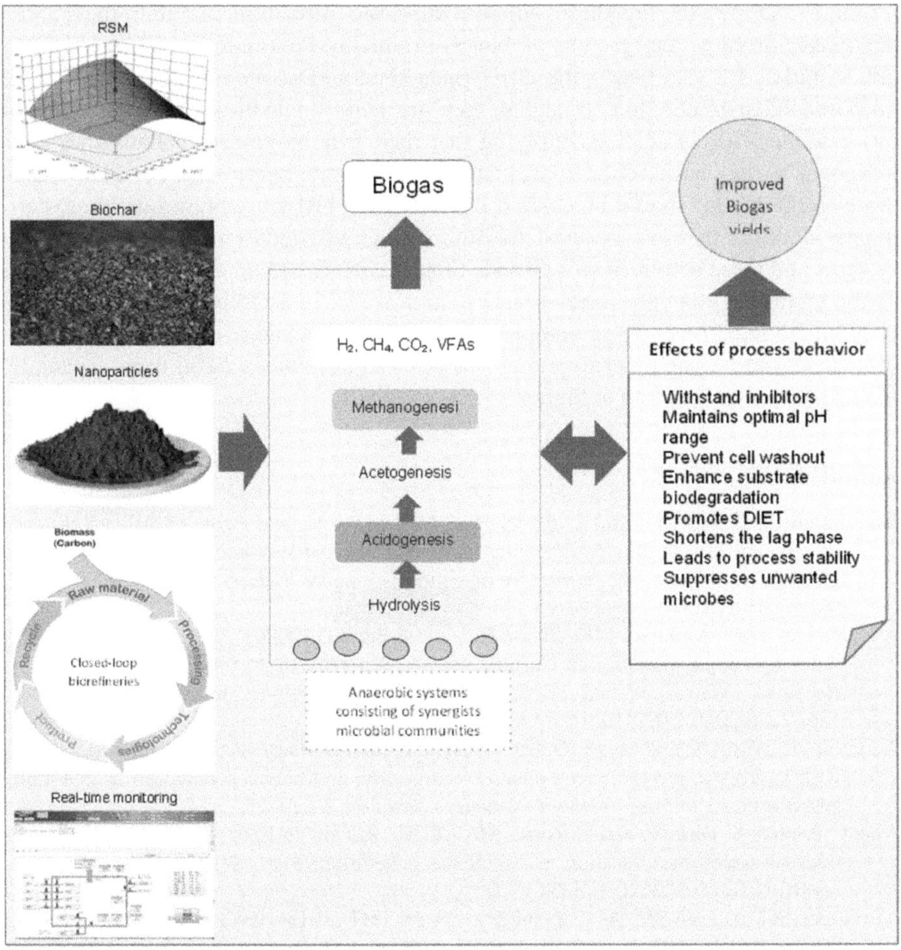

FIGURE 5.2 Schematic of bioaugmentation strategies for biogas-producing microbes.

biochemical pathways that are associated with the biogas production process. From an economic standpoint, these computer web-enabled tools are beneficial as they not only maintain the desirable setpoint variables but also allow high-throughput actionable information as the process data are measured at every millisecond during the entire anaerobic digestion process (Kana et al., 2013). Figure 5.2 illustrates some of the proposed bioaugmentation strategies applicable in anaerobic digestion systems, resulting in elevated/stable biogas operations.

5.5 CONCLUDING REMARKS AND FUTURE OUTLOOK

This chapter explored the emerging bioaugmentation strategies applicable in anaerobic digestion systems targeting the production of biogas. As highlighted in this study, these bioaugmentation methods allow microbes to withstand various inhibitors, shorten the lag phase, enhance substrate utilization, maintain the desired pH range, promote the growth of biogas-producing communities, improve CEC, and stimulate EC that boosts the DIET phenomenon. Despite the advantages of the proposed bioaugmentation strategies, they are mainly conducted at lower technology readiness levels (TRLs), implying that their true bioprocess behavior remains unknown as they have not been evaluated at large-scale operations. For this reason, more R&D studies should be carried out at higher TRLs to obtain reliable experimental datasets that can be used for future scale-up studies in biogas production systems and these activities will provide biogas experts with more realistic outcomes as the technology of biogas has reached large-scale. In addition, the technical and economic competitiveness in comparison to existing technologies needs to be scrutinized so that we can determine the methods that have meaningful uses, especially those that can be operated at higher TRLs.

REFERENCES

Abdelsalam E, Samer M, Attia Y, Abdel-Hadi M, Hassan H, and Badr Y (2017). Influence of zero valent iron nanoparticles and magnetic iron oxide nanoparticles on biogas and methane production from anaerobic digestion of manure. *Energy*. 120: 842–853. https://doi.org/10.1016/j.energy.2016.11.137

Abdelwahab TAM, Mohanty MK, Sahoo PK, and Behera D (2020). Application of nanoparticles for biogas production: Current status and perspectives. *Energy Sources, Part A: Recovery, Utilization, and Environmental Effects*. 46: 1–13. https://doi.org/10.1080/15567036.2020.1767730

Aderonke A, Wasiu A, and Moses O (2017). Isolation and characterization of microorganisms with hydrolytic profile during anaerobic digestion and biogas production of cow dung and rice husk. *Journal of Natural Sciences Research*. 7: 20–30.

Ajay C, Mohan S, Dinesha P, and Rosen MA (2020). Review of impact of nanoparticle additives on anaerobic digestion and methane generation. *Fuel*. 277: 118234. https://doi.org/10.1016/j.fuel.2020.118234

Alrumman S, Mostafa YSM, Al-Qahtani S, and Taha THT (2018). Hydrolytic enzyme production by thermophilic bacteria isolated from Saudi hot springs. *Open Life Sciences*. 13: 470–480.

Aung T, Kim S-J, and Eun J-B (2022). A hybrid RSM-ANN-GA approach on optimisation of extraction conditions for bioactive component-rich laver (Porphyra dentata) extract. *Food Chemistry*. 366: 130689. https://doi.org/10.1016/j.foodchem.2021.130689

Awode AE, Babatola JO, and Adelodun AA (2022). Effect of biochar on biogas yield of poultry wastes. *Environmental Protection Research*. 2(2): 62–74. https://doi.org/10.37256/epr.2220221349

Battista F, Fino D, and Mancini G (2016). Optimization of biogas production from coffee production waste. *Bioresource Technology*. 200: 884–890. https://doi.org/10.1016/j.biortech.2015.11.020

Borth PLB, Perin JKH, Torrecilhas AR, Lopes DD, Santos SC, Kuroda EK, and Fernandes F (2022). Pilot-scale anaerobic co-digestion of food and garden waste: Methane potential, performance and microbial analysis. *Biomass and Bioenergy*. 157: 106331. https://doi.org/10.1016/j.biombioe.2021.106331

Bruno M and de Oliveira RA (2013). Performance of UASB reactors in two stages followed by post-treatment with activated sludge in wastewater batch of wet-processed coffee. *Engenharia Agrícola*. 33: 808–819. https://doi.org/10.1590/S0100-69162013000400020

Čater M, Fanedl L, Malovrh Š, and Logar RM (2015). Biogas production from brewery spent grain enhanced by bioaugmentation with hydrolytic anaerobic bacteria. *Bioresource Technology*. 186: 261–269. https://doi.org/10.1016/j.biortech.2015.03.029

Cavinato C, Fatone F, Bolzonella D, and Pavan P (2010). Thermophilic anaerobic co-digestion of cattle manure with agro-wastes and energy crops: Comparison of pilot and full scale experiences. *Bioresource Technology*. 101: 545–550. https://doi.org/10.1016/j.biortech.2009.08.043

Chen B, Zhou D, and Zhu L (2008). Transitional adsorption and partition of nonpolar and polar aromatic contaminants by biochars of pine needles with different pyrolytic temperatures. *Environmental Science and Technology*. 42: 5137–5143. https://doi.org/10.1021/es8002684

Conrad R (2007). Microbial ecology of methanogens and methanotrophs. *Advances in Agronomy*. 96: 1–63. https://doi.org/10.1016/S0065-2113(07)96005-8

Daud M, Rizvi H, Akram MF, Ali S, Rizwan M, Nafees M, and Jin ZS (2018). Review of upflow anaerobic sludge blanket reactor technology: Effect of different parameters and developments for domestic wastewater treatment. *Journal of Chemistry*. 2018: 1–13. https://doi.org/10.1155/2018/1596319

de Barros VG, Duda RM, da Silva Vantini J, Omori WP, Ferro MIT, and de Oliveira RA (2017). Improved methane production from sugarcane vinasse with filter cake in thermophilic UASB reactors, with predominance of Methanothermobacter and Methanosarcina archaea and Thermotogae bacteria. *Bioresource Technology*. 244: 371–381. https://doi.org/10.1016/j.biortech.2017.07.106

Demirel B and Scherer P (2008). The roles of acetotrophic and hydrogenotrophic methanogens during anaerobic conversion of biomass to methane: A review. *Reviews in Environmental Science and Bio/Technology*. 7: 173–190. https://doi.org/10.1007/s11157-008-9131-1

Dikshit PK, Kumar J, Das AK, Sadhu S, Sharma S, Singh S, Gupta PK, and Kim BS (2021). Green synthesis of metallic nanoparticles: Applications and limitations. *Catalysts*. 11: 902. https://doi.org/10.3390/catal11080902

Djimtoingar SS, Derkyi NSA, Kuranchie FA, and Yankyera JK (2022). A review of response surface methodology for biogas process optimization. *Cogent Engineering*. 9: 2115283. https://doi.org/10.1080/23311916.2022.2115283

Elfghi F (2016). A hybrid statistical approach for modeling and optimization of RON: A comparative study and combined application of response surface methodology (RSM) and artificial neural network (ANN) based on design of experiment (DOE). *Chemical Engineering Research and Design*. 113: 264–272. https://doi.org/10.1016/j.cherd.2016.05.023

Emezirinwune MU, Adejumobi IA, Adebisi OI, and Akinboro FG (2024). Synergizing hybrid renewable energy systems and sustainable agriculture for rural development in Nigeria. *e-Prime-Advances in Electrical Engineering, Electronics and Energy*. 7: 100492. https://doi.org/10.1016/j.prime.2024.100492

Enders A, Hanley K, Whitman T, Joseph S, and Lehmann J (2012). Characterization of biochars to evaluate recalcitrance and agronomic performance. *Bioresource Technology*. 114: 644–653. https://doi.org/10.1016/j.biortech.2012.03.022

España-Gamboa EI, Mijangos-Cortés JO, Hernández-Zárate G, Maldonado JAD, and Alzate-Gaviria LM (2012). Methane production by treating vinasses from hydrous ethanol using a modified UASB reactor. *Biotechnology for Biofuels*. 5: 1–9. https://doi.org/10.1186/1754-6834-5-82

Fia FR, Matos AT, Borges AC, Fia R, and Cecon PR (2012). Treatment of wastewater from coffee bean processing in anaerobic fixed bed reactors with different support materials: performance and kinetic modeling. *Journal of Environmental Management*. 108: 14–21. https://doi.org/10.1016/j.jenvman.2012.04.033

Frisvad JC, Møller LL, Larsen TO, Kumar R, and Arnau J (2018). Safety of the fungal workhorses of industrial biotechnology: update on the mycotoxin and secondary metabolite potential of Aspergillus niger, Aspergillus oryzae, and Trichoderma reesei. *Applied Microbiology and Biotechnology*. 102: 9481–9515. https://doi.org/10.1007/s00253-018-9354-1

Germec M, Demirci A, and Turhan I (2020). Biofilm reactors for value-added products production: An in-depth review. *Biocatalysis and Agricultural Biotechnology*. 27: 101662. https://doi.org/10.1016/j.bcab.2020.101662

Gkotsis P, Kougias P, Mitrakas M, and Zouboulis A (2023). Biogas upgrading technologies–Recent advances in membrane-based processes. *International Journal of Hydrogen Energy*. 48: 3965–3993. https://doi.org/10.1016/j.ijhydene.2022.10.228

Glivin G, Kalaiselvan N, Mariappan V, Premalatha M, Murugan P, and Sekhar J (2021). Conversion of biowaste to biogas: A review of current status on techno-economic challenges, policies, technologies and mitigation to environmental impacts. *Fuel*. 302: 121153. https://doi.org/10.1016/j.fuel.2021.121153

Habchi S, Lahboubi N, Sallek B, and El Bari H (2023). Response surface methodology for anaerobic digestion of waste from poultry slaughterhouse: Optimization of load and hydraulic retention time. *Results in Engineering*. 18: 101215. https://doi.org/10.1016/j.rineng.2023.101215

Harirchi S, Wainaina S, Sar T, Nojoumi SA, Parchami M, Parchami M, Varjani S, Khanal SK, Wong J, and Awasthi MK (2022). Microbiological insights into anaerobic digestion for biogas, hydrogen or volatile fatty acids (VFAs): A review. *Bioengineered*. 13: 6521–6557. https://doi.org/10.1080/21655979.2022.2035986

Hughes SR and Jones MA (2020). The global demand for biofuels and biotechnology-derived commodity chemicals: Technologies, markets, and challenges. *Green energy to sustainability: strategies for global industries* (Vertès AA, Qureshi N, Blaschek HP & Yukawa H, eds.), pp. 137–156. Wiley.

Ingabire H, M'arimi MM, Kiriamiti KH, and Ntambara B (2023). Optimization of biogas production from anaerobic co-digestion of fish waste and water hyacinth. *Biotechnology for Biofuels and Bioproducts*. 16: 110. https://doi.org/10.1186/s13068-023-02360-w

Ivanova V, Petrova P, and Hristov J (2011). Application in the ethanol fermentation of immobilized yeast cells in matrix of alginate/magnetic nanoparticles, on chitosan-magnetite microparticles and cellulose-coated magnetic nanoparticles. *International Review of Chemical Engineering*. 3: 289–299. https://doi.org/10.48550/arXiv.1105.0619

Jensen MB, Strübing D, de Jonge N, Nielsen JL, Ottosen LDM, Koch K, and Kofoed MVW (2019). Stick or leave–Pushing methanogens to biofilm formation for ex situ biomethanation. *Bioresource Technology*. 291: 121784. https://doi.org/10.1016/j.biortech.2019.121784

Ji M, Wang X, Usman M, Liu F, Dan Y, Zhou L, Campanaro S, Luo G, and Sang W (2022). Effects of different feedstocks-based biochar on soil remediation: A review. *Environmental Pollution*. 294: 118655. https://doi.org/10.1016/j.envpol.2021.118655

Jung K-W, Kim D-H, Lee M-Y, and Shin H-S (2012). Two-stage UASB reactor converting coffee drink manufacturing wastewater to hydrogen and methane. *International Journal of Hydrogen Energy*. 37: 7473–7481. https://doi.org/10.1016/j.ijhydene.2012.01.150

Kana EG, Schmidt S, and Kenfack RA (2013). A web-enabled software for real-time biogas fermentation monitoring–Assessment of dark fermentations for correlations between medium conductivity and biohydrogen evolution. *International Journal of Hydrogen Energy*. 38: 10235–10244. https://doi.org/10.1016/j.ijhydene.2013.06.019

Kao-Kniffin J, Freyre DS, and Balser TC (2010). Methane dynamics across wetland plant species. *Aquatic Botany*. 93: 107–113. https://doi.org/10.1016/j.aquabot.2010.03.009

Karray F, Ben Abdallah M, Kallel N, Hamza M, Fakhfakh M, and Sayadi S (2018). Extracellular hydrolytic enzymes produced by halophilic bacteria and archaea isolated from hypersaline lake. *Molecular Biology Reports*. 45: 1297–1309. https://doi.org/10.1007/s11033-018-4286-5

Kasinath A, Fudala-Ksiazek S, Szopinska M, Bylinski H, Artichowicz W, Remiszewska-Skwarek A, and Luczkiewicz A (2021). Biomass in biogas production: Pretreatment and codigestion. *Renewable and Sustainable Energy Reviews*. 150: 111509. https://doi.org/10.1016/j.rser.2021.111509

Kemausuor F, Adaramola MS, and Morken J (2018). A review of commercial biogas systems and lessons for Africa. *Energies*. 11: 2984. https://doi.org/10.3390/en11112984

Khalid MJ, Waqas A, and Nawaz I (2019). Synergistic effect of alkaline pretreatment and magnetite nanoparticle application on biogas production from rice straw. *Bioresource Technology*. 275: 288–296. https://doi.org/10.1016/j.biortech.2018.12.051

Khan SZ, Zaidi AA, Naseer MN, and AlMohamadi H (2022). Nanomaterials for biogas augmentation towards renewable and sustainable energy production: A critical review. *Frontiers in Bioengineering and Biotechnology*. 10: 868454. https://doi.org/10.3389/fbioe.2022.868454

Khayum N, Anbarasu S, and Murugan S (2018). Biogas potential from spent tea waste: A laboratory scale investigation of co-digestion with cow manure. *Energy*. 165: 760–768. https://doi.org/10.1016/j.energy.2018.09.163

Kougias PG, Treu L, Benavente DP, Boe K, Campanaro S, and Angelidaki I (2017). Ex-situ biogas upgrading and enhancement in different reactor systems. *Bioresource Technology*. 225: 429–437. https://doi.org/10.1016/j.biortech.2016.11.124

Kumar V, Koul B, Taak P, Yadav D, and Song M (2023). Journey of Trichoderma from pilot scale to mass production: A review. *Agriculture*. 13: 2022. https://doi.org/10.3390/agriculture13102022

Kushkevych I, Kobzová E, Vítězová M, Vítěz T, Dordević D, and Bartoš M (2019). Acetogenic microorganisms in operating biogas plants depending on substrate combinations. *Biologia*. 74: 1229–1236. https://doi.org/10.2478/s11756-019-00283-2

Lapponi MJ, Méndez MB, Trelles JA, and Rivero CW (2022). Cell immobilization strategies for biotransformations. *Current Opinion in Green and Sustainable Chemistry*. 33: 100565. https://doi.org/10.1016/j.cogsc.2021.100565

Leng L, Xiong Q, Yang L, Li H, Zhou Y, Zhang W, Jiang S, Li H, and Huang H (2021). An overview on engineering the surface area and porosity of biochar. *Science of the Total Environment*. 763: 144204. https://doi.org/10.1016/j.scitotenv.2020.144204

Li C, Li J, Pan L, Zhu X, Xie S, Yu G, Wang Y, Pan X, Zhu G, and Angelidaki I (2020). Treatment of digestate residues for energy recovery and biochar production: From lab to pilot-scale verification. *Journal of Cleaner Production*. 265: 121852. https://doi.org/10.1016/j.jclepro.2020.121852

Li C, Wang X, Zhang G, Li J, Li Z, Yu G, and Wang Y (2018). A process combining hydrothermal pretreatment, anaerobic digestion and pyrolysis for sewage sludge dewatering and co-production of biogas and biochar: Pilot-scale verification. *Bioresource Technology*. 254: 187–193. https://doi.org/10.1016/j.biortech.2018.01.045

Li H, Chen J, Li X, Gan J, Liu H, Jian Z, Xu S, Zhang A, Li G, and Chen K (2024a). Artificial neural network and genetic algorithm coupled fermentation kinetics to regulate L-lysine fermentation. *Bioresource Technology*. 393: 130151. https://doi.org/10.1016/j.biortech.2023.130151

Li R, Jiang Y, Huang J, Luo K, Fan X, Guo R, Liu T, Zhang Y, and Fu S (2024b). Simultaneous biogas upgrading and single cell protein production using hydrogen oxidizing bacteria. *Chemical Engineering Journal*. 490: 151576. https://doi.org/10.1016/j.cej.2024.151576

Lim JW, Park T, Tong YW, and Yu Z (2020). The microbiome driving anaerobic digestion and microbial analysis. *Advances in Bioenergy*. 5: 1–61.

Lin L (2022). Bottom-up synthetic ecology study of microbial consortia to enhance lignocellulose bioconversion. *Biotechnology for Biofuels and Bioproducts*. 15: 14. https://doi.org/10.1186/s13068-022-02113-1

Maleki Shahraki Z and Mao X (2022). Biochar application in biofiltration systems to remove nutrients, pathogens, and pharmaceutical and personal care products from wastewater. *Journal of Environmental Quality*. 51: 129–151. https://doi.org/10.1002/jeq2.20331

Mao C, Feng Y, Wang X, and Ren G (2015). Review on research achievements of biogas from anaerobic digestion. *Renewable and Sustainable Energy Reviews*. 45: 540–555. https://doi.org/10.1016/j.rser.2015.02.032

María De la Rosa J, Espejo De Miguel M, Knicker H, and Fernández Boy E (2018). Testing established method for the determination of the cation exchange capacity in soils for the characterization of biochars, p. 5071.

Mazzucotelli CA, Ponce AG, Kotlar CE, and Moreira Md R (2013). Isolation and characterization of bacterial strains with a hydrolytic profile with potential use in bioconversion of agroindustial by-products and waste. *Food Science and Technology*. 33: 295–303. https://doi.org/10.1590/S0101-20612013005000038

Menzel T, Neubauer P, and Junne S (2020). Role of microbial hydrolysis in anaerobic digestion. *Energies*. 13: 5555. https://doi.org/10.3390/en13215555

Mohammadzadeh V, Barani M, Amiri MS, Yazdi MET, Hassanisaadi M, Rahdar A, and Varma RS (2022). Applications of plant-based nanoparticles in nanomedicine: A review. *Sustainable Chemistry and Pharmacy*. 25: 100606. https://doi.org/10.1016/j.scp.2022.100606

Monga D, Shetti NP, Basu S, Reddy KR, Badawi M, Bonilla-Petriciolet A, and Aminabhavi TM (2022). Engineered biochar: A way forward to environmental remediation. *Fuel*. 311: 122510. https://doi.org/10.1016/j.fuel.2021.122510

Moon HC and Song I (2011). Enzymatic hydrolysis of foodwaste and methane production using UASB bioreactor. *International Journal of Green Energy*. 8: 361–371. https://doi.org/10.1080/15435075.2011.557845

Moraes B, Triolo JM, Lecona VP, Zaiat M, and Sommer SG (2015). Biogas production within the bioethanol production chain: Use of co-substrates for anaerobic digestion of sugar beet vinasse. *Bioresource Technology*. 190: 227–234. https://doi.org/10.1016/j.biortech.2015.04.089

Nagata Y and Chu KH (2003). Optimization of a fermentation medium using neural networks and genetic algorithms. *Biotechnology Letters*. 25: 1837–1842. https://doi.org/10.1023/A:1026225526558

Nie W, He S, Lin Y, Cheng JJ, and Yang C (2023). Functional biochar in enhanced anaerobic digestion: Synthesis, performances, and mechanisms. *Science of the Total Environment*. 490: 167681. https://doi.org/10.1016/j.scitotenv.2023.167681

Nzila A (2017). Mini review: Update on bioaugmentation in anaerobic processes for biogas production. *Anaerobe*. 46: 3–12. https://doi.org/10.1016/j.anaerobe.2016.11.007

Odejobi OJ, Odekanle EL, Bamimore A, Falowo OA, and Akeredolu F (2022). Anaerobic digestion of abattoir wastes for biogas production: Optimization via performance evaluation comparison. *Cogent Engineering*. 9: 2122150. https://doi.org/10.1080/23311916.2022.2122150

Osterberg M, Henn KA, Farooq M, and Valle-Delgado JJ (2023). Biobased nanomaterials—The role of interfacial interactions for advanced materials. *Chemical Reviews*. 123: 2200–2241. https://doi.org/10.1021/acs.chemrev.2c00492

Otieno EO, Kiplimo R, and Mutwiwa U (2023). Optimization of anaerobic digestion parameters for biogas production from pineapple wastes co-digested with livestock wastes. *Heliyon*. 9: e14041. https://doi.org/10.1016/j.heliyon.2023.e14041

Pan X, Zhao L, Li C, Angelidaki I, Lv N, Ning J, Cai G, and Zhu G (2021). Deep insights into the network of acetate metabolism in anaerobic digestion: Focusing on syntrophic acetate oxidation and homoacetogenesis. *Water Research*. 190: 116774. https://doi.org/10.1016/j.watres.2020.116774

Park H-S, Jun S-C, Han K-H, Hong S-B, and Yu J-H (2017). Diversity, application, and synthetic biology of industrially important Aspergillus fungi. *Advances in Applied Microbiology*. 100: 161–202. https://doi.org/10.1016/bs.aambs.2017.03.001

Park J-H, Kang H-J, Park K-H, and Park H-D (2018). Direct interspecies electron transfer via conductive materials: A perspective for anaerobic digestion applications. *Bioresource Technology*. 254: 300–311. https://doi.org/10.1016/j.biortech.2018.01.095

Pomdaeng P, Kongthong O, Tseng C-H, Dokmaingam P, and Chu C-Y (2024). An immobilized mixed microflora approach to enhancing hydrogen and methane productions from high-strength organic loading food waste hydrolysate in series batch reactors. *International Journal of Hydrogen Energy*. 52: 160–169. https://doi.org/10.1016/j.ijhydene.2023.09.187

Prakash O, Dewala SR, Nimonkar Y, Patil SK, Chauhan A, Yadav A, Dhotre DP, and Ranade DR (2024). Culture-based and culture-independent approach for the study of the methanogens and obligate anaerobes from different landfill sites. *Frontiers in Microbiology*. 14: 1273037. https://doi.org/10.3389/fmicb.2023.1273037

Rachbauer L, Voitl G, Bochmann G, and Fuchs W (2016). Biological biogas upgrading capacity of a hydrogenotrophic community in a trickle-bed reactor. *Applied Energy*. 180: 483–490. https://doi.org/10.1016/j.apenergy.2016.07.109

Rahimieh A, Nosrati M, and Zamir SM (2024). Anaerobic digestion of a curious VFA complex feed for biomethane production; A study on ANN modeling optimized with genetic algorithm. *Desalination and Water Treatment*. 317: 100257. https://doi.org/10.1016/j.dwt.2024.100257

Rajendran N and Han J (2022). Techno-economic analysis of food waste valorization for integrated production of polyhydroxyalkanoates and biofuels. *Bioresource Technology*. 348: 126796. https://doi.org/10.1016/j.biortech.2022.126796

Ram Talib NS, Halmi MIE, Abd Ghani SS, Zaidan UH, and Shukor MYA (2019). Artificial neural networks (ANNs) and response surface methodology (RSM) approach for modelling the optimization of chromium (VI) reduction by newly isolated *Acinetobacter* radioresistens strain NS-MIE from agricultural soil. *BioMed Research International*. 2019: 1–14. https://doi.org/10.1155/2019/5785387

Rusanowska P, Zieliński M, and Dębowski M (2023). Removal of CO_2 from biogas during mineral carbonation with waste materials. *International Journal of Environmental Research and Public Health*. 20: 5687. https://doi.org/10.3390/ijerph20095687

Satar I, Daud WRW, Kim BH, Somalu MR, and Ghasemi M (2017). Immobilized mixed-culture reactor (IMcR) for hydrogen and methane production from glucose. *Energy*. 139: 1188–1196. https://doi.org/10.1016/j.energy.2017.08.071

Sekoai PT and Daramola MO (2018). Effect of metal ions on dark fermentative biohydrogen production using suspended and immobilized cells of mixed bacteria. *Chemical Engineering Communications*. 205: 1011–1022. https://doi.org/10.1080/00986445.2018.1428958

Sekoai PT and Kana EBG (2013). Fermentative biohydrogen modelling and optimization research in light of miniaturized parallel bioreactors. *Biotechnology & Biotechnological Equipment*. 27: 3901–3908. https://doi.org/10.5504/BBEQ.2013.0046

Sekoai PT and Kana EBG (2015). Semi-pilot scale production of hydrogen from organic fraction of solid municipal waste and electricity generation from process effluents. *Biomass and Bioenergy*. 60: 156–163. https://doi.org/10.1016/j.biombioe.2013.11.008

Sekoai PT, Awosusi AA, Yoro KO, Singo M, Oloye O, Ayeni AO, Bodunrin M, and Daramola MO (2018a). Microbial cell immobilization in biohydrogen production: A short overview. *Critical Reviews in Biotechnology*. 38: 157–171. https://doi.org/10.1080/07388551.2017.1312274

Sekoai PT, Chunilall V, and Ezeokoli O (2023a). Creating value from acidogenic biohydrogen fermentation effluents: An innovative approach for a circular bioeconomy that is acquired via a microbial biorefinery-based framework. *Fermentation*. 9: 602. https://doi.org/10.3390/fermentation9070602

Sekoai PT, Chunilall V, Msele K, Buthelezi L, Johakimu J, Andrew J, Zungu M, Moloantoa K, Maningi N, and Habimana O (2023b). Biowaste biorefineries in South Africa: Current status, opportunities, and research and development needs. *Renewable and Sustainable Energy Reviews*. 188: 113870. https://doi.org/10.1016/j.rser.2023.113870

Sekoai PT, Chunilall V, Sithole B, Habimana O, Ndlovu S, Ezeokoli OT, Sharma P, and Yoro KO (2022). Elucidating the role of biofilm-forming microbial communities in fermentative biohydrogen process: An overview. *Microorganisms*. 10: 1924. https://doi.org/10.3390/microorganisms10101924

Sekoai PT, Daramola MO, Mogwase B, Engelbrecht N, Yoro KO, du Preez SP, Mhlongo S, Ezeokoli OT, Ghimire A, and Ayeni AO (2020a). Revising the dark fermentative H_2 research and development scenario–An overview of the recent advances and emerging technological approaches. *Biomass and Bioenergy*. 140: 105673. https://doi.org/10.1016/j.biombioe.2020.105673

Sekoai PT, Engelbrecht N, du Preez SP, and Bessarabov D (2020b). Thermophilic biogas upgrading via ex situ addition of H_2 and CO_2 using codigested feedstocks of cow manure and the organic fraction of solid municipal waste. *ACS Omega*. 5: 17367–17376. https://doi.org/10.1021/acsomega.0c01725

Sekoai PT, Ghimire A, Ezeokoli OT, Rao S, Ngan WY, Habimana O, Yao Y, Yang P, Fung AHY, and Yoro KO (2021). Valorization of volatile fatty acids from the dark fermentation waste Streams-A promising pathway for a biorefinery concept. *Renewable and Sustainable Energy Reviews*. 143: 110971. https://doi.org/10.1016/j.rser.2021.110971

Sekoai PT, Mhlongo SI, and Ezeokoli OT (2019a). Progress in the development of methods used for the abatement of microbial contaminants in ethanol fermentations: A review. *Reviews in Environmental Science and Bio/Technology*. 18: 795–821. https://doi.org/10.1007/s11157-019-09511-2

Sekoai PT, Ouma CNM, Du Preez SP, Modisha P, Engelbrecht N, Bessarabov DG, and Ghimire A (2019b). Application of nanoparticles in biofuels: An overview. *Fuel*. 237: 380–397. https://doi.org/10.1016/j.fuel.2018.10.030

Sekoai PT, Roets-Dlamini Y, O'Brien F, Ramchuran S, and Chunilall V (2024). Valorization of food waste into single-cell protein: An innovative technological strategy for sustainable protein production. *Microorganisms*. 12: 166. https://doi.org/10.3390/microorganisms12010166

Sekoai PT, Yoro KO, Bodunrin MO, Ayeni AO, and Daramola MO (2018b). Integrated system approach to dark fermentative biohydrogen production for enhanced yield, energy efficiency and substrate recovery. *Reviews in Environmental Science and Bio/Technology*. 17: 501–529. https://doi.org/10.1007/s11157-018-9474-1

Sekoai PT, Yoro KO, and Daramola MO (2016). Batch fermentative biohydrogen production process using immobilized anaerobic sludge from organic solid waste. *Environments*. 3: 38. https://doi.org/10.3390/environments3040038

Sekoai PT, Yoro KO, and Daramola MO (2018c). Effect of nitrogen gas sparging on dark fermentative biohydrogen production using suspended and immobilized cells of anaerobic mixed bacteria from potato waste. *Biofuels*. 9: 595–604. https://doi.org/10.1080/17597269.2018.1432275

Senko O, Gladchenko M, Maslova O, and Efremenko E (2019). Long-term storage and use of artificially immobilized anaerobic sludge as a powerful biocatalyst for conversion of various wastes including those containing xenobiotics to biogas. *Catalysts*. 9: 326. https://doi.org/10.3390/catal9040326

Sewsynker-Sukai Y, Faloye F, and Kana EBG (2017). Artificial neural networks: An efficient tool for modelling and optimization of biofuel production (a mini review). *Biotechnology & Biotechnological Equipment*. 31: 221–235. https://doi.org/10.1080/13102818.2016.1269616

Shah FA, Mahmood Q, Rashid N, Pervez A, Raja IA, and Shah MM (2015). Co-digestion, pretreatment and digester design for enhanced methanogenesis. *Renewable and Sustainable Energy Reviews*. 42: 627–642. https://doi.org/10.1016/j.rser.2014.10.053

Sharma R, Garg P, Kumar P, Bhatia SK, and Kulshrestha S (2020). Microbial fermentation and its role in quality improvement of fermented foods. *Fermentation*. 6: 106. https://doi.org/10.3390/fermentation6040106

Singh A, Moestedt J, Berg A, and Schnürer A (2021a). Microbiological surveillance of biogas plants: Targeting acetogenic community. *Frontiers in Microbiology*. 12: 700256. https://doi.org/10.3389/fmicb.2021.700256

Singh A, Müller B, and Schnürer A (2021b). Profiling temporal dynamics of acetogenic communities in anaerobic digesters using next-generation sequencing and T-RFLP. *Scientific Reports*. 11: 13298. https://doi.org/10.1038/s41598-021-92658-2

Singh AK, Pal P, Rathore SS, Sahoo UK, Sarangi PK, Prus P, and Dziekański P (2023). Sustainable utilization of biowaste resources for biogas production to meet rural bioenergy requirements. *Energies*. 16: 5409. https://doi.org/10.3390/en16145409

Tamura K, Stecher G, and Kumar S (2021). MEGA11: Molecular evolutionary genetics analysis version 11. *Molecular Biology and Evolution*. 38: 3022–3027. https://doi.org/10.1093/molbev/msab120

Tan Z, Zou J, Zhang L, and Huang Q (2018). Morphology, pore size distribution, and nutrient characteristics in biochars under different pyrolysis temperatures and atmospheres. *Journal of Material Cycles and Waste Management*. 20: 1036–1049. https://doi.org/10.1007/s10163-017-0666-5

Tarelho L, Hauschild T, Vilas-Boas A, Silva D, and Matos M (2020). Biochar from pyrolysis of biological sludge from wastewater treatment. *Energy Reports*. 6: 757–763. https://doi.org/10.1016/j.egyr.2019.09.063

Tian G, Yang B, Dong M, Zhu R, Yin F, Zhao X, Wang Y, Xiao W, Wang Q, and Zhang W (2018). The effect of temperature on the microbial communities of peak biogas production in batch biogas reactors. *Renewable Energy*. 123: 15–25. https://doi.org/10.1016/j.renene.2018.01.119

Timmers PH, Vavourakis CD, Kleerebezem R, Damsté JSS, Muyzer G, Stams AJ, Sorokin DY, and Plugge CM (2018). Metabolism and occurrence of methanogenic and sulfate-reducing syntrophic acetate oxidizing communities in haloalkaline environments. *Frontiers in Microbiology*. 9: 425991. https://doi.org/10.3389/fmicb.2018.03039

Ungerfeld E and Pitta D (2024). Biological consequences of the inhibition of rumen methanogenesis. *Animal*. 19: 101170. https://doi.org/10.1016/j.animal.2024.101170

Valentino F, Moretto G, Lorini L, Bolzonella D, Pavan P, and Majone M (2019). Pilot-scale polyhydroxyalkanoate production from combined treatment of organic fraction of municipal solid waste and sewage sludge. *Industrial & Engineering Chemistry Research*. 58: 12149–12158. https://doi.org/10.1021/acs.iecr.9b01831

Wang C, Hong F, Lu Y, Li X, and Liu H (2017). Improved biogas production and biodegradation of oilseed rape straw by using kitchen waste and duck droppings as co-substrates in two-phase anaerobic digestion. *PLoS One*. 12: e0182361. https://doi.org/10.1371/journal.pone.0182361

Wang T, Zhang D, Dai L, Chen Y, and Dai X (2016). Effects of metal nanoparticles on methane production from waste-activated sludge and microorganism community shift in anaerobic granular sludge. *Scientific Reports*. 6: 25857. https://doi.org/10.1038/srep25857

Wang Y, Baral NR, Yang M, and Scown CD (2023). Co-processing agricultural residues and wet organic waste can produce lower-cost carbon-negative fuels and bioplastics. *Environmental Science & Technology*. 57: 2958–2969. https://doi.org/10.1021/acs.est.2c06674

Ward AJ, Bruni E, Lykkegaard MK, Feilberg A, Adamsen AP, Jensen AP, and Poulsen AK (2011). Real time monitoring of a biogas digester with gas chromatography, near-infrared spectroscopy, and membrane-inlet mass spectrometry. *Bioresource Technology*. 102: 4098–4103. https://doi.org/10.1016/j.biortech.2010.12.052

Westerholm M, Moestedt J, and Schnürer A (2016). Biogas production through syntrophic acetate oxidation and deliberate operating strategies for improved digester performance. *Applied Energy*. 179: 124–135. https://doi.org/10.1016/j.apenergy.2016.06.061

Wiedner K, Rumpel C, Steiner C, Pozzi A, Maas R, and Glaser B (2013). Chemical evaluation of chars produced by thermochemical conversion (gasification, pyrolysis and hydrothermal carbonization) of agro-industrial biomass on a commercial scale. *Biomass and Bioenergy*. 59: 264–278. https://doi.org/10.1016/j.biombioe.2013.08.026

Wu D, Peng X, Li L, Yang P, Peng Y, Liu H, and Wang X (2021). Commercial biogas plants: Review on operational parameters and guide for performance optimization. *Fuel*. 303: 121282. https://doi.org/10.1016/j.fuel.2021.121282

Yang H, Huan B, Chen Y, Gao Y, Li J, and Chen H (2016). Biomass-based pyrolytic polygeneration system for bamboo industry waste: evolution of the char structure and the pyrolysis mechanism. *Energy & Fuels*. 30: 6430–6439. https://doi.org/10.1021/acs.energyfuels.6b00732

Yang P, Cai T, and Wang M-L (1988). Immobilized mixed microbial cells for wastewater treatment. *Biological Wastes*. 23: 295–312. https://doi.org/10.1016/0269-7483(88)90017-1

Yellezuome D, Zhu X, Liu X, Liu X, Liu R, Wang Z, Li Y, Sun C, Abd-Alla MH, and Rasmey A-HM (2023). Integration of two-stage anaerobic digestion process with in situ biogas upgrading. *Bioresource Technology*. 369: 128475. https://doi.org/10.1016/j.biortech.2022.128475

Zhao S, Guo H, Chen Z, Chen L, Wei G, and Yu H (2024). The key metabolic pathway for enhanced anaerobic digestion of chicken manure with coal slime for methane production. *Journal of Environmental Chemical Engineering*. 12: 111922. https://doi.org/10.1016/j.jece.2024.111922

Zhu X, Vrachioli M, Baldoni E, M'Barek R, and Sauer J (2024). Is biorefinery establishment enhancing the EU labour market on a regional level? *Regional Studies*. 58: 1–15. https://doi.org/10.1080/00343404.2023.2292251

6 Multi-Criteria Benchmarking of Bioproducts from Sugarcane Molasses

Perspectives on Production Costs, Market Prices, and Greenhouse Gas Emissions

Eunice Sefakor Dogbe, Janus Louw, Jeanne Louw, and Johann F. Görgens

6.1 INTRODUCTION AND BACKGROUND INFORMATION

The integrated sugarcane biorefinery concept, adopted in this study for the comparison of 20 potential products on an equal basis, aims to overcome the challenges facing the global sugar industry, including weak sugar prices in the international (dumped) market leading to unattractive export opportunities, and increasing energy prices (Farzad et al., 2017). In many countries, such as South Africa, India, and Vietnam, the price of imported sugar is frequently lower than that of locally produced sugar, a situation often exacerbated by unfair subsidies in certain exporting countries (SA Cane Growers, 2025). The biorefinery concept generally seeks to diversify the sugar industry by co-producing sugar alongside bioproducts using sugar streams and/or sugar mill by-products to enhance the economic sustainability of the industry.

The challenge of climate change has motivated the global transition towards a circular economy, in which biobased chemicals and materials form one of the key pillars of sustainability (de Jong et al., 2020). Bioethanol is the most dominant biobased product presently produced from sugarcane, with Brazil being a key player, supported by its mature sugar industry (Santos et al., 2017). However, bioethanol-only biorefineries are constrained in the economic benefits that they offer due to ongoing subsidy requirements and international agreements, providing motivation for increasing attention towards a next generation of biobased products, such as chemicals, materials, and advanced fuels, owing to their high market values and attractive growth prospects (Sanford et al., 2016). Biochemicals and biomaterials have higher

DOI: 10.1201/9781003745150-6

profit margins than conventional biofuels and when co-produced with bioethanol, have the potential to improve the economics of the bioethanol biorefinery (Axelsson et al., 2012; Lynd et al., 2022; Mandegari et al., 2017b).

Higher-value biobased chemicals, materials, and food ingredients produced from sugar may altogether replace commodity products such as ethanol in commercial biorefineries to maximise the viability and market relevance of these new investments. Conducting techno-economic assessments (TEAs) of biorefineries dedicated solely to producing biobased chemicals, materials, or food ingredients is therefore essential for their rapid development and scale-up (Sanford et al., 2016). However, the raw materials used and the methodologies applied in individual product assessments vary widely across the research landscape, limiting comparability. These variations arise from differences in research context, such as the country of study, which influences economic input data, as well as from the use of raw versus simulated data, the choice and cost of feedstocks, and the selection of assessment tools (de Jong et al., 2020; Mandegari et al., 2017a).

The diversity in bioproduct TEA studies poses a significant challenge for comparing the financial attractiveness of different bioproducts, thereby hindering the prioritisation of bioproducts for further research, development, and potential commercialisation. The type and cost of raw material have a significant impact on the economic feasibility of bioproducts (Lynd et al., 2022). Sugarcane-based raw materials are among the many biorefinery raw materials presented in the literature, as the global sugar industry is shifting towards product diversification (Eggleston and Lima, 2015). To assess which bioproducts may be more suitable or attractive for a particular biorefinery configuration, it is imperative to assess their production on the same technical, financial, and market basis, rather than rely on the divergent TEA studies that have been reported. The economic performance of alternative products depends on process technology performance, input requirements, and the scale of production. Therefore, market assessments, based on market reports for bioproducts, are included in this work to account for varying market demands for different products and the economies-of-scale implications for the techno-economic performances of each product.

In addition to economic viability, greenhouse gas (GHG) emissions play a crucial role in the market attractiveness of bioproducts. Similar to TEAs, GHG assessments are often conducted using varying system boundaries, allocation methods, and data sources, making direct comparison between studies challenging (Kendall and Yuan, 2013). This lack of methodological consistency can obscure the true environmental benefits of different bioproducts and limit the ability to prioritise options based on their climate impact.

This study, therefore, aims to benchmark the techno-economic performance of 20 potential bioproducts on the same manufacturing basis with consistent process simulation, financial, and GHG emission methodologies, to prioritise them for biorefinery integration into an existing sugar mill. Additionally, the effect of carbon credits on the economic performance of bioproducts is investigated to determine whether their required green price premiums are in line with their decarbonisation potentials/benefits. Ultimately, the research presents novel perspectives on the prioritisation of products for the sustainable use of sugarcane bioresources (sugars), which are likely to apply to other sugar-based biorefineries as well.

6.1.1 Twenty Bioproducts Selected for the Benchmarking and Their Production Processes

Twenty bioproducts derived from the sugars present in one common feedstock, sugarcane A-molasses, were selected for this benchmarking study. To revitalise the sector, sugar industries worldwide are increasingly adopting product diversification strategies. Among these, diverting A-molasses (rich in readily fermentable sugars) from conventional sugar production to biorefineries has been identified as one of the most energy-efficient and economically viable options (Dogbe et al., 2020). A-molasses is a particularly attractive feedstock, as it contains a higher concentration of sugars than B- and C-molasses, with fewer impurities (Dogbe et al., 2020). Moreover, its diversion to biorefinery applications eliminates two crystallisation steps during sugar production, thereby reducing operating costs (Dogbe et al., 2020) and potentially reducing the cost per unit of fermentable/monomeric sugar. This also frees up additional boiler capacity and sugarcane bagasse, facilitating greater energy integration and improving the potential for bioenergy self-sufficiency within the biorefinery (Dogbe et al., 2020).

The 20 bioproducts were selected based on their attractive markets and the availability of technical data to enable process simulation. Figure 6.1 lists these products, grouped into four categories: food ingredients, biochemicals, bioplastics, and advanced biofuels. Corresponding flow diagrams for each product are provided in the Supplementary material.

Biobased products such as ethanol, lactic acid, succinic acid, and 1,4-butanediol (1,4-BDO) are most commonly produced through microbial fermentation. However, chemical catalytic routes for converting sugars to bioproducts, such as levulinic acid and glucaric acid, have also been established. In addition, the production of several biobased products, including ethyl lactate, ethyl acetate, jet fuel, and biopolymers, involves hybrid pathways that combine fermentation (to key intermediates or monomers) with subsequent catalytic upgrading to the final product. Table 6.1 provides an overview of the production routes for the bioproducts considered in this study, categorised as biological, chemical, catalytic, or hybrid processes.

6.2 FOOD PRODUCTS

6.2.1 Glutamic Acid

Glutamic acid ($C_5H_6NO_4$), chemically known as (2S)-2-aminopentanedioic acid, is a non-essential, odourless, and water-soluble solid amino acid. It is one of the most important amino acids produced globally, owing to its variety of food, pharmaceutical, and industrial applications. The polymer derived from glutamic acid, poly-γ-glutamic acid (γ-PGA), has been successfully commercialised due to its desirable properties, including edibility, biodegradability, solubility, and non-toxicity

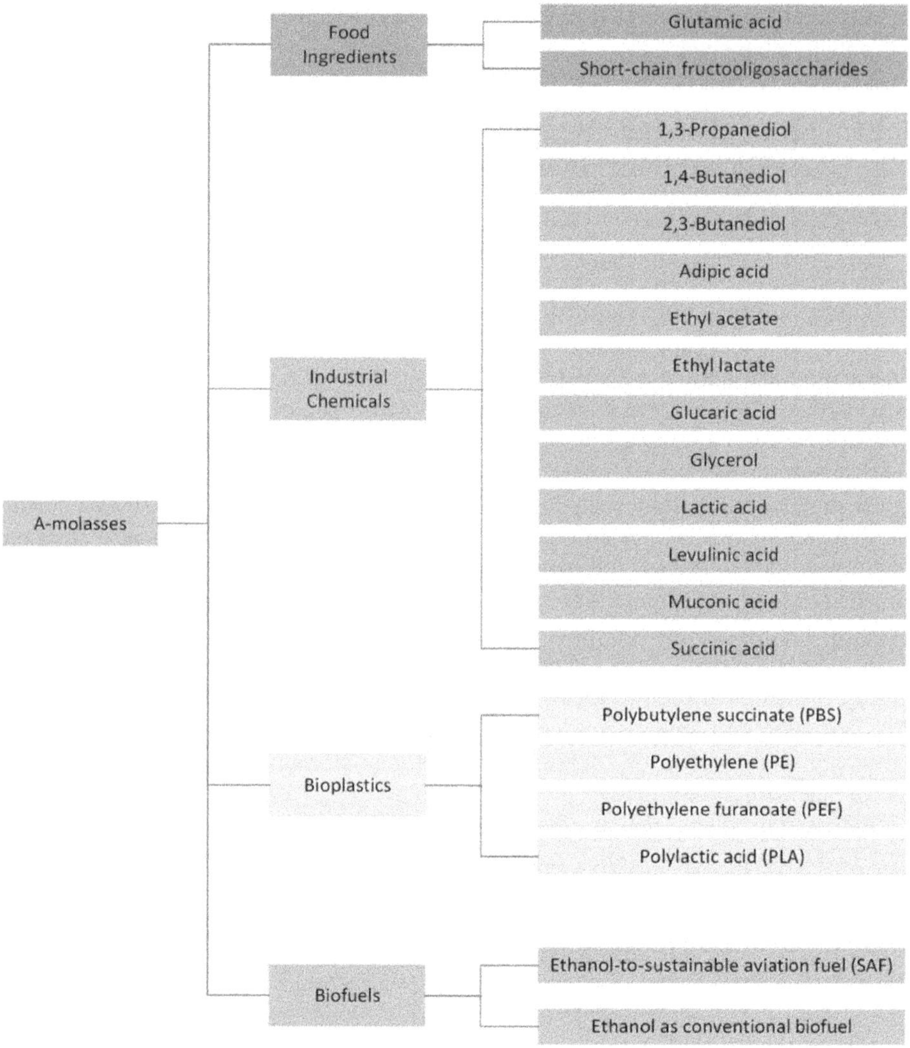

FIGURE 6.1 Overview of the 20 selected biorefinery products considered in this study for benchmarking, derived from A-molasses as feedstock.

(D'Este et al., 2018). Glutamic acid, particularly its sodium salt (monosodium glutamate; MSG), is used largely as a flavour enhancer in food processing (Zhang et al., 2012), representing its largest market segment. It is also used in the pharmaceutical industry, where it serves as a building block for protein synthesis and as a precursor to glutamine, which plays a vital role in immune function and intestinal health (Williams et al., 2005). In addition, it is used as an additive in animal feed, particularly in poultry production. The fermentative process for the production of glutamic acid was developed in 1957, where it was produced via the direct fermentation of carbohydrates and ammonia using *Corynebacterium glutamicum* (Kinoshita, 1959). Since the discovery of this microbial process, the industrial production of glutamic

TABLE 6.1
Classification of biobased products according to their production pathway

Product	Selected production pathway from sugars
	Direct biological conversion
Glutamic acid	Direct fermentation
Short-chain fructooligosaccharides	Direct enzymatic conversion
1,3-Propanediol	Direct fermentation
1,4-Butanediol	Direct fermentation
2,3-Butanediol	Direct fermentation
Adipic acid	Direct fermentation
Glycerol	Direct fermentation
Lactic acid	Direct fermentation
Muconic acid	Direct fermentation
Succinic acid	Direct fermentation
Ethanol	Direct fermentation
	Chemical catalytic conversion
Glucaric acid	Catalytic oxidation of sugars to glucaric acid
Levulinic acid	Catalytic dehydration of sugars to 5-hydroxymethyl furfural, followed by rehydration to levulinic acid
	Hybrid conversion
Ethyl lactate	Intermediates (ethanol and lactic acid) produced through fermentation, followed by their esterification to ethyl lactate
Ethyl acetate	Intermediate (ethanol) produced through fermentation, followed by dehydrogenation to ethyl acetate
Polybutylene succinate	Monomers (1,4-butanediol and succinic acid) produced through fermentation, followed by their co-polymerisation to PBS
Polyethylene	Intermediate (ethanol) produced through fermentation, followed by catalytic dehydration to ethylene, and polymerisation of ethylene to PE
Polyethylene furanoate	Polymerisation of two monomers: 2,5-furandicarboxylic acid (produced through two-step catalytic conversion of sugars) and monoethylene glycol (produced through fermentation of sugars to ethanol, followed by catalytic dehydration, oxidation and hydration steps)
Polylactic acid	Monomer (lactic acid) produced via fermentation, followed by polymerisation to PLA
Sustainable aviation fuel	Intermediate (ethanol) produced through fermentation, followed by catalytic dehydration, oligomerisation, and hydrogenation steps to produce SAF

acid has shifted from chemical synthesis to fermentation, owing to its lower process cost and environmental advantages (Sano, 2009). In 2014, its global market size was estimated to be 2,900 ktpa (GMI, 2021), with a market price of around 1,800–2,500 US$/ton in 2026 (Statsnex, 2026). Figure S6.1 in the Supplementary material shows the process steps involved in the biobased production of glutamic acid through the fermentation of a 1G feedstock (A-molasses).

6.2.2 Short-Chain Fructooligosaccharides (scFOS)

Short-chain fructooligosaccharides (scFOS) are oligomers consisting of a terminal glucose group linked by β (1 → 2)-glycosidic bonds to between 2 and 4 fructose units, namely 1-kestose (GF2), nystose (GF3), and fructofuranosyl nystose (GF4), respectively (Ur Rehman et al., 2016). They can be produced either by the degradation of inulin or by the transformation of sucrose using enzymes derived from bacterial, fungal, or yeast sources (Bedzo et al., 2019). In the present study, the conversion of sucrose in A-molasses to scFOS was considered. The global production of scFOS in 2015 was reported at 134 kt/y and was projected to nearly double by 2024, driven by increasing demand for functional foods, the prebiotic and low-calorie sweetening properties, and the health benefits of scFOS (Dominguez et al., 2014). These health benefits include the prevention of colon cancer, the reduction of cholesterol levels (thereby lowering obesity risks and being safe for diabetic patients), and improved mineral absorption in the gut (Dominguez et al., 2014). The market price of scFOS was reported at 3,200 US$/t (Dogbe et al., 2020). Figure S6.2 in the Supplementary material illustrates the manufacturing process from A-molasses, including the production of a fungal enzyme called ß-D-fructofuranosidase (FFase), which has trans-fructosilase activity for the transformation of sucrose, and the subsequent production and purification of scFOS.

6.3 INDUSTRIAL CHEMICALS AND SOLVENTS

6.3.1 Adipic Acid

Adipic acid (ADA) is a dicarboxylic acid primarily used for the production of nylon-6,6 fibres and engineered resins. Among the many building block chemicals reviewed, biobased ADA exhibits one of the lowest market penetrations relative to its fossil-based equivalent, showing substantial potential for future growth. At the end of 2018, DSM N.V. was the only producer of biobased ADA, with an annual production capacity of 3 kt/y (Skoczinski et al., 2020), despite the United States Department of Energy identifying ADA as the most important dicarboxylic acid with the greatest market potential in 2004 (Zhao et al., 2018). The global fossil-based ADA market was nearly 3,000 kt/a, with a value of nearly $6 billion in 2016, and was projected to grow at a compound annual growth rate of 4.7% (Skoog et al., 2018).

Rennovia and Verdezyne have developed processes for ADA production from glucose via chemo-catalytic conversion and direct fatty acid conversion, respectively. Both companies reached pilot-scale production, but ceased operations in 2018 following a sharp decline in oil prices driven by the expansion of shale gas supply (Lane, 2018). Several companies own intellectual property related to ADA production, including INVISTA, Amyris, Genomatica, BioAmber Inc., Celexion, ZymoChem, LEAF Technologies, and Aemetis (Skoczinski et al., 2020). One of the primary drivers towards the development of biobased ADA is the high environmental impact of its fossil-based production route. It has been estimated that the replacement of fossil-derived ADA with its biobased form could reduce GHG emissions by up to 95% (Skoog et al., 2018).

Figure S6.3 in the Supplementary material shows a block flow diagram for biobased ADA production from a 1G feedstock (A-molasses) by means of fermentation for direct conversion of sugars, followed by vacuum evaporation and reactive distillation. Alternative process routes for its production via intermediates such as glucaric acid and muconic acid are also under consideration, and in fact, are likely to offer better economic outcomes than the direct fermentation of sugars considered here (Cronjé et al., 2023).

6.3.2 2,3-Butanediol

2,3-Butanediol (2,3-BDO) is a promising biobased platform chemical, with its microbial production dating back to 1906 (Maina et al., 2019). 2,3-BDO has a wide range of applications, including use as an antifreeze agent and in the production of printing inks, moistening and softening agents, and fumigants. This versatile compound can also be converted to other valuable compounds such as methyl ethyl ketone and 1,3-butadiene; the former is used as a solvent for resins and lacquers, and the latter is used for the production of synthetic rubbers (Celińska and Grajek, 2009; Ishii et al., 2018). Although 2,3-BDO has a small reported market valuation of 270.4 million US$ in 2022, it has a significant future market potential, owing to its derivatives having an estimated global market value of 43 billion US$/y at a market volume of 32,000 kt/y, while having a current selling price of around 10,000–50,000 US$/t (Rosales-Calderon and Arantes, 2019; TMR, 2024).

The biobased production of 2,3-BDO has been established, although no information was found related to the size of the biobased market compared with that of the fossil-based product. While 2,3-BDO is currently being produced commercially through the hydrogenation of corn sugar by Global Biochem, it is unclear whether its commercial production from sugarcane-derived sugars has been established (Rosales-Calderon and Arantes, 2019). A block flow diagram for the biobased production of 2,3-BDO from 1G feedstock (A-molasses) via fermentation and vacuum flash distillation is shown in Figure S6.4 in the Supplementary material.

6.3.3 1,4-Butanediol

1,4-BDO is a commodity chemical used as a solvent, food additive, and in the pharmaceutical industry. 1,4-BDO is a precursor for the production of the biodegradable polymers polybutylene (PBS) and polybutylene adipate terephthalate (PBAT) and is also used as an intermediate chemical in the manufacturing of polyurethanes. 1,4-BDO was traditionally produced from petrochemicals, typically through the Reppe process, by the reaction of acetylene with formaldehyde (Satam et al., 2019). However, more recently, Genomatica (USA) developed a patented process (US8067214B2) for biobased 1,4-BDO production by direct microbial conversion of sugars derived from renewable biomass (Dani et al., 2017). In 2016, Novamont in Italy became the first company to produce 1,4-BDO at an industrial scale (30 kt/y) directly from sugars, using Genomatica's patented technology (Novamont, n.d.). Polymer-grade 1,4-BDO is recovered by a filtration system, evaporators, and distillation columns as shown in Figure S6.5 of the Supplementary material. While the fossil-based market for 1,4-BDO was approximately 3 million t/y in 2021, the biobased market remains comparatively small, estimated at around 50 kt/y (Skoczinski et al., 2020). However, it is projected to exceed 100 kt/y, supported by the commissioning of a reported 66 kt/y facility jointly owned by Cargill and HELM, which began operations in 2025 (The Lycra Company, 2025). In addition, Hyosung plans to commence operations at a 50 kt/y 1,4-BDO plant in Vietnam in 2026, with expansion to 200 kt/y anticipated by 2035 (Blois, 2024). The biobased selling price of 1,4-BDO was around 2,500 US$/t in 2019 (Satam et al., 2019).

6.3.4 Ethyl Acetate

Ethyl acetate is an organic ester and oxygenated solvent with a molecular formula of $C_4H_8O_2$. It is primarily used as a solvent for cleaning, paint removal, and coatings (Rosales-Calderon and Arantes, 2019). The production of ethyl acetate is typically achieved through the classic Fischer esterification reaction of acetic acid and ethanol. However, other routes include the dimerisation of acetaldehyde, the direct addition of acetic acid to ethylene, and the dehydrogenation of ethanol (Phasha, 2025). Hence, for ethyl acetate production to be biobased, the reactants, such as ethanol, acetic acid, acetaldehyde, and ethylene, should be biobased (Mandegari et al., 2017b). Figure S6.6 in the Supplementary material shows the production of ethyl acetate through the dehydrogenation of bioethanol, produced by the fermentation of A-molasses. In 2018, the global ethyl acetate market size was about 4,750 kt/y, with projected global annual growth rate of over 4%–5% in terms of volume per year (Market Research Future, 2021). The fossil-based selling price ranged from 0.86 to 1.21 US$/t in different countries in 2026 (Imarc Group, 2026d).

6.3.5 Ethyl Lactate

Ethyl lactate is a monobasic ester with the molecular formula $C_5H_{10}O_{3.}$ It is a non-volatile, biodegradable solvent produced through the esterification of (bio) ethanol and lactic acid (Phasha, 2025). It is primarily used as a functional replacement for metal cleaning solvents, as its properties and performance have been found to meet or exceed those of traditional solvents such as toluene, methyl ethyl ketone, and N-methyl-pyrrolidone (Biddy et al., 2016). It is considered to be environmentally friendly and less toxic relative to conventional solvents with similar applications (Pereira et al., 2009). Ethyl lactate is typically produced through the esterification of lactic acid with ethanol, in the presence of an acid catalyst (Figure S6.7 in the Supplementary material). Because almost all global lactic acid production is via fermentation of sugars, currently all the ethyl lactate produced globally, is either partially biobased or fully biobased, depending on the source of ethanol. The main producers of ethyl lactate are ADM, Galactic (15 kt/y) and Cargill Dow (Biddy et al., 2016). The total global production of ethyl lactate surpassed 65 kt/y in 2023 (Market Reports World, 2026a). The price of ethyl lactate was reportedly in the range of 3,300 to 4,400 US$/t in 2019, with high expectations that it will drop in the future due to economies-of-scale and supply chain improvements that are competitive with some fossil-derived solvents selling at 2,000 US$/t (Rosales-Calderon and Arantes, 2019).

6.3.6 Glucaric Acid

Glucaric acid has a wide range of industrial applications, but the most important is its use as a replacement for phosphates or citric acid in detergent builders. Phosphates are now banned from use in dishwasher detergents, and glucaric acid is used as an alternative. Another major application of glucaric acid is as a precursor to renewable ADA (Cronjé et al., 2023). While glucaric acid production from sugar by means of chemical catalytic oxidation is currently the most common method, it can also be produced by means of microbial fermentation. The global glucaric acid market was valued at $1.12 billion in 2023, with the Asia Pacific region accounting for the largest market share (35%) (Zion Market Research, 2023). Pure glucaric acid is expected to dominate the market in the future, but potassium-, sodium-, and calcium-D-glucarate, together with D-glucaric acid-1,4-lactone, also form part of the market. The global production capacity of biobased glucaric acid was 50.5 kt/y in 2015 (IAR, 2015), with Rennovia and Rivertop Renewables being the major producers. There have been significant challenges in producing pure, biobased glucaric acid due to the formation of lactones when the product is dried, even at a temperature of 50°C (Armstrong et al., 2017). Thaore et al. (2020) conducted a TEA on the production of pure glucaric acid, but with a different production strategy from the one employed here. The process developed here (Figure S6.8 in the Supplementary material) would be better suited for glucaric acid as a precursor to ADA production, as the major impurity in the final product stream is the solvent used in ADA bioproduction. If it is considered a precursor to ADA production, then vacuum distillation may not be required, depending on the purity of glucaric acid needed for ADA production. This would reduce the production costs of the glucaric acid intermediate used for ADA production.

6.3.7 Glycerol

Glycerol is a commodity chemical with more than 2,000 known applications in several different markets, including cosmetics, foodstuffs, fine chemicals, and emerging markets such as hydrogen, methanol, ethanol, and several other chemical industry products. Glycerol is produced as a by-product of oleochemical fatty acid production, soap production, and biodiesel production. With biodiesel production increasing internationally, crude glycerol from biodiesel has become abundant, causing market price volatility. However, the purification of crude glycerol is expensive, and the wide variety of applications of this product make it an attractive option for production in a sugarcane biorefinery. A process diagram for the production of glycerol from a 1G feedstock (A-molasses) is shown in Figure S6.9 in the Supplementary material. The global glycerol market was estimated to be 2,800 kt/y in 2020, of which the biobased market is 1,580 kt/y. It is projected that the global total glycerol market will reach 4,000 kt/y by 2027 (Businesswire, 2021). The reported selling price of glycerol was 859–1,410 US$/t in 2025 across different regions (Intratec, 2026).

6.3.8 Lactic Acid

Lactic acid as a purified monomer is an industrial chemical used in the food, bioplastic, pharmaceutical, cosmetic, and chemical industries. Lactic acid has received significant attention as a building block for the polylactic acid (PLA) polymer, as discussed in Section 6.4.5. Lactic acid can be produced via chemical or biobased routes, although the biobased pathway is responsible for over 90% of the global lactic acid production capacity. Fossil-derived lactic acid produces a racemic mixture of D/L isomers, which has limited applications. The L-isomer is commonly produced on an industrial scale as it has a wide range of applications.

Conventional lactic acid fermentation requires large amounts of neutralisers, typically $CaCO_3$ or $Ca(OH)_2$, which lead to the formation of gypsum during downstream reacidification, posing significant environmental and waste disposal challenges (Wang et al., 2026). In recent years, alternative neutralising agents, such as $Mg(OH)_2$, have been explored to avoid gypsum formation and improve the environmental performance of lactic acid production (Wang et al., 2026; Brobbey et al., 2025). Another approach involves using acid-tolerant yeast, which eliminates the need for neutralisation agents (Wang et al., 2026).

The gypsum-free process for biobased production of the lactic acid monomer from sugars was considered in the present study, using $Mg(OH)_2$ to neutralise the fermentation broth, whereafter Mg-lactate is complexed with trimethylamine (TMA) in an exchange reactor and later released by decomposition. Both the $Mg(OH)_2$ and the TMA are recovered and reused. Polymer-grade purity of lactic acid is achieved by reactive distillation (esterification and hydrolysis – Figure S6.10 in the Supplementary material).

The estimated global market size for lactic acid in 2026 exceeded 1.8 million t/y in 2025 (Market Reports World, 2026b), with production largely dominated by China (nova-Institut GmbH, 2020), and a market selling price of 1,450 US$/t (Mandegari et al., 2017b). Downstream development opportunities for PLA production are the main driver of the lactic acid market (de Jong et al., 2020), where substantial growth in demand has continuously outstripped supply. The current techno-economics for lactic acid production consider its final sale as a monomer, for comparison to its conversion to PLA before sale.

6.3.9 Levulinic Acid

Levulinic acid is a short-chain molecule with a chemical formula of $C_5H_8O_3$. It contains both carbonyl (C=O) and carboxylic acid (COOH) functionalities, resulting in a variety of reaction pathways with other functional groups. In 2004, it was listed by the US Department of Energy Biomass Program as one of the top 12 promising biobased chemicals (Signoretto et al., 2019). Over the past several years, levulinic acid has become a chemical of interest, as it can be produced with relatively high yield from acid pretreatment of C6 sugars (Taylor et al., 2015), and it provides a feasible chemical overlap between biomass and petroleum refining. A number of levulinic acid derivatives are used for fuel applications, including chemicals such as γ-valerolactone (GVL), ethyl levulinate (EL), and methyl tetrahydrofuran (MTHF) (Signoretto et al., 2019). Moreover, it can be used as an additive in fuels such as diesel and gasoline, as a member of the valerate ester family. Additionally, other markets for levulinic acid exist in the solvent, cosmetic, plasticiser, and pharmaceutical industries. However, the current global market remains rather small at an estimated

6 kt/y, with market prices up to 17,000 US$/t reported in 2025 (Gérardy et al., 2020; ChemAnalyst, 2025a, 2025b). Figure S6.11 in the Supplementary material shows the assumed process flow to produce levulinic acid from sugarcane A-molasses.

6.3.10 Muconic Acid

Muconic acid is a C6, naturally occurring, unsaturated dicarboxylic acid. The two terminal carboxylic groups in the muconic acid structure provide the capability to participate in several chemical reactions, enabling muconic acid to serve as a precursor and platform chemical for the production of several bioplastics and products (Gorden et al., 2016). Muconic acid is an important compound for the production of ADA, which subsequently serves as a precursor for the synthesis of polymers such as plastics, coatings, and nylons (Cronjé et al., 2023). Muconic acid is also applied as a starting material for the synthesis of value-added derivatives such as caprolactam and in polymer reactions to produce muconic homo- and co-polymers (Khalil et al., 2020). Production processes for intermediates derived from muconic acid, including ADA, are presently petroleum-based. The current market size for muconic acid is estimated at 79 kt/y at a selling price of 1,500 US$/t in 2023 (Blaga et al., 2023). Figure S6.12 in the Supplementary material shows the process flow diagram for the production of muconic acid via fermentation of A-molasses.

6.3.11 1,3-Propanediol

1,3-Propanediol (1,3-PDO) is a useful building block that can be converted into a wide variety of products. The main application of 1,3-PDO is the production of polytrimethylene terephthalate (PTT), a polymer that is used in textiles due to its superior durability and stain resistance (Biddy et al., 2016). Other applications of 1,3-PDO include polyurethane production, personal care products, solvents, and lubricants. The total global production of 1,3-PDO was estimated to be between 200 and 300 kt/y in 2024 (Marr, 2024), with a biobased production rate of 83.5 kt/y and a market price of 2,052 euro/t in 2015 (IAR, 2015).

Biobased 1,3-PDO has been jointly produced by DuPont and Tate & Lyle since 2004 from corn-derived glucose, with the manufacturing facility recently incorporated into CovationBio Inc. Other biobased producers include METabolic Explorer, Shenghong Group, and Zhanjiagang Glory Biomaterial, which produce 1,3-PDO from biodiesel-derived crude glycerol (Biddy et al., 2016; de Jong et al., 2020). Compared to petrochemically derived 1,3-PDO, biobased 1,3-PDO has a lower cost of production, making biobased production very attractive and the preferred source (IAR, 2015). Figure S6.13 in the Supplementary material shows the flowsheet for the production of 1,3-PDO via fermentation of A-molasses.

6.3.12 Succinic Acid

Biobased succinic acid is produced by the microbial conversion of plant-derived sugars (Biddy et al., 2016). It is a platform chemical widely used for high-value, niche applications in the food and beverage industries, as well as for high-volume products such as plasticisers, polyurethanes, resins, coatings, and more than 30 commercially important products (Vaswani, 2010). The potential for succinic acid as a platform chemical to replace maleic anhydride for the 1,4-BDO production (Mazière et al., 2017) is the main driver for its the growing global market, although this has mostly been replaced by direct production of 1,4-BDO from sugars (see the previous section). Succinic acid is also a monomer required for the production of the biodegradable plastic polybutylene succinate (PBS; see Section 6.4.1). The estimated biobased market size for succinic acid in 2025 was 70 kt/y (ChemAnalyst, 2025b). In 2019, the market prices of fossil- and biobased succinic acid were 2,500 US$/t and 2,860 US$/t, respectively (Gérardy et al., 2020).

Despite attractive techno-economics, four commercial facilities for succinic acid production have failed due to (i) much higher production costs than anticipated, apparently due to purification requirements, (ii) competition from Genomatica's process for direct conversion of sugars into 1,4-BDO, and (iii) rapid changes in market prices for competing fossil-based chemicals around 2015, which negatively affected market prices for biobased chemicals as well (Li and Mupondwa, 2021). In short, a product must demonstrate robust process performance assumptions and techno-economic viability substantially superior to that of succinic acid to be prioritised for biorefinery implementation. Figure S6.14 in the Supplementary material presents the process flowsheet for bio-succinic acid production from A-molasses.

6.4 BIOBASED POLYMERS

6.4.1 Polybutylene Succinate (PBS)

PBS is a biodegradable polyester that can be produced from 1,4-BDO and SA monomers obtained either from fossil fuels or renewable biomass, although the biobased monomers offer superior environmental benefits. PBS is used in packaging materials, transportation, electric appliances, and electronics applications. Biobased PBS has recently attracted intense attention as a green substitute for poly(ethylene)terephthalate (PET), polyethylene (PE), polypropylene (PP), and polystyrene (PS) owing to its excellent processability and good biodegradability. The first commercial-scale PBS manufacturing was established by Showa Highpolymer and Showa Denko in 1993 under the trademark Bionolle, utilising petroleum-derived monomers. Mitsubishi Chemical Corporation has also successfully commercialised 1,4-BDO and established a joint venture with PTT Global in 2017 to produce 1,4-BDO in Thailand (Kato et al., 2023).

Fossil-derived PBS is commercially produced from succinic acid and 1,4-BDO with a capacity of 91 kt/y in 2019, while the capacity is expected to remain the same until 2024 (Skoczinski et al., 2020). The biobased PBS market value was estimated to be around 24 million US$/y with market volume ranging from 5 to 6 kt/y, and it is expected to grow by 37% compounded annual growth rate (CAGR) by 2020 (Taylor et al., 2015). Although the PBS market volume is currently low, it is expected to maintain high growth due to its superior biodegradability when produced from either fossil-derived or biobased monomers, compared to other polymers with similar applications. It does not require any expensive equipment to facilitate its biodegradability, as it can easily degrade in water as well as soil at ambient conditions. PBS has also favourable market prices ranging from 4,500 to 5,000 US$/t in 2014 (Taylor et al., 2015) and a fossil-based market price of 2,380 US$/t (Skoczinski et al., 2020).

Figure S6.15 in the Supplementary material shows the PBS polymerisation and recovery from its monomers (1,4-BDO and SA produced from sugar). PBS is produced by polymerisation of monomers in two steps, the first of which is esterification at temperatures around 200°C and under substantial vacuum. Subsequent to esterification, transesterification is started by increasing the reactor temperature to 230°C at high vacuum to form a high-molecular-weight PBS. PBS is further purified by dissolving it in chloroform and precipitating it in cold methanol. The PBS biorefinery considered in the present study had both succinic acid and 1,4-BDO production facilities onsite, with the common A-molasses feedstock split between the monomers such that the stoichiometric ratio of 1:1.27, necessary for efficient PBS polymerisation, is maintained. The process therefore produces a fully bio-derived PBS.

6.4.2 Polyethylene (PE)

Biobased PE is a replacement for the conventional fossil-based PE but also considered an alternative to other biobased polymers (bioplastics). PE is one of the most widely used plastics on Earth, with a wide range of applications ranging from food packaging to construction and use in medical devices. PE is conventionally produced through the polymerisation of fossil-based ethylene. In recent years, the commercial production of biobased PE from Brazilian bioethanol has been realised by first converting ethanol into monomeric ethylene. Braskem is the largest player (about 75% market share) in the bio-PE market with an annual production capacity of 200 kt/y from sugarcane (nova-Institut GmbH, 2020) and plans to expand this. Biobased PE forms only a small percentage (<0.5%) of the total PE market, with substantial growth potential in future. The current biobased market price of PE is 1,830 US$/t (European Bioplastics, 2025).

Figure S6.16 in the Supplementary material shows the biobased process for PE production from bioethanol, with ethanol derived from A-molasses. The process considered here considers all three steps, that is, ethanol, ethylene, and PE production, as add-ons to an existing sugarcane mill. This is a conservative approach, since there is a global overcapacity of polymerisation capacity (Greenwood, 2017), indicating that further cost savings may be possible relative to the production method considered here through the polymerisation of biobased ethylene in existing polymerisation plants.

6.4.3 Polyethylene Furanoate (PEF)

Polyethylene furanoate (PEF) is a thermoplastic that is considered to be a functional, biobased replacement for fossil-based PET, with a cost of manufacturing that is potentially lower than that of biobased PET (Louw et al., 2023). PEF has a wide range of potential applications, including fabrics, food packaging, and the manufacturing of automotive parts, and is produced through the polycondensation of biobased monoethylene glycol (MEG; typically derived from bioethanol) and 2,5-furandicarboxylic acid (FDCA). FDCA is chemically analogous to terephthalic acid (TPA), which is combined with MEG for the production of PET (Taylor et al., 2015).

Compared to fossil-based PET, the production of PEF can decrease GHG emissions by 45%–55% and non-renewable energy use by 40%–50% (Eerhart et al., 2012). The lower melting point, higher glass transition temperature, and greater CO_2 and O_2 barrier properties of PEF make it a superior packaging material compared to PET (Papageorgiou et al., 2016). Although biobased (ethanol-derived) MEG is presently used for the manufacture of partially biobased PET (e.g., Coca-Cola PlantBottle®), there are substantial challenges in the manufacture of fully biobased TPA, to replace the fossil-based TPA presently used for PET manufacture; PEF avoids these challenges by utilising FDCA instead of TPA.

Currently, PEF manufacturing is yet to be commercialised, but Avantium® has successfully produced PEF on a pilot scale with plans to complete the production of a 5 kt/y facility by the end of 2023 (Avantium, 2020). The current production of MEG for PlantBottle® is done using Brazilian ethanol that is shipped to India for conversion. UPM Biochemicals recently announced the start of construction of its lignocellulose-based Leuna Biorefinery in Europe, which will produce bio-MEG from lignocelluloses as its primary product. Figure S6.17 in the Supplementary material provides the flowsheet for the production of PEF, where both the ethanol and FDCA required for its production are derived from A-molasses.

6.4.5 Polylactic Acid (PLA)

PLA is a biobased and compostable polymer used in packaging materials, transportation, electric appliances, and electronics applications. PLA is exclusively produced from polymer-grade, biobased lactic acid, via ring-opening polymerisation. The prepolymers are first formed at reduced pressure and 150°C, then the prepolymers are fed into a lactide reactor to form cyclic dimers, which are further purified through multiple distillations. Purified lactide is polymerised to form a high-molecular-weight PLA, which is further purified by dissolving it in chloroform and precipitating it in cold methanol (Figure S6.18 in the Supplementary material). The global production capacity for PLA was estimated at 611 kt/y (European Bioplastics, 2025) and the global PLA market price is currently around 2,500 US$/t (Imarc group, 2026c).

6.5 BIOFUELS

6.5.1 Conventional Bioethanol

Bioethanol is conventionally considered a renewable fuel for decarbonising the road transport fuel sector by blending it with gasoline, where its production from sugarcane crops is a well-established industrial technology. However, bioethanol is also considered as both a critical platform chemical for the growing bio-economy as well as an industrial solvent, offering several technical advantages in terms of yield and final product titre compared to alternative platform chemicals derived from sugar (Lynd et al., 2022). Ethanol is a feedstock to produce industrial chemicals, solvents, and polymers such as ethyl lactate, ethyl acetate, PE, MEG (as precursor for PEF and PET plastics), and butadiene, as well as sustainable aviation fuel (SAF) (alcohol-to-jet conversion) (Dagle et al., 2020; Godavari, n.d.). Several of these derivative products are considered elsewhere in this report, representing market sizes that may be comparable to that of fuel ethanol (Rosales-Calderon and Arantes, 2019). The current price for biobased ethanol is around 687 US$/t (Imarc Group, 2026b). Figure S6.19 in the Supplementary material shows the production of bioethanol from A-molasses *via* fermentation. Although for several developing countries there is an opportunity for tariff-free export to the EU of fuel-grade ethanol, the amount of so-called 'crop-based' bioethanol is capped and already fulfilled by imports from Brazil and the USA. The bioethanol considered for production from A-molasses in the present study would also be considered as crop-based, according to EU-REDII. For the future, there is substantial potential to increase the export of so-called waste-based biofuels to the EU, where these are considered as advanced biofuels that fetch premium prices in the market. Ethanol produced from sugarcane bagasse and harvest residues (2G ethanol) may be considered as waste-based advanced biofuels, for example, in a 1G2G facility for the co-production of ethanol from A-molasses and lignocelluloses. However, such facilities were not further considered in the present study.

6.5.2 Sustainable Aviation Fuel (From Bioethanol)

The market for SAF is driven primarily by international agreements and markets, such as the EU, USA, and the International Air Transport Association (IATA), which will continue to stimulate global demand based on agreed targets for substantial reductions in GHG emissions (Petersen et al., 2021; Transport & Environment, 2016). Therefore, with the EU consuming about 75 billion litres of aviation fuel per annum, this could translate into a SAF demand of at least 21.5 billion litres per annum (17.2 million kt/y), although it should be kept in mind that the same regulations related to crop-based and waste-based (advanced) biofuels as relevant to ethanol export to the EU, will apply here. Currently, only around 1,000 kt/y of SAF is produced, which is far below the agreed-upon targets (IATA, 2024).

Due to international agreements and regulations such as these, there is a strong global market demand for SAF, which presently outstrips supply (KPMG, 2022). LanzaJet and its partners have started to commercialise a process for ethanol conversion to SAF, which was co-developed with the Pacific Northwest National Laboratory (PNNL) in the USA. LanzaJet launched the world's first commercial-scale ethanol-to-jet SAF plant in 2024, with a production capacity of more than 90 kt/y of SAF and biodiesel (LanzaJet, 2024). Other competitors in alcohol-to-jet technologies, such as Gevo, use isobutanol as the starting alcohol, although the latter is substantially more expensive to produce than ethanol (Geleynse et al., 2018). The current market prices for SAF are therefore highly variable and difficult to determine. A typical price of 1,870 US$/t is assumed here (S&P Global, 2020). However, this price does not account for the distinction between first-generation crop-based SAF and second-generation waste-based SAF, the former of which will not be accepted under EU regulations (Climate Catalyst, 2026). The production of SAF from A-molasses via bioethanol is shown in Figure S6.20 in the Supplementary material.

6.6 BENCHMARKING METHODOLOGY

This study employed a standardised benchmarking approach to both the process simulations and TEAs for the production of the aforementioned 20 bioproducts. In each scenario, the bioproduct is produced in a biorefinery integrated into an existing sugar mill by utilising A-molasses from the sugar mill as feedstock, thus providing the same amount of feedstock to each potential product. Aspen Plus® chemical process simulator was used to simulate each of the processes to generate the mass and energy balance data required for the TEAs. All biorefineries were designed to be energy self-sufficient based on sugarcane materials only, including the supply of the required process energy to the existing sugar mill. These simulations were based on process technical data obtained from the literature with technical assumptions aimed at industrial practices. The techno-economic results of the bioproducts, first at full production scale using all A-molasses and, secondly, at a scaled-down capacity of 10% of the total global market size, were compared to rank the products in order of economic feasibility based on the minimum selling price (MSP) compared to the market price of each product. Furthermore, the global warming potential (GWP) of each product was calculated on a cradle-to-gate basis using SimaPro LCA software.

6.6.1 The Benchmarking Strategy

The simulation-based approach that is used in the present TEA starts with an existing, diffuser-based sugarcane mill as described in the Aspen Plus® process simulations developed by Guest et al. (2019) and refined by Dogbe et al. (2020, 2018).

6.6.1.1 The Integrated Biorefinery Design Approach

The annexation of molasses-based biorefineries to existing sugarcane mills requires suitable energy integration strategies, with additional boilers and/or upgrades to combined heat and power (CHP) facilities (where needed), to ensure that the combined facilities remain bioenergy self-sufficient without the need to import external biomass- or fossil-based energy sources. This consistent basis for integration of the biorefinery with the sugar mill will allow the differences in capital and operating costs and market prices of the different products to influence the apparent attractiveness of each product independently of the sugar processing performance. The resulting comparisons therefore offer a robust, conservative view of the commercial viability and environmental (GHG) benefits of various products on an equal technical basis. The standard simulation modelling approach applied for each of the 20 biorefinery scenarios is presented in Table 6.2. The table shows the mass and energy balances of a typical existing sugar mill with three stages of crystallisation (ABC boils) and those of the proposed single-stage crystallisation (A boil only).

6.6.1.1.1 Mass Integration

The crystallisation unit of a typical sugarcane mill is designed for maximum sucrose extraction and operates in three stages (A, B, and C). The A boiling stage recovers approximately half of the sugar, with the B and C stages designed to recover the remaining crystallisable sugar for reprocessing in the A stage. Diverting all of the A-molasses to an annexed biorefinery implies that only the first (A) stage crystallisation unit is operated, decreasing the amount of crystalline sucrose and final C-molasses produced for sale. However, the processing costs associated with this sugar will be lower than those associated with the traditional approach, since the reprocessing of B and C sugar is avoided (Dogbe et al., 2018).

TABLE 6.2
Overall performance of sugar mill (Louw et al., 2025)

Parameter	Unit	ABC boils[a] (normal sugar mill – 45% SoC)	ABC boils[a] (improved sugar mill – 40% SoC)	A boil only[b] (A-molasses extraction)
Cane crushing capacity	t/h	300	300	300
Raw sugar flowrate	t/h	38.37	38.37	27.66
Steam demand	t/h	135	120	104.5
Electricity demand	kW	8,100	8,100	8,100
Bagasse combusted (DM)	t/h	29.37	26.03	22.67
Bagasse leftover (DM)	t/h	14.91	18.25	21.61
Total bagasse produced (DM)	t/h	44.28	4.28	44.28

[a] A, B, and C pans are used, hence steam is required for A, B, and C pans (SoC = steam-on-cane).
[b] Only A-pans are used, hence no steam required for the B and C pans.

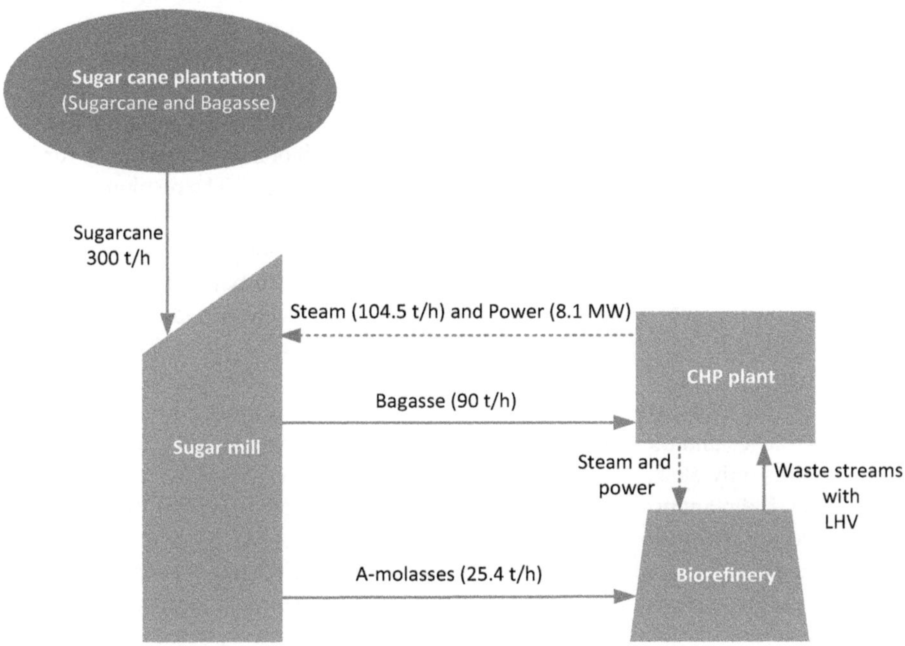

FIGURE 6.2 Schematic diagram of the 1G integrated biorefinery annexed to a typical South African sugar mill.

A-molasses has a higher sugar content than B- and C-molasses and may replace industrial sugar products as biorefinery feedstock, while yielding biorefinery products with less complex purification requirements than B- and C-molasses. In Figure 6.2, the schematic of a molasses-based biorefinery annexed to a sugar mill is shown. Each biorefinery is designed to be energy self-sufficient using only sugarcane materials obtained from current agricultural activities. Depending on the energy demand of the biorefinery, the existing CHP plant or, where necessary, either a new utility boiler or a new CHP plant (preferably a medium-pressure boiler), is used to supply the heat and power demands of both the sugar mill and biorefinery. The fuel supply for boiler(s) would be the existing bagasse together with the brown leaves component of harvesting residues (if required), with no external sources of energy, either bio- or fossil-based.

6.6.1.1.2 Energy Integration

The existing boiler in a typical sugarcane mill was assumed to produce 135 t/h of steam for a sugar factory processing 300 t cane/h, resulting in a sugar mill steam efficiency of 45% steam-on-cane, representing an average of several mills operating at lower or higher efficiencies. It was assumed that the sugar mill steam economy would be improved from the current average of 45% to 40% steam-on-cane, together with biorefinery development, with the latter corresponding to 120 t/h sugar mill steam requirement as shown in Table 6.2. It was further estimated that an additional 15 t/h steam could be supplied to the biorefinery from the current boiler, since 6.8 t/h wet bagasse would be liberated as a result of improving the steam efficiency from 45% to 40%. The biorefinery should pay for the investment costs of generating the steam and bagasse savings (the liberation cost). The base case steam demand for the 300 t/h cane sugar mill, that is, 120 t/h, included 96.5 t/h of low-pressure (LP) steam (130°C and 2.62 bar) and 23.5 t/h of high-pressure (HP) steam (360°C and 28.6 bar). The HP steam demand remains the same irrespective of biorefinery integration, while the LP steam demand reduces to 81.0 t/h when A-molasses is diverted to the biorefinery, leaving a further excess steam capacity of 15.5 t/h to be utilised in the biorefinery. The resultant condensate returned from the sugar mill (90°C and 1.01 bar) for the biorefinery is 104.5 t/h.

When the existing CHP unit was not able to supply enough steam for both the sugar mill and the biorefinery, a small, medium-pressure (10 bar; subject to the biorefinery steam pressure and electricity demand) boiler was installed in the biorefinery to supplement the 15 t/h surplus capacity in the existing sugar mill CHP. The maximum theoretical power demand for a typical diffuser mill is 27 kWh/ton of cane (8,100 kW), including the cogeneration system, of which almost 50% is utilised by boiler auxiliaries and the powerhouse turbine. Hence, for a new boiler, the sugar mill itself requires only 50%, which is 4,050 kW for 300 t/h cane processed. The existing sugar mill CHP plant produces 1,857.5 kW surplus electricity (over 8,100 kW) that is available to the biorefinery.

The present TEA considers only the 1G-biorefineries where A-molasses is used as feedstock for conversion to primary products, while surplus bagasse and/or harvest residues are used to supplement existing energy supply systems. The price of A-molasses provided to the biorefinery was calculated as the sum of annual revenue losses to the sugar mill from producing less sugar and no C-molasses, divided by the flowrate of A-molasses, thus representing a robust estimate of the opportunity cost for the sugar mill.

6.7 TECHNO-ECONOMIC ANALYSIS OF THE BIOREFINERY

A TEA for each biorefinery scenario was conducted to inform process/product selection and investment decision-making for implementation (Mandegari et al., 2017a). The simulation of the biorefinery scenarios in Aspen Plus® generated the mass and energy balance and costing data for some common process units, such as pumps, heat exchangers, and separation columns, while other process units were based on costs provided in the literature, adjusted for the capacity and timing of the study (Mandegari et al., 2018). The calculation of the total capital investment (TCI), operation costs (OPEX), and the cash flow analysis was performed in Microsoft Excel spreadsheet (Humbird et al., 2011). Table 6.3 presents the set of economic parameters used for the techno-economic analyses. Discounted cash flow analysis (DCFA) was conducted in real terms to obtain the MSP of the various products based on the

TABLE 6.3
Economic parameters used for the economic analysis (Louw et al., 2025)

Parameter	Value
Cost year analysis	2,023
Term	Real term
Desired internal rate of return (IRR)	20%
Plant operating hours, based on sugarcane harvesting	5,000 h/year
Income tax rate	28%
Plant life	25 years
Depreciation:	
Depreciation method	Straight line
Depreciation period	5 years (20%/ year)
Salvage value	0%
Construction period:	
% Spent in −2	10%
% Spent in −1	60%
% Spent in 0	30%
Startup time	1 year
Revenues during startup	50%
Variable costs incurred during startup	75%
Fixed costs incurred during startup	100%
Working capital	5% of FCI
Land cost	R0
Prices	US$/t
C-molasses price	145.7
Sugar price	303.0
Bagasse price	29.6
Price of brown leaf (collection and transport)	31.6
Calculated A-molasses price	193.9

minimum desired internal rate of return (IRR) of 20% (real terms, not nominal). The latter desired IRR was set at a conservative value to accommodate cost increases in further detailed engineering design and process scale-up performance. The MSP therefore provides a conservative, realistic view of the cost of manufacturing for different products, which can be directly compared to the wholesale market prices for these products to determine financial viability.

6.7.1 Markets Analysis

It is important to compare the production rates of the products, as well as their cost of production indicated by the MSPs from the DCFA to their market sizes and selling prices to ascertain their profitability and ease of access to the market. The market sizes and selling prices used in this study were sourced from the open literature (Imarc Group, 2026a–d; market.us, 2026; Facts and Factors, 2022; IATA, 2024; ChemAnalyst, 2025b; Phasha et al., 2025; Blaga et al., 2023; European Bioplastics, 2025) and market reports (Future Market Insights, 2026; Grand View Research, 2021a–e; Maia Research, 2024; MAResearch, 2022); prices were adjusted for the 2023 costing year as a common basis for comparison.

6.7.2 Benchmarked Production Capacity for 10% of Market Size

The proportion of an existing market for a particular product that may be captured by a new producer/entrant should be based on a realistic assumption for market penetration, set at 10% in the present study; production rates for all products were scaled such that this indicator was not exceeded in the pursuit of economy-of-scale advantages (Lees, 2019). This conservative approach ensured that supply would not exceed demand (oversupply), causing price crashes and the eventual demise of new businesses (Li and Mupondwa, 2021). The assumed limit of 10% penetration of the products' global market sizes was deemed to be a reasonable limit to keep the product price fairly stable for new entrants (Nieder-Heitmann et al., 2019).

Based on the total amount of A-molasses available from the existing sugarcane mill, some products would exceed the market penetration target of 10%. For these products the amount of A-molasses diverted for their production was therefore reduced, with the CAPEX estimates scaled down by the 6–10th rule. Variable operation costs were scaled down proportionally to the production scale, while some fixed operating costs such as labour remained the same.

6.7.3 Greenhouse Gas Emissions of the Biorefinery

6.7.3.1 Goal, Scope, System Boundary, and Allocation

The goal of this study is to quantify and compare the GHG emissions associated with each product from integrated A-molasses biorefineries. The functional unit is defined as kilograms of CO_2-equivalent emissions per kilogram of bioproduct. The bioproducts are derived from biorefineries annexed to existing South African sugar

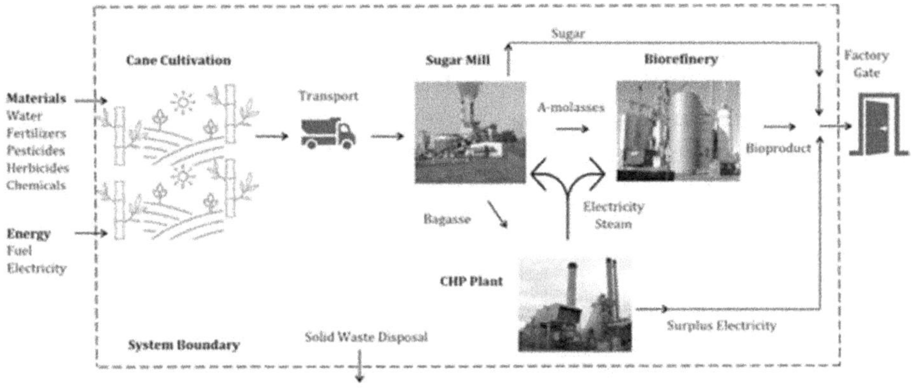

FIGURE 6.3 System boundary for the calculation of greenhouse gas emissions in this study.

mills, using A-molasses as the primary feedstock. In this study, the GHG assessment was performed on a cradle-to-gate basis, meaning that the system boundary included all processes from sugarcane cultivation to the production of final bioproducts at the biorefinery gate, as illustrated in Figure 6.3, but excluded downstream stages such as product distribution, use and end-of-life treatment. Furthermore, the GHG analysis focused exclusively on non-biogenic CO_2 emissions, that is, emissions resulting from fossil-derived inputs such as electricity from non-renewable grids, chemical reagents, and transportation fuels.

Biogenic CO_2 emissions, which result from the combustion or degradation of biomass-derived materials (e.g., sugarcane, bagasse, or A-molasses), were not included in the GHG totals, following standard practice for cradle-to-gate assessments. These emissions are typically considered carbon-neutral under the assumption that the carbon released during biogenic combustion was recently captured from the atmosphere during biomass growth. Non-CO_2 biogenic emissions, such as methane and carbon monoxide, were included in the carbon accounting, as these are not reabsorbed during sugarcane growth. Net carbon sequestration was excluded from the scope of the analysis. Including sequestration would require extending the system boundary beyond the biorefinery gate to account for long-term carbon storage in products or soils, or for land-use change dynamics – factors that were not considered in this study. This approach enables consistent and conservative benchmarking of different bioproducts within the defined system boundary. However, it should be noted that excluding carbon sequestration may underestimate the climate benefits of products that ultimately sequester carbon over extended periods (e.g., durable bioplastics). Future cradle-to-grave analyses could provide a more comprehensive understanding of such benefits by including end-of-life scenarios and carbon storage dynamics. An economic allocation approach was applied based on the relative selling prices of co-products and raw materials. The environmental impacts of sugarcane cultivation and milling were allocated among A-molasses, bagasse, raw sugar, and filter cake according to their respective selling prices and mass flowrates.

6.7.3.2 Life-Cycle Inventory and Methods

The inventories for each bioproduct were developed based on the simulated mass and energy balances in Aspen Plus®, supplemented by literature data for sugarcane cultivation and typical sugar milling operations, as shown in Tables 6.4 and 6.5, respectively. A midpoint life-cycle impact assessment (LCIA) approach was applied, using the CML-IA baseline methodology (Version 3.08, EU25) developed by the Centre of Environmental Sciences (CML) at Leiden University in collaboration with Technische Universität Berlin. The analysis focused specifically on GHG emissions.

To obtain emission factors (EFs) for raw materials not included in the Ecoinvent database in SimaPro, data were sourced from the Roundtable on Sustainable Biomaterials (RSB) GHG calculator tool or relevant literature. Based on the inputs presented in Tables 6.4 and 6.5, the combined non-CO_2 biogenic and non-biogenic CO_2-equivalent EFs for A-molasses and sugarcane bagasse were calculated in SimaPro as 0.19 kg CO_2eq/kg and 0.03 kg CO_2eq/kg, respectively. EFs for all other chemicals used in this study are provided in Table S6.5 of the Supplementary document.

TABLE 6.4
Life-cycle inventory for sugarcane cultivation (Mashoko et al., 2010)

Sugar cane agriculture	Value	Unit
Cultivation area	400,000	ha
Average cane harvest	60	t/ha
Irrigation water requirements	8,000	m^{3}/ha
Electricity consumption for irrigation	108	kWh/ha
N_2O emissions from soil	1.5 (1.25% of N input)	kg
NO_x emissions from soil	0.6 (0.5% of N input)	kg
Fertiliser application rate/ha		
Urea, N	120	kg
Diammonium phosphate (DAP)	30	kg
Potassium oxide, KCl	125	kg
Herbicides use	1,614 (26.9 g/t sugarcane)	g
Herbicides loss in water bodies	3.228 (0.2%)	g
Nitrogen loss in water bodies	161.4 (10%)	g
Phosphorus loss in surface runoff/ha	1	kg
Pesticide use	132.6	g
Cane transportation		
Transportation, average distance, by:		
Road (94% of cane)	25	km
Rail (6% of cane)	50	km
Diesel consumption:		
Truck	1.08	MJ/tkm
Rail	0.68	MJ/tkm
Fertilisers and herbicides transport distance	60	km

TABLE 6.5
Life-cycle inventory of a typical sugar mill when A-molasses is diverted for bioconversion (Louw et al., 2024)

Input	
Sugarcane (t/h)	300
Steam consumed (t/h)	104.5
Bagasse consumed (t/h)	59.52
Electricity consumed (kWh)	6,242.5
Coal consumption (kg/t cane)	8.4
Water consumed (m^3/t cane processed)	0.6
Output	
Raw sugar produced (t/h)	27.66
Bagasse produced (t/h)	90.00
Excess bagasse (t/h) (output–input)	30.48
Filter cake produced (t/h)	1.13
Steam generated (t/h)	135
Electricity generated (kWh)	8,100
Carbon dioxide emissions to air (t/h)	51.23
Ash emissions to soil (t/h)	1.37
COD/t cane	3,320
BOD5/t cane	1,590

6.8 BENCHMARKING RESULTS AND DISCUSSION

6.8.1 Mass and Energy Balance Results

The overall mass and energy balance results (product yields as well as heating, cooling, and electricity demands) of the 20 products are presented in Figure 6.4, with a more detailed analysis provided in Tables S6.1–S6.4 in the Supplementary material. Products with high total mass yields (t product per t A-molasses) of more than 40% were lactic acid (68%), ethyl lactate (50%), succinic acid (46%), and glutamic acid (43%), while PE had the lowest yield, at 20%. Product yield affects the economics of the product, since the feedstock cost contributes substantially to biorefinery operation costs (Mandegari et al., 2017b).

In general, products produced via chemical catalytic or hybrid pathways exhibited higher specific energy demands than those derived from direct biological conversion, with four of the five most energy-intensive processes following hybrid routes (Figure 6.4). An exception was ADA, a fermentation-based product, which showed the highest overall energy demand (22.1 $MW/ton_{product}$). This was primarily due to its relatively low product titre (50 g/L), necessitating energy-intensive downstream processing steps such as evaporation, reactive distillation, and crystallisation (Cronjé et al., 2023).

Ethyl lactate, produced via a hybrid pathway, had the second-highest energy demand (18.7 $MW/ton_{product}$). This was attributed to its multi-step production route,

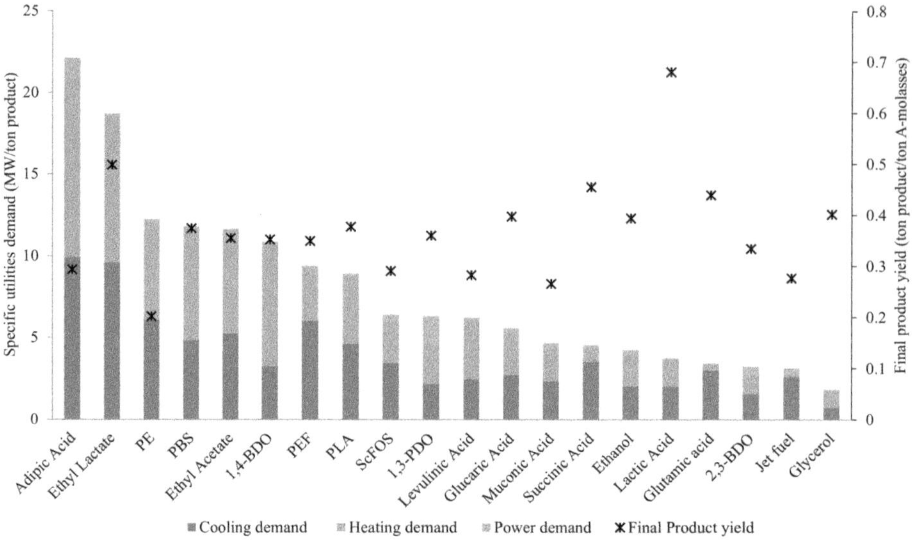

FIGURE 6.4 Summary of mass and energy balance results for all of the products considered.

involving the separate production and purification of ethanol and lactic acid, followed by esterification and extensive downstream separation. The latter requires four distillation columns to isolate ethyl lactate from ethanol, lactic acid, water, and oligomers (Phasha, 2025). Similar trends were observed across most catalytic and hybrid pathways, which typically involve multiple reaction steps and more complex purification systems. Bioplastic production pathways are notable examples, with relatively high utility demands observed for PE (12.2 $MW/ton_{product}$), PBS (11.8 $MW/ton_{product}$), PEF (9.3 $MW/ton_{product}$), and PLA (8.9 $MW/ton_{product}$).

In contrast, biological processes generally exhibited lower energy demands due to simpler production pathways, in particular those with direct microbial or enzymatic conversion of sugars into final products, and less intensive downstream processing. Among all products evaluated, glycerol had the lowest energy demand (1.8 MW/ton), attributed to its high product titre (219 g/L) and the use of energy-efficient multi-effect evaporation and vacuum distillation for water removal (Cronjé et al., 2023). Other low-energy biological processes included 2,3-BDO (3.26 MW/ton), glutamic acid (3.44 MW/ton), lactic acid (3.73 MW/ton), ethanol (4.24 MW/ton), succinic acid (4.54 MW/ton), and muconic acid (5.06 MW/ton), all ranking among the eight least energy-intensive scenarios. ScFOS (6.39 MW/ton) and 1,3-PDO (6.34 MW/ton) showed moderately higher energy demands, ranking ninth and tenth overall. In contrast, 1,4-BDO (10.80 MW/ton) was the second most energy-intensive biological process, after ADA, and ranked sixth overall, largely due to the energy-intensive distillation required to achieve polymer-grade purity (99.99 wt%) from large volumes of water and heavy impurities (Ratshoshi et al., 2021).

In an energy-sufficient biorefinery scenario such as the one considered in this study, high process energy intensity could limit the utilisation of all available feedstock for product synthesis due to insufficient feedstock for energy supply and, hence, worsen the financial viability owing to the economies of scale at lower production rates. However, it is worth noting that, with the current sugar mill energy supply assumptions, the combination of bagasse with harvest residues (brown leaves), supplemented by waste products from the biorefinery, such as biogas from a wastewater treatment plant, provided adequate feedstock to meet the energy demands of all the integrated sugar mill and biorefinery scenarios. No external energy sources, such as grid-based electricity, coal, natural gas, or external biomass sources, were used to supplement these for process energy supply. In practice, the use of harvest residues for energy supply would incur additional feedstock costs compared to conventional cane harvesting and would best be avoided, if possible. Although some improvements may be proposed to the process configurations of the mass and energy balance models for the sugar mill and the new product facilities, it is important to note that the simulations have been performed consistently, such that any bias introduced by specific assumptions should similarly influence all similar biorefinery models. Thus, the results are suitable for comparison and benchmarking of the technical feasibility of each product based on mass and energy balances, independently of the sugar processing performance. The mass and energy balance results, therefore, provide a realistic perspective on the manufacturing requirements for the different products, according to the best-reported process performances, which may also apply to sugar-based feedstocks other than sugarcane A-molasses. Weaker process performances also highlight specific opportunities for technology improvements in the future.

6.9 TECHNO-ECONOMIC RESULTS

6.9.1 Techno-economic Results of Full-Scale Production Using All of the A-molasses

The benchmarked techno-economic results of the 20 products, expressed in terms of specific OPEX, specific amortised CAPEX, and MSP, are presented in Figure 6.5. In general, there is a strong positive correlation between the specific amortised CAPEX and the MSP. For example, SAF had the highest specific amortised CAPEX (1,476 US$/t) and MSP (5,528 US$/t), while glycerol, with the lowest specific amortised CAPEX (196 US$/t), had the third lowest MSP (1,238 US$/t). The scFOS, which had the second lowest specific amortised CAPEX of 224 US$/t, had the lowest MSP of 1,022 US$/t. The economic performance of scFOS was strongly dependent on the sale of invert sugars as a by-product of the scFOS production process. In the absence of invert sugar recovery and valorisation, the MSP of scFOS increases substantially to 1,800 US$/t, exceeding the MSPs of ethyl acetate (1,749 US$/t), 1,3-PDO (1,667 US$/t), succinic acid (1,576 US$/t), ethanol (1,379 US$/t), glycerol (1,238 US$/t), and lactic acid (1,205 US$/t). Co-production of saleable by-products, therefore, has a significant effect on the economics of the biorefineries.

The correlation between process complexity, defined in terms of the number of unit operations and the nature of processing routes (biological, catalytic, or hybrid), and the specific amortised CAPEX of bioproducts is generally positive, but not uniform. For instance, several products derived via direct fermentation of A-molasses,

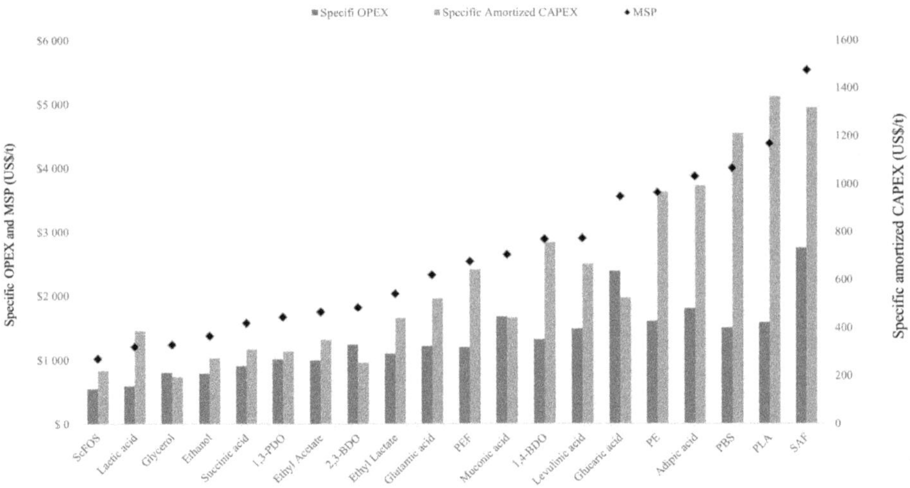

FIGURE 6.5 Techno-economic results of bioproducts produced at full capacity from all available A-molasses.

including ADA (995 US$/t), 1,4-BDO (757 US$/t), glutamic acid (552 US$/t), and muconic acid (443 US$/t), exhibit higher specific amortised CAPEX values than certain products produced through chemical catalysis, such as levulinic acid (667 US$/t) and glucaric acid (526 US$/t), as well as some multi-step hybrid processes, including ethyl lactate (440 US$/t) and ethyl acetate (352 US$/t). Consequently, it cannot be assumed that biological processes are inherently less capital-intensive than catalytic or hybrid processes. This deviation from an expected trend can be attributed to factors such as low yields, titres, and productivities in direct microbial conversions, as well as the complexity of downstream separation and purification, the type and cost of process equipment, and the technological maturity of the production pathway. These factors collectively exert a strong influence on specific CAPEX and overall process economics (Lynd et al., 2022; Mancini et al., 2022; Pothakos et al., 2018).

However, it is evident that biopolymers, which are typically associated with higher process complexity and costly polymerisation steps, consistently exhibited relatively high specific amortised CAPEX values (PLA: 1,363 US$/t; PBS: 1,211 US$/t; PE: 968 US$/t; PEF: 644 US$/t). Conversely, biological processes dominated the lower end of the CAPEX spectrum. Well-established enzymatic and fermentation-based processes, such as scFOS and glycerol, displayed the lowest specific amortised CAPEX across all scenarios (scFOS: 224 US$/t; glycerol: 196 US$/t), followed by 2,3-BDO (257 US$/t), ethanol (274 US$/t), 1,3-PDO (303 US$/t), and succinic acid (311 US$/t). This trend reflects the relative simplicity, scalability, and technological maturity of these processes.

The relationship between specific OPEX and MSP is positive but relatively weaker than that of CAPEX, as shown in Figure 6.5. This can largely be attributed to the uniformity of raw material costs across all scenarios, given that feedstock constitutes the dominant component of OPEX. Nevertheless, elevated specific OPEX values for certain products, such as glucaric acid (2,390 US$/t) and muconic acid (1,679 US$/t), had a pronounced impact on their respective MSPs (3,560 US$/t and 2,648 US$/t), despite their comparatively low specific amortised CAPEX values (526 US$/t and 443 US$/t, respectively). This behaviour is attributed to the extensive use of external chemicals, including solvents (acetic acid and acetonitrile), and oxygen in glucaric acid production (Cronjé et al., 2023), and sodium hydroxide in muconic acid production (Phasha et al., 2025).

In general, purely biological processes tend to exhibit lower specific OPEX due to minimal chemical input requirements. This was evident for scFOS (549 US$/t), lactic acid (597 US$/t), ethanol (792 US$/t), glycerol (805 US$/t), and succinic acid (909 US$/t), which rank among the lowest OPEX values across all scenarios. Notable exceptions include ADA (1,802 US$/t) and muconic acid (1,679 US$/t), which display comparatively high OPEX values among biological routes. This is primarily due to increased external chemical inputs, including enzymes for sucrose hydrolysis (in the case of ADA), fermentation nutrients, and neutralising agents (e.g., sodium hydroxide and ammonium hydroxide), in combination with relatively low process yields (29.3% and 26.5% for ADA and muconic acid, respectively).

In addition to ADA, several other products, such as glucaric acid and 1,3-PDO, require costly enzymatic inversion of sucrose prior to fermentation. To mitigate this expense, potential strategies could include genetically engineering microorganisms to produce the necessary enzymes themselves or enabling them to metabolise sucrose directly. Implementing such approaches could eliminate the need to purchase or produce enzymes onsite, thereby significantly reducing the production costs of these bioproducts (Löwe et al., 2017).

Accordingly, simple biological processes achieve the lowest MSPs across all scenarios, driven by their relatively low specific OPEX and CAPEX. Products such as scFOS (1,022 US$/t), lactic acid (1,205 US$/t), glycerol (1,238 US$/t), ethanol (1,379 US$/t), succinic acid (1,576 US$/t) and 1,3-PDO (1,667 US$/t) rank as the most economically favourable scenarios, ranking from the lowest to the sixth-lowest MSP among all scenarios, respectively. These scenarios are notably all biological-based processes, reflecting their low production costs relative to chemical and hybrid production routes. In contrast to the expected trend, several hybrid processes, combining biological conversion with subsequent chemical catalytic upgrading, also exhibited comparatively low MSPs. Ethyl acetate (1,749 US$/t) and ethyl lactate (2,034 US$/t) achieved lower MSPs than a number of purely biological processes, including glutamic acid (2,329 US$/t), muconic acid (2,648 US$/t), 1,4-BDO (2,892 US$/t), and ADA (3,875 US$/t). This can be attributed to the low MSPs of their key intermediates (ethanol for ethyl acetate, and both ethanol and lactic acid for ethyl lactate), as well as the relatively simple, solvent-free conversion of these intermediates to the final products.

In contrast, more complex hybrid processes associated with biopolymer production exhibit high MSPs, with PEF (2,543 US$/t), PE (3,620 US$/t), PBS (3,999 US$/t), and PLA (4,383 US$/t) ranking among the highest-cost scenarios, with the 10th, 5th, 3rd, and 2nd highest MSPs, respectively, across all products considered. These elevated costs are primarily associated with additional processing complexity and capital-intensive polymerisation steps.

Notably, levulinic acid and glucaric acid, both produced via chemical catalytic pathways, also exhibit relatively high MSPs, ranking as the 7th and 6th highest at 2,904 US$/t and 23,560 US$/t, respectively. For both products, this is largely attributed to high make-up solvent requirements and co-feedstock inputs (e.g., oxygen in the case of glucaric acid), as well as elevated CAPEX associated with downstream recovery and wastewater treatment (Cronjé et al., 2023; Morakile et al., 2022). In addition, levulinic acid is characterised by a relatively low process yield (28.3%), further contributing to its high MSP.

6.9.2 Market Analysis of Full-Scale Production Using All A-molasses

The MSPs and annual production capacities of all 20 products were compared to their respective market prices and global market sizes to evaluate their potential for market entry. A ranking criterion was developed based on the linear demand curve equation (Q = a − b(P)), which describes the relationship between product price (P) and quantity demand (Q). In this formulation, *a* represents all non-price factors affecting demand (assumed constant), while *b* is the slope of the demand curve. This criterion captures the combined effect of production rate and MSP on market entry potential under prevailing market conditions. A trade-off exists between MSP and production rate: the product rate/market size ratio (Q) reflects how easily a product can be absorbed into the market, whereas the MSP/market price ratio (P) reflects the competitiveness and ease of sale. Lower percentages are preferred in both cases, indicating more favourable market entry conditions.

Table 6.6 presents the results of this analysis, including the ratios of MSP to market price (P) and annual production rate to market size (Q), as well as the combined ranking criterion (R = Q + 2P), all expressed as percentages and calculated assuming a demand curve slope of 2. Q and P values below 100% reflect production rates and MSPs below market sizes and prices, respectively. The ranking criterion was subsequently normalised using min–max normalisation to calculate a product score (S) ranging from 100% (representing the highest economic viability) to 0% (representing the lowest economic feasibility).

Based on the combined evaluation, the five most attractive products were identified as scFOS (Q = 54.1%; P = 13.8%), 1,3-PDO (Q = 16.6%; P = 62.8%), lactic acid (Q = 4.6%; P = 81.3%), glycerol (Q = 1.6%; P = 82.9%), and glucaric acid (Q = 29.3%; P = 99.7%). In contrast, the five least attractive products were levulinic acid (Q = 318.3%; P = 30.2%), ethanol (Q = 0.05%; P = 201.2%), PE (Q = 8.7%; P = 201.3%), PEF (Q = 0.1%; P = 223.0%), and SAF (Q = 1.4%; P = 224.7%). The low market score of ethanol was somewhat surprising, given that it is produced via a relatively simple and mature biological process that is widely implemented globally. However, the high market demand is matched by equally large supply volumes, resulting in ethanol having the lowest market price among the products considered. This is primarily driven by the strong cost competitiveness and large-scale production of sugarcane-derived ethanol in Brazil and corn-derived ethanol in the United States. Consequently, emerging producers in developing regions, such as sugarcane-based ethanol production in South Africa, face significant challenges in competing with Brazilian ethanol, where feedstock costs and overall production economics are more favourable

TABLE 6.6
Market analysis of food products, chemicals, polymers, and fuels produced at full capacity from all available A-molasses

Rank	Bioproduct	Product type	MSP (US$/t)	MSP/market price (P) (%)	Production rate/market size (Q) (%)	Ranking criteria (R = 2P + Q) (%)	Normalised score (S) (%)	Reference for market size and market price
1	ScFOS	FI	$1,022	13.8	54.1	81.7	100.0	FMI (2026)
2	1,3-PDO[a]	IC	$1,667	62.8	16.6	142.3	83.6	FMI (2026)
3	Lactic acid[a]	IC	$1,205	81.3	4.6	167.1	76.9	FMI (2026)
4	Glycerol	IC	$1,238	82.9	1.6	167.3	76.8	GVR (2021c)
5	Glucaric acid[a]	IC	$3,560	99.7	29.3	228.6	60.2	FMI (2026) and Maia Research (2024)
6	Succinic acid[a]	IC	$1,576	91.6	82.7	265.8	50.1	ChemAnalyst (2025b) and GVR (2021b)
7	2,3-BDO	IC	$1,821	15.4	243	273.7	48.0	Maia Research (2024)
8	Ethyl acetate[b]	IC	$1,749	141.2	0.9	283.3	45.4	GVR (2021d)
9	PBS[a]	BP	$3,999	124.9	38.3	288.1	44.1	FMI (2026)
10	Ethyl lactate	IC	$2,034	110.5	72.0	292.9	42.8	FMI (2026)
11	1,4-BDO[a]	IC	$2,892	146.4	1.4	294.3	42.4	FMI (2026)
12	Muconic acid	IC	$2,648	155.8	42.7	354.3	26.1	Blaga et al. (2023) and Phasha (2025)
13	PLA	BP	$4,383	174.2	7.8	356.3	25.6	European Bioplastics (2025) and GVR (2021b)
14	Glutamic acid	FI	$2,329	177.9	0.8	356.6	25.5	MAResearch (2022)
15	Adipic acid[b]	IC	$3,875	186.1	1.0	373.2	21.0	GVR (2021a)
16	Levulinic acid	IC	$2,904	30.2	318.3	378.7	19.5	Maia Research (2024)
17	Ethanol[a]	BF	$1,379	201.2	0.0	402.4	13.1	market.us (2026) and Imarc Group (2026b)
18	PE	BP	$3,620	201.3	8.7	411.3	10.7	European Bioplastics (2025) and GVR (2021b)
19	PEF[b]	BP	$2,543	223.0	0.1	446.2	1.2	Facts and Factors (2022) and Imarc Group (2026a)
20	SAF	BF	$5,528	224.7	1.4	450.7	0.0	IATA (2024) and SP Global (2020)

FI = food ingredient, IC = industrial chemical, BP = biobased polymers (bioplastics), BF = biofuels.

[a] Overall market size (biobased + fossil-based) used, due to similar biobased and fossil-based market prices.

[b] Fossil-based market data used, due to no existing commercial biobased market or lack of info on biobased market.

(Crago et al., 2010). For similar reasons, PE and SAF also struggle to compete in their respective markets, as their production relies on ethanol as a key intermediate.

Figure 6.6 further illustrates the market positioning of the lower-ranked products. Most are located either outside or near the boundary of an inclusion region defined by two criteria: (i) a production rate-to-market size ratio below 100% and (ii) an MSP-to-market price ratio below 150%. The latter reflects the assumption that consumers may be willing to pay up to a 50% price premium for biobased alternatives (Gaffey et al., 2021). Products that fall outside or close to this boundary are therefore considered less competitive under the current and anticipated future market conditions.

As shown in Figure 6.6, 2,3-BDO and levulinic acid fall outside the exclusion zone and are therefore deemed unattractive, despite achieving MSPs well below their respective market prices (P = 15.4% and 30.2%, respectively), because their production rates exceed the corresponding market sizes by 243% and 318.3%, respectively, indicating limited market absorption capacity. Ethyl acetate, 1,4-BDO, muconic acid, PLA, glutamic acid, ADA, PE, ethanol, PEF, and SAF lie outside the inclusion region; although their production rates are substantially lower than their market sizes, their MSPs exceed their market prices by more than 50%, reducing their economic competitiveness. PEF and SAF emerge as the least attractive products due to their MSPs exceeding their market prices by 223% and 225%, respectively.

As shown in Table 6.6, utilisation of all available A-molasses results in production volumes exceeding 10% of the respective global market sizes for nine products, including 1,3-PDO (16.6%), glucaric acid (29.3%), PBS (38.3%), muconic acid (42.7%), scFOS (54.1%), ethyl lactate (72.0%), succinic acid (82.7%), 2,3-BDO (243%), and levulinic acid (318%). To maintain market stability and avoid oversupply, it is assumed that the introduction of a new biobased product should not exceed 10% of the prevailing market size.

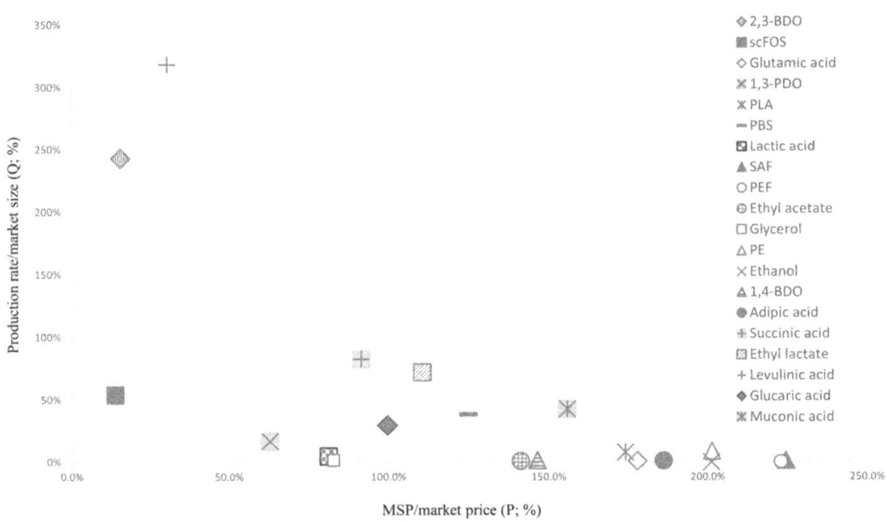

FIGURE 6.6 Summary of screening results of all products based on techno-economic results and market information.

This threshold is therefore used as a benchmark to assess the feasibility of large-scale market entry for each product (Nieder-Heitmann et al., 2019).

Many of these chemicals are platform molecules that can be converted into a wide range of higher-value derivatives, often associated with significantly larger market sizes. Consequently, future studies could focus on upgrading these intermediates into downstream products to circumvent market size limitations. Promising pathways to consider in future studies include the conversion of muconic acid and glucaric acid to ADA or polymer precursors; 1,3-PDO to PTT; 2,3-BDO to methyl ethyl ketone or 1,3-butadiene; ethyl lactate to 1,2-propanediol; and succinic acid to tetrahydrofuran, γ-butyrolactone, or 1,4-BDO.

Diversion of A-molasses from more than one sugar mill to an annexed biorefinery could be explored in future research to increase production capacity for the eleven products that did not reach the 10% market size threshold, thereby further reducing the cost of manufacturing and improving financial viability. Brobbey et al. (2024) demonstrated the benefits of increased production scale through the aggregation of A-molasses from multiple sugar mills, where six mills resulted in an MSP of 849 US$/t compared to 1,312 US$/t for a single mill. This highlights the advantages of economies of scale, particularly for high-volume commodity chemicals, and further incentivises increased sugar diversion. Interestingly, six of the products considered, that is, 1,4-BDO, glutamic acid, ADA, ethanol, PEF (with PET as its fossil analogue), and SAF, are high-volume commodities with production rates well below this limit, yet they are ranked among the bottom ten products. This is primarily due to their relatively high MSP-to-market price ratios. Therefore, combining feedstock from multiple sugar mills could potentially improve the attractiveness of these products by providing economies-of-scale benefits, without risking oversaturation of their respective markets.

High market premiums do not inherently correspond to high production costs, as ethanol and glycerol are among the least expensive products to produce. Rather, this highlights the challenge of competing in mature, high-volume markets characterised by low margins and strong price competition. In such markets, production from A-molasses, particularly in regions where sugar feedstock costs are relatively high compared to those of major producers such as Brazil, India, and Thailand, is less competitive. In these contexts, it may be more advantageous to prioritise the production of higher-value, lower-volume chemicals, even if their production costs are comparatively higher. This is demonstrated by the high ranking of products such as 1,3-PDO, scFOS, glucaric acid, and 2,3-BDO, which have market prices exceeding 2,000 US$/t.

Contrary to the findings of a previous techno-economic study that focused on lignocellulosic feedstock, processes, and products (Lynd et al., 2022), 16 of the products benchmarked in this study offer better economics than ethanol using A-molasses as the feedstock. This could be attributed to their relatively low MSP versus selling price, and comparable market size (for some of the products) that are in the same order as the current global markets for ethanol as a biofuel. This makes market assessment an important consideration in determining the economic feasibility of new biorefineries, as indicated by Li and Mupondwa (2021) in their study of succinic acid. It is also noteworthy that several products present better economics than succinic acid, meaning a higher probability of commercial success, compared to the four biobased succinic acid producers that closed down in recent years (Li and Mupondwa, 2021).

In general, the relative ranking of products should not change significantly for a biorefinery integrated into any sugar mill, anywhere in the world, as the economic viability mostly depends on production processes and global markets. This assumption is also likely to hold for other sugar-based feedstocks, such as corn dextrose, cassava starch, and sugar beet, provided that the biorefinery is implemented as an annexed facility that leverages existing infrastructure. However, rankings may differ for stand-alone facilities that do not benefit from integrated energy systems. In such cases, energy costs can have a more pronounced impact on overall economics, disproportionately affecting processes with high energy demand.

The following section presents a further assessment of smaller production scales for products with limited market sizes, and their effect on the economic feasibility of bioproducts.

6.9.3 Sensitivity Analysis of the Production Scale of at Most 10% of the Market Size

A techno-economic and market analysis was conducted for all products based on scaled-down production capacities at a single biorefinery that was no more than 10% of their respective global market sizes. This was done as a sensitivity analysis to validate the robustness of the proposed ranking criterion in this study, by assuming that only a portion of the available A-molasses from a typical sugar mill would be utilised (see Table 6.7), with all the products ranked based on the MSP/market price index. It can be observed that the ranking based on MSP-to-market price ratios at <10% market shares closely matches the initial ranking based on the demand–price relationship, with both the top- and bottom-performing products remaining largely unchanged. This confirms that the demand–price relationship is a sufficient and robust criterion for preliminary economic screening of bioproducts, regardless of production scale relative to market size.

As expected, reducing the production scale increased the MSP, with more pronounced effects at low scaling factors and for capital-intensive processes, due to higher specific capital costs, as described by the six-tenths rule. For example, 2,3-BDO and levulinic acid were scaled down by 95.9% and 96.9%, respectively; however, levulinic acid, which has a higher specific amortised CAPEX, experienced a substantially larger MSP increase (227% total) compared to 2,3-BDO (98.6%). Despite these significant reductions in production scale, both products show notable improvements in ranking, with 2,3-BDO rising from 7th to 2nd and levulinic acid from 16th to 6th. This can be largely attributed to their exceptionally high market prices (>9,000 US$/t), reflecting their status as niche chemicals, which allows them to remain economically competitive even at reduced scales. However, single-product biorefineries that utilise only a fraction of the available A-molasses are inherently inefficient. This is particularly evident for 2,3-BDO and levulinic acid, which utilise only 4.1% and 3.1% of the available feedstock, respectively. As a result, rankings based solely on MSPs relative to market prices are insufficient to fully capture the feasibility of bioproducts. Products with very small market sizes, despite appearing

TABLE 6.7

Techno-economic results of food products, chemicals, polymers, and fuels with production capacity of at most 10% of the product global market size

Base scale rank	Scaled rank	Bioproduct	Production rate (ktpa)	Scaled % of market size	A-Molasses utilisation (%)	Base case MSP	Scaled MSP (US$/t)	% MSP increase	Scaled MSP/ market price (%)
1	1	ScFOS	6.83	10.0	18.5	$1,022	$1,897	85.6	25.6
7	2	2,3-BDO	1.78	10.0	4.1	$1,821	$3,617	98.6	30.6
2	3	1,3-PDO	27.51	10.0	60.1	$1,667	$1,811	8.7	68.3
3	4	Lactic acid	89.60	4.6	100.0	$1,205	$1,205	0.0	81.3
4	5	Glycerol	51.03	1.6	100.0	$1,238	$1,238	0.0	82.9
16	6	Levulinic acid	1.13	10.0	3.1	$2,904	$9,484	226.6	98.7
5	7	Glucaric acid	17.21	10.0	34.1	$3,560	$4,223	18.6	118.2
8	8	Ethyl acetate	45.07	0.93	100.0	$1,749	$1,749	0.0	141.2
6	9	Succinic acid	7.00	10.0	12.1	$1,576	$2,519	59.9	146.4
11	10	1,4-BDO	44.76	1.4	100.0	$2,892	$2,892	0.0	146.4
13	11	PLA	47.94	7.8	100.0	$4,383	$4,383	0.0	174.2
10	12	Ethyl lactate	8.80	10.0	13.9	$2,034	$3,242	59.3	176.0
14	13	Glutamic acid	55.90	0.8	100.0	$2,329	$2,329	0.0	177.9
9	14	PBS	12.34	10.0	26.1	$3,999	$5,811	45.3	181.5
15	15	Adipic acid	38.62	1.0	100.0	$3,875	$3,875	0.0	186.1
17	16	Ethanol	49.84	0.05	100.0	$1,379	$1,379	0.0	201.2
18	17	PE	25.53	8.7	100.0	$3,620	$3,620	0.0	201.3
12	18	Muconic acid	7.90	10.0	23.4	$2,648	$3,666	38.4	215.7
19	19	PEF	44.37	0.08	100.0	$2,543	$2,543	0.0	223.0
20	20	SAF	13.78	1.4	100.0	$5,528	$5,528	0.0	224.7

economically attractive, do not offer a viable pathway for meaningful sugar industry diversification. A potential solution lies in the development of multiproduct biorefineries, where two or more high-value products with small market sizes are co-produced to enable more complete feedstock utilisation. For instance, scFOS and 1,3-PDO could be co-produced using 18.5% and 60.1% of A-molasses, respectively, with each meeting 10% of its relative market demands. The remaining 21.4% of A-molasses could then be directed towards another high-performing product such as lactic acid, or distributed across multiple lower-volume products, including 2,3-BDO and levulinic acid. In such configurations, higher-margin products like scFOS (MSP = 25.6% of the selling price) can be produced at their maximum market entry capacity, while other products such as lactic acid are produced at a reduced scale.

Notably, the top five bioproducts are all derived from biological processes, which present an opportunity for shared bioprocessing equipment and unit operations to allow efficient utilisation of equipment in production campaigns for different products. This could significantly reduce capital investment compared to co-producing products that require fundamentally different processing pathways. Amanful et al. (2024) demonstrated that a multiproduct facility is a viable approach for producing six high-value, low-volume sweeteners within a single refinery. This configuration achieved both economic viability and complete A-molasses utilisation, while limiting the production of each sweetener to 10% or less of total output. Furthermore, the study showed that employing a single large bioreactor operated in a scheduled manner was more economically attractive, yielding an IRR of 60%, compared to 50% when using six smaller, dedicated reactors. These co-production strategies warrant further investigation for the selected low-volume products considered in this study.

In summary, after adjusting production capacities according to market sizes, the top five bioproducts based on financial viability were scFOS, 2,3-BDO, 1,3-PDO, lactic acid, and glycerol, in that order. Although all five are produced via direct biological processes, this does not imply that biological routes are inherently superior to chemical or hybrid pathways. Notably, levulinic acid and glucaric acid (chemical processes), along with ethyl acetate (a hybrid process), rank 6th–8th overall and exhibit lower price premiums than several biological products, including succinic acid, 1,4-BDO, glutamic acid, ADA, ethanol, and muconic acid. In contrast, SAF and the four bioplastics, produced via more complex hybrid processing routes, rank within the bottom ten, with scaled MSPs exceeding 150% of their respective market prices. Overall, the most attractive investment opportunities lie in producing high-value products via direct biological conversion, which generally offers lower CAPEX and OPEX than more complex chemical and hybrid routes, thereby reducing investment risk, while higher market prices help offset production costs. Such products also benefit from rapid advances in synthetic biology as a means to develop more efficient production processes that will further enhance financial viability and environmental benefits (Sikazwe et al., 2025).

6.10 GREENHOUSE GAS EMISSIONS RESULTS

6.10.1 Overview of Total and Non-biogenic GHG Emissions

Figure 6.7 presents both the non-biogenic and total (biogenic + non-biogenic) GHG emissions for each scenario. Non-CO_2 biogenic emissions, such as methane and

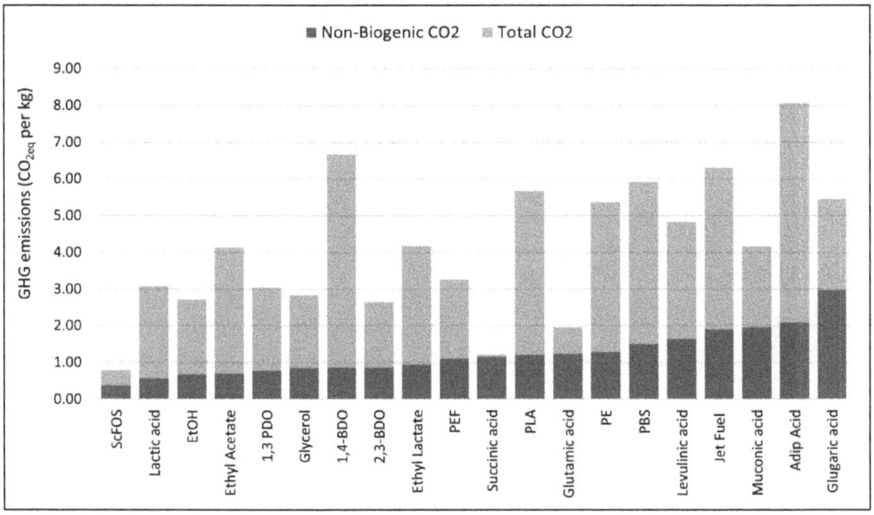

FIGURE 6.7 Non-biogenic and total GHG emissions of each biorefinery.

carbon monoxide, are also accounted for as part of the non-biogenic fraction, as they are not reabsorbed from the atmosphere during sugarcane growth. This is because the controlled combustion of sugarcane bagasse produces low volumes of CO, CH_4, NO_2, and other volatile non-CO_2 gases (Costa et al., 2023), while biogas generated during wastewater treatment is captured and combusted, thereby converting methane to CO_2. The results highlight the critical role of integrated biorefineries utilising biomass as an energy source in achieving low GHG emissions for biobased products. This is particularly important for products with energy-intensive processing requirements, such as 1,4-BDO, PLA, PBS, and ADA. In contrast, if non-renewable energy sources such as coal or natural gas were used, the GHG emissions associated with most biobased products would increase substantially. Notably, certain products, including scFOS, succinic acid, and glutamic acid, maintained total GHG emissions below 2 kg CO_2eq/kg of product, due to their relatively low energy demands.

Table 6.8 presents an overview of the total non-biogenic CO_2-equivalent emissions for each scenario, identifying the primary contributors to their respective GHG footprints. Overall, A-molasses was the dominant contributor to GHG emissions across most scenarios, with contributions ranging from 16.2% to 91.2%. Despite this, the use of A-molasses remains advantageous from an environmental perspective. Only 35.3% of the impacts associated with sugarcane cultivation and milling were economically allocated to A-molasses, with the bulk of the burden attributed to raw sugar production. This allocation supports the selection of A-molasses as a favourable feedstock for integrated biorefineries.

The lowest GHG emissions (i.e., less than 1 kg CO_2eq/kg of product) were observed in scenarios involving the production of scFOS (0.38 kg CO_2eq/kg product), lactic acid (0.58 kg CO_2eq/kg product), ethanol (0.68 kg CO_2eq/kg product), ethyl acetate (0.70 kg CO_2eq/kg product), 1,3-PDO (0.77 kg CO_2eq/kg product), glycerol (0.86 kg CO_2eq/kg product), 1,4-BDO (0.87 kg CO_2eq/kg product), 2,3-BDO (0.87 kg CO_2eq/

TABLE 6.8
Non-biogenic CO_2eq emissions for each scenario, highlighting the most significant process

Product	Non-biogenic CO_2eq (kg/h)	Non-biogenic CO_2eq (kg/kg)	Individual process contribution				
ScFOS	2,793.7	0.38	A-Molasses	Glycerol	Bagasse	Ash	NH_4OH
			91.2%	3.8%	3.7%	1.0%	0.2%
Lactic acid	9,586.3	0.58	A-Molasses	K_2HPO_4	Triethylamine	Bagasse	Ethanol
			50.7%	17.5%	16.2%	7.3%	2.9%
Ethanol	6,771.2	0.68	A-Molasses	Glucose	Bagasse	Urea	Ash
			71.8%	21.7%	4.2%	1.0%	1.0%
Ethyl acetate	6,285.5	0.70	A-Molasses	Glucose	Bagasse	Ammonia	Ash
			68.6%	12.6%	10.9%	3.0%	2.6%
1,3-PDO	7,074.5	0.77	A-Molasses	KH_2PO_4	Ammonia	Enzymes	Bagasse
			68.7%	7.5%	5.4%	4.1%	4.0%
Glycerol	8,793.3	0.86	A-Molasses	KH_2PO_4	Urea	K_2SO_4	Bagasse
			55.3%	29.0%	6.8%	3.6%	2.2%
1,4-BDO	7,774.0	0.87	A-Molasses	Bagasse	Ammonia	NaOH	Ash
			62.5%	15.1%	12.6%	4.6%	3.5%
2,3-BDO	7,408.8	0.87	A-Molasses	Oxygen	Bagasse	Electricity	DAP
			65.6%	24.7%	2.7%	2.3%	1.9%
Ethyl lactate	11,961.1	0.94	A-Molasses	NaOH	Bagasse	Glucose	H_2SO_4
			37.8%	35.6%	9.7%	7.0%	5.5%
PEF	9,924.4	1.11	A-Molasses	Platinum	Oxygen	Glucose	K_2CO_3
			49.0%	19.6%	17.3%	5.0%	2.8%
Succinic acid	13,317.3	1.15	$NaHCO_3$	A-Molasses	1-Octanol	Ammonia	Tryoctylamine
			54.9%	36.5%	4.6%	2.1%	0.8%

(Continued)

TABLE 6.8
(Continued)

Product	Non-biogenic CO_2eq (kg/h)	Non-biogenic CO_2eq (kg/kg)	Individual process contribution				
PLA	11,613.5	1.21	A-Molasses	Chloroform	K_2HPO_4	Triethylamine	Bagasse
			41.8%	18.5%	14.5%	13.4%	5.9%
Glutamic acid	13,815.7	1.24	Urea	A-Molasses	KH_2PO_4	$MnSO_4 \cdot H_2O$	$MgSO_4 \cdot 7H_2O$
			44.0%	35.2%	19.8%	0.6%	0.5%
PE	6,959.4	1.28	A-Molasses	Glucose	Bagasse	NaOH	Ash
			69.8%	21.1%	4.2%	1.8%	1.0%
PBS	14,204.6	1.50	A-Molasses	$NaHCO_3$	Chloroform	Bagasse	Ammonia
			34.2%	27.4%	15.2%	7.5%	5.6%
Levulinic acid	11,803.1	1.64	MIBK	A-Molasses	Ammonia	Bagasse	H_2SO_4
			42.5%	41.2%	13.2%	1.7%	1.1%
SAF	15,758.5	1.90	A-Molasses	Glucose	Bagasse	Hydrogen	Urea
			69.9%	21.1%	4.1%	2.6%	1.0%
Muconic acid	14,321.1	1.96	A-Molasses	NaOH	Ammonia	Glucose	H_2SO_4
			33.9%	28.7%	26.0%	6.8%	4.0%
Adipic acid	15,571.9	2.09	NH_4OH	A-Molasses	Bagasse	Enzymes	Ash
			55.7%	31.2%	7.5%	1.9%	1.8%
Glucaric acid	29,959.3	2.97	Acetonitrile	Acetic acid	A-Molasses	Oxygen	Bagasse
			51.2%	16.5%	16.2%	9.7%	2.8%

kg product), and ethyl lactate (0.94 kg CO_2eq/kg product). These products were primarily generated via biological processes such as fermentation or enzymatic conversion, which generally require fewer and less impactful chemical inputs, compared to chemical-catalytical conversion processes. Ethyl acetate and ethyl lactate, although involving both biological and chemical conversion steps, were notable exceptions, with low emissions of 0.70 and 0.94 kg CO_2eq/kg product, respectively. This is attributed to the absence of solvents in their processes and the use of catalysts with low EFs. Additionally, both scenarios exhibited high production rates and generated valuable co-products, which absorbed a portion of the environmental burden and further reduced emissions per kilogram of the main product.

Moderate emissions, ranging between 1.00 and 1.50 kg CO_2eq/kg product, were associated with PEF (1.11 kg CO_2eq/kg product), succinic acid (1.15 kg CO_2eq/kg product), PLA (1.21 kg CO_2eq/kg product), glutamic acid (1.24 kg CO_2eq/kg product), PE (1.28 kg CO_2eq/kg product), and PBS (1.50 kg CO_2eq/kg product). PEF, PLA, PE, and PBS are biobased polymers produced through a combination of biological and chemical processes. In the case of PEF, the use of a platinum catalyst and oxygen contributed 19.6% and 17.3% of its total emissions, respectively. For PLA and PBS, the use of chloroform during purification resulted in 18.5% and 15.2% of their respective emissions.

Succinic acid and PBS scenarios were particularly impacted by the use of sodium bicarbonate ($NaHCO_3$) as a neutralising agent, which contributed 54.9% and 27.4% of their GHG burdens, respectively. The use of fermentation nutrients, such as ammonia, urea, dipotassium phosphate (K_2HPO_4), potassium dihydrogen phosphate (KH_2PO_4), potassium sulphate (K_2SO_4), and diammonium phosphate (DAP), also significantly influenced the carbon footprints of several products. For instance, urea alone was responsible for 44.0% of the emissions associated with glutamic acid production. These findings emphasise the importance of optimising nutrient usage and minimising the reliance on neutralising agents such as $NaHCO_3$ to enhance the environmental sustainability of fermentation-based processes.

PE represents a notable outlier among the moderate-emission products. Its GHG emissions (1.28 kg CO_2eq/kg product) were not driven by extensive use of chemicals or catalysts but were instead the result of its low process yield (20.1%). While the PE biorefinery ranked fourth lowest in terms of total CO_2 emissions per hour, it fell to the 14th place when emissions were normalised per kilogram of product because it had the lowest production rate among all scenarios (5.42 t/h). The low mass yield of PE is inherently associated with its production pathway from ethanol, which involves dehydration to its monomer, ethylene. During this step, approximately 0.391 kg of product mass is lost as water per kilogram of ethanol converted, as dictated by reaction stoichiometry. This case clearly illustrates the significant impact that process yield can have on both the environmental and economic viability of biobased products.

The highest GHG emissions, in kg CO_2eq/kg product, were observed for levulinic acid (1.64 kg CO_2eq/kg product), jet fuel (1.90 kg CO_2eq/kg product), muconic acid (1.96 kg CO_2eq/kg product), ADA (2.09 kg CO_2eq/kg product), and glucaric acid (2.97 kg CO_2eq/kg product). For levulinic acid and glucaric acid, chemical inputs had a particularly large impact. In the levulinic acid scenario, the use of methyl isobutyl ketone (MIBK) during solvent extraction accounted for 42.5% of total emissions. Given MIBK's high emission factor (4.43 kg CO_2eq/kg), exploring alternative green solvents or more efficient recovery strategies is strongly recommended.

The GHG emissions associated with jet fuel were predominantly attributed (97.4%) to the use of bioethanol, which was derived from A-molasses with an emission factor of 0.68 kg CO_2eq/kg, as established in this study. Although ethanol itself exhibited a relatively low carbon footprint, the conversion of ethanol to jet fuel involved a low process yield (27.6%) and a comparatively high market value for jet fuel relative to that of its co-products, namely gasoline and diesel. As a result, economic allocation assigned 75.4% of the upstream GHG emissions to jet fuel. A key limitation of this benchmarking study is the application of a single allocation approach (economic allocation) to compare products with fundamentally different end uses. While this method allows for consistent comparison across diverse bioproducts, it may not be the most representative for energy carriers such as jet fuel. For energy products, burden distribution is more conventionally based on energy content, rather than the economic allocation applied here. For example, when applying an energy allocation approach, jet fuel accounts for only 50% of the total GHG emissions of the fuel slate, corresponding to a much lower GWP of 1.27 kg CO_2eq/kg of jet fuel, with the remaining emissions allocated to the diesel and gasoline co-products.

Muconic acid also exhibited high GHG emissions (1.96 kg CO_2eq/kg product), despite being produced via a relatively simple fermentation process. This was due to substantial consumption of sodium hydroxide (3.24 t/h) and ammonia (1.14 t/h) as a neutralising agent and a nutrient, respectively. Additionally, its relatively low yield (54.0%) and low production rate (6.75 t/h) further contributed to its high emissions. ADA, another fermentation-derived product, also ranked poorly due to the high usage of ammonium hydroxide (5.45 t/h) required to maintain pH during production. Glucaric acid exhibited the highest emissions among all scenarios evaluated, at 2.97 kg CO_2eq/kg product. The most significant contributor was acetonitrile, used during product recovery, which accounted for 51.2% of the total emissions, followed by acetic acid (16.5%), employed as a solvent during synthesis. These findings collectively underscore the importance of feedstock allocation, process yields, chemical input optimisation, and product recovery strategies in improving the environmental performance of integrated biorefineries.

The choice of molasses type can significantly influence the GHG emissions of bioproducts through its effect on economic allocation. In this study, A-molasses was used as the primary feedstock, contributing substantially to the non-biogenic CO_2eq emissions in most scenarios. However, using final (C-) molasses could result in even lower GHG emissions due to its substantially lower market price, which translates to a smaller allocation of the environmental burden from sugarcane cultivation and milling. For example, bioethanol produced from final molasses in Kenya and Nepal (Mbothu et al., 2019; Khatiwada and Silveira, 2011) demonstrated GWPs of 0.34 kg CO_2eq/kg (cradle-to-gate) and 0.55 kg CO_2eq/kg (cradle-to-grave), respectively. These low emissions were primarily due to final molasses being priced 81.8% and 91.7% lower than sugar in Kenya and Nepal, respectively – far greater than the 36.0% difference assumed for A-molasses in this study. As a result, only 5.2% and 4.3% of the upstream sugarcane impacts were allocated to final molasses, compared to

35.3% for A-molasses in this work. This highlights the potential of final molasses as a highly sustainable feedstock for integrated biorefineries.

6.10.2 Comparative GHG Reductions vs. Fossil-Based Equivalents

Figure 6.8 presents the comparative reduction in GHG emissions of biobased products relative to their fossil-based counterparts. The analysis includes only 13 bioproducts, as scFOS, glutamic acid, ethyl lactate, levulinic acid, glucaric acid, muconic acid, and lactic acid are predominantly or exclusively produced from biobased feedstocks, thus having no fossil-based equivalents to compare to. It's also worth noting that, while fossil-based production routes for lactic acid (and PLA) exist, the majority of the global supply is derived from bioresources (FMI, 2026). The biobased plastics PEF and PLA are compared with fossil-based PET, the most widely used thermoplastic that they are intended to replace.

All evaluated bioproducts demonstrated substantial reductions in non-biogenic CO_2eq emissions, ranging from 43.9% to 88.5%, when compared to conventional fossil-derived alternatives. Among these, 1,3-PDO exhibited the highest reduction (88.5%), followed by ADA (84.1%), and 1,4-BDO (80.9%). These findings suggest that even bioproducts with relatively higher cradle-to-gate emissions, such as ADA and PBS, can offer considerable climate benefits due to the displacement of carbon-intensive fossil-based production pathways. In contrast, certain bioproducts with lower absolute emissions, such as ethanol and PE, may deliver comparatively smaller GHG savings if their fossil-based alternatives already have relatively lower footprints.

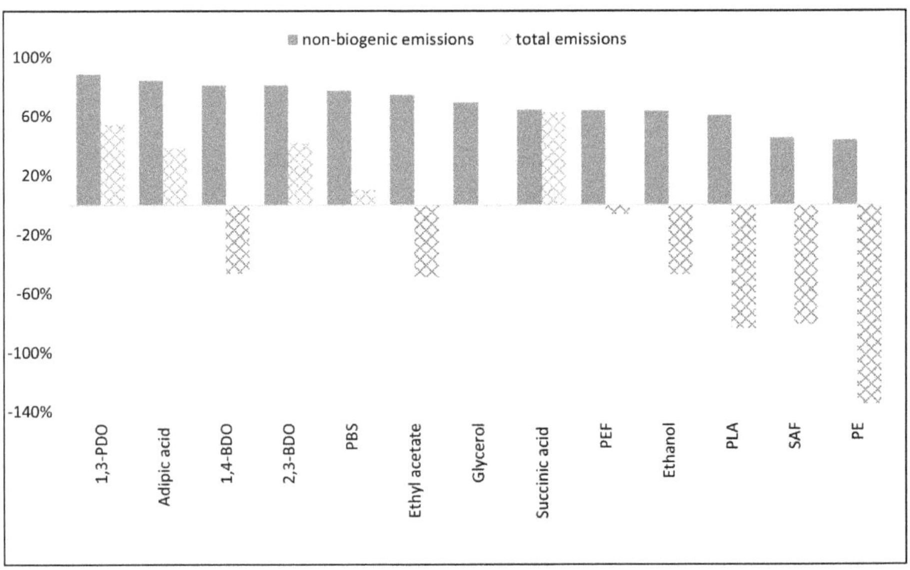

FIGURE 6.8 GHG emission reductions of biobased products compared to their fossil-based counterparts (positive values indicate a reduction in emissions, while negative values indicate an increase).

Notably, several bioproducts, including 1,3-PDO, 1,4-BDO, ethyl acetate, and glycerol, achieved both low absolute GHG emissions (<1 kg CO_2eq/kg of product) and significant non-biogenic GHG reductions (>69.0%), reinforcing their potential as leading candidates for sustainable biochemical production. Furthermore, Figure 6.8 underscores the importance of utilising biogenic carbon sources for process energy. While some bioproducts, including 1,4-BDO, ethyl acetate, PEF, ethanol, jet fuel, and PE, exhibited higher total (biogenic + non-biogenic) CO_2eq emissions than their fossil-based counterparts, this is largely due to the inclusion of biogenic emissions. These emissions, however, are generally considered neutral in long-term carbon accounting, as they originate from recent atmospheric carbon uptake during biomass growth. Thus, the integrated design of biorefineries and the use of sugarcane bagasse as a renewable energy source are critical to realising the full climate benefits of bio-based production systems and enhancing their viability as alternatives to fossil-based products.

6.11 MULTI-CRITERIA ANALYSIS

Each product was assigned an economic (E) and GWP score using min–max normalisation, as shown in the footnote of Table 6.9. The economic and GWP scores were calculated based on the scaled MSP-to-market price ratio (P) and total CO_2-equivalent emissions, respectively. As lower MSPs and GWPs are preferred, the normalisation was structured such that the lowest value corresponds to a score of 100%. The multi-criteria ranking of the 20 products, based on their economic and GWP scores, is presented in Table 6.9. The overall ranking was calculated by normalising the weighted sum of the economic and GWP scores (2E + GWP), thereby assigning greater importance to economic performance. This weighting reflects the observation that most products exhibited relatively low GHG emissions and comparable savings relative to their fossil-based counterparts.

The results in Table 6.9 indicate that GHG emissions have a limited influence on the overall attractiveness of biorefinery investments. The six most economically attractive products, that is, scFOS, 2,3-BDO, 1,3-PDO, lactic acid, glycerol, and levulinic acid, retain their rankings after incorporating GHG emissions into the assessment, owing to their low to moderate GHG values (0.38, 0.87, 0.77, 0.58, 0.86, and 1.64 kg CO_2eq/kg product, respectively). ScFOS, produced via a relatively simple biological process, achieves both the lowest GWP and the lowest market price premium across all scenarios. Although 2,3-BDO ranks only eighth in terms of GWP, its strong economic performance enables it to maintain its position as the second most attractive product overall. 1,3-PDO, lactic acid, and glycerol arguably emerge as the strongest candidates for near-term implementation alongside existing sugar mills, combining competitive MSP-to-market price ratios (ranked 3rd, 4th, and 5th, respectively) with low GHG emissions (ranked 5th, 2nd, and 6th, respectively), while fully utilising the available A-molasses feedstock.

In general, a moderate correlation is observed between the economic and GWP rankings of the bioproducts. The ten most economically attractive processes achieve a mean GWP score of 81.0%, which is 20.1% better than that of the ten least attractive products (60.9%). This trend can largely be attributed to the poorer-performing

TABLE 6.9
Multi-criteria ranking of all bioproducts according to scaled economic viability and greenhouse gas emissions

Combined rank	Product	Economic score (E)[a] (%)	GWP score (G) (%)	Multi-criteria rank (2E + G) (%)	Economic rank (%)	GWP rank
1	ScFOS	26	100	100	100	1
2	2,3-BDO	31	97	81	92	2
3	1,3-PDO	68	79	85	81	3
4	Lactic acid	81	72	92	79	4
5	Glycerol	83	71	81	75	5
6	Levulinic acid	99	63	51	59	6
7	Ethyl acetate	141	42	88	57	8
8	1,4-BDO	146	39	81	53	10
9	Succinic acid	146	39	70	50	9
10	Ethyl lactate	176	24	78	42	12
11	PLA	174	25	68	40	11
12	Glutamic acid	178	23	67	38	13
13	Ethanol	201	12	88	37	16
14	Glucaric acid	118	53	0	36	7
15	PBS	181	22	57	33	14
16	PE	201	12	65	29	17
17	PEF	223	1	72	25	19
18	Adipic acid	186	19	34	24	15
19	Muconic acid	216	5	39	16	18
20	SAF	225	0	41	14	20

[a] $E_i = \frac{P_{max} - P_i}{P_{max} - P_{min}}$; $G_i = \frac{CO_{2max} - CO_{2i}}{CO_{2max} - CO_{2min}}$

products typically relying on more complex and chemically intensive processing routes and/or exhibiting lower process yields, both of which contribute to higher GHG emissions. Similarly, purely biological processes achieve a higher average GWP score (74.4%) compared to chemical and hybrid routes (57.8%), reflecting their generally lower energy and chemical input requirements.

However, notable exceptions to these trends highlight the importance of considering both economic and environmental metrics simultaneously. The most prominent outlier is glucaric acid, which, despite achieving the 6th best economic performance, ranks 20th in terms of GWP, resulting in a drop to the 14th place in the combined assessment. This discrepancy is primarily due to its chemical production pathway, which relies on fossil-based solvents such as acetonitrile and acetic acid, leading to the highest GHG emissions among all scenarios. In contrast, ethanol demonstrates the opposite behaviour: although it ranks 16th economically, largely due to the highly competitive and saturated nature of its market, it improves to 13th overall when GHG emissions are considered, owing to its strong environmental performance (3rd lowest GWP). These deviations underscore that while general trends exist, individual process characteristics and input choices can significantly influence the balance between economic viability and environmental sustainability.

It is important to note that the rankings presented thus far are based on the absolute GWP of each product, rather than their potential to reduce GHG emissions through the displacement of fossil-based alternatives. The latter is arguably a more relevant metric, as it reflects the broader climate benefit of biorefinery systems. Accordingly, this impact was assessed and incorporated into the multi-criteria ranking, with the results presented in Table 6.10. In this analysis, the economic and environmental performances were integrated by assigning a monetary value to GHG reductions, based on current and projected European carbon prices[1] (ICAP, 2022; Enerdata, 2025; Intereconomics, 2021). These avoided emissions, calculated relative to fossil-based equivalents, were treated as an additional revenue stream in the DCFA. Consequently, the MSP of each product was recalculated to reflect the financial value of the decarbonisation that its biobased production achieves. Products with an MSP-to-market price ratio below 100% after incorporating carbon credits remain economically unfeasible, demonstrating that carbon reductions alone are insufficient to justify their price premiums. The results presented in Table 6.10 demonstrate that current 2024 EU carbon prices lead to modest reductions in the MSP-to-market price ratios for most of the 13 bioproducts that displace fossil-based alternatives; however, these reductions do not substantially alter the overall multi-criteria rankings. A notable exception is ADA, whose MSP-to-market price ratio decreases by 37.5% to 149%, improving its ranking from the 15th to 11th place, although it remains economically unfeasible since its adjusted selling price still exceeds the market price. Several other products exhibit reductions greater than 10%, including 1,3-PDO (15.8% reduction, maintaining the 3rd place), 1,4-BDO (13.2% reduction, improving from 10th to 9th), PEF (12.2% reduction, shifting from 19th to 18th), ethanol (12.1% reduction, remaining in 16th position), ethyl acetate (11.9% reduction, retaining the 8th place), and PBS (11.3% reduction, moving from 14th to 13th). In contrast, some products with significantly lower GWPs relative to their fossil counterparts, such as 2,3-BDO and succinic acid, show only marginal MSP reductions. This is primarily due to their limited market sizes, which constrain the total volume of fossil-based CO_2 displacement. By comparison, products that utilise a larger fraction, or the entirety, of the available A-molasses, such as ADA, achieve greater absolute emissions reductions and therefore benefit more substantially from carbon pricing mechanisms.

By 2040, the impact of carbon pricing on the economic performance of bioproducts becomes significantly more pronounced. Most notably, ADA exhibits a 77.7% reduction in its MSP-to-market price ratio (corresponding to a decrease in MSP from 3,875 US$/t to 2,257 US$/t), resulting in a substantial improvement in ranking from 15th in the no-credit scenario to the 7th place in the multi-criteria assessment. Similarly, 1,3-PDO experiences a 32.8% reduction in MSP/market price under 2040 carbon pricing, although it retains its 3rd-place ranking, indicating that its strong economic position is already well established.

Importantly, the leading bioproducts, that is, scFOS, 2,3-BDO, 1,3-PDO, glycerol, and lactic acid, largely maintain their rankings from the no-credit scenario through to 2040, with only a minor interchange between glycerol and lactic acid in the 4th and 5th positions. Likewise, the lowest-performing products, that is, ethanol, PE, muconic acid, PEF, and SAF, remain largely unchanged, highlighting the limited influence of carbon pricing on fundamentally uncompetitive pathways. The only notable shift within this group is ethanol, which improves slightly to the 15th place, while glutamic acid drops from 13th to 16th. Overall, these results suggest that while by 2040 carbon prices can significantly enhance the competitiveness of certain

TABLE 6.10
Benchmarking of products based on economic and GHG considerations. Benefits of GHG reductions quantified by incorporating carbon sales (based on EU carbon prices) into the economic model. Analysis performed for reduced scales, where the maximum production rate is smaller than 10% of the market size

	MSP/market price				Multi-criteria product ranking			
	No credits (%)	2,024 credits (%)	2,040 credits (%)	2,050 credits (%)	No credits	2,024 credits	2,040 credits	2,050 credits
ScFOS	26	26	26	26	1	1	1	5
2,3-BDO	31	28	26	13	2	2	2	4
1,3-PDO	68	52	35	−59	3	3	3	2
Lactic acid	81	81	81	81	4	5	5	9
Glycerol	83	74	64	9	5	4	4	3
Levulinic acid	99	99	99	99	6	6	6	11
Glucaric acid	118	118	118	118	7	7	9	13
Ethyl acetate	141	129	117	46	8	8	8	7
Succinic acid	146	138	129	78	9	10	11	8
1,4-BDO	146	133	119	41	10	9	10	6
PLA	174	169	163	132	11	12	13	15
Ethyl lactate	176	176	176	176	12	14	14	17
Glutamic acid	178	178	178	178	13	15	16	18
PBS	181	170	158	91	14	13	12	10
Adipic acid	186	149	108	−114	15	11	7	1
Ethanol	201	189	176	105	16	16	15	12
PE	201	197	193	170	17	17	17	16
Muconic acid	216	216	216	216	18	19	20	20
PEF	223	211	198	125	19	18	18	14
SAF	225	220	215	188	20	20	19	19

emissions-intensive products, particularly those with large-scale feedstock utilisation, they do not fundamentally alter the relative attractiveness of the majority of bioproduct pathways.

Surprisingly, even at projected carbon prices through 2040, only scFOS, 2,3-BDO, 1,3-PDO, lactic acid, glycerol, and levulinic acid remain economically viable, with adjusted selling price-to-market price ratios below 100% (Table 6.10). The selling prices of all other products continue to exceed their market prices, demonstrating that carbon credits alone are insufficient to offset their higher production costs relative to conventional alternatives. This highlights the importance of additional factors, such as sustainability branding, regulatory incentives, and subsidies, in supporting market adoption of higher-priced biobased chemicals.

Looking ahead to 2050, the multi-criteria ranking shifts substantially, with products demonstrating significant carbon reduction benefits occupying the top positions. ADA emerges as the leading product, followed by 1,3-PDO and glycerol. At this stage, the revenue from carbon reductions becomes so pronounced that ADA and 1,3-PDO achieve negative MSPs (–2,384 US$/t, –1,555US$/t, respectively), indicating that carbon credits alone are sufficient to offset production costs. However, these projections should be interpreted with caution, given the uncertainty associated with long-term carbon price trajectories and the potential for improvements in the carbon intensity of fossil-based production routes. Despite this, the results provide valuable insights into the potential impact of carbon credits on the economic and environmental performance of bioproducts.

In contrast, products with exclusively biobased markets, such as muconic acid and glucaric acid, do not benefit under the current framework, as they do not directly displace fossil-based alternatives. However, these products could potentially be positioned to replace a range of fossil-derived chemicals depending on their end-use case. For example, ethyl lactate could replace many conventional solvents as a green alternative; muconic and glucaric acids could substitute fossil-based cyclohexane as precursors to ADA; levulinic acid can be converted to GVL, a green solvent that could replace multiple fossil-based solvents; and lactic acid could compete with several fossil-based monomers, such as ethylene. This underscores the importance of developing standardised methods to quantify GWP reduction potential across all chemicals, enabling a more comprehensive assessment of their environmental and economic attractiveness in the future.

6.12 CONCLUSIONS AND FUTURE PERSPECTIVES

A benchmarking framework was developed to rank 20 biobased products, including food products, industrial chemicals, polymers, and fuels, produced via biological, chemical catalytic, and hybrid pathways, based on integrated techno-economic, market, and GHG analyses. The results demonstrate that manufacturing all of the evaluated products from A-molasses can be integrated into a conventional sugar mill, while maintaining energy self-sufficiency, with bagasse and harvest residues providing sufficient energy for combined sugar and biorefinery operations. The ranking methodology effectively captures key trade-offs between production scale, process economics, and market constraints by incorporating MSP relative to market size, thereby providing a more realistic assessment of market entry potential. The study further highlights the critical role of market analysis in biorefinery feasibility assessments, particularly in avoiding overestimation of opportunities for products with limited demand. Overall, the benchmarking approach enables the identification and prioritisation of the most promising bioproducts for investment and suggests that the relative ranking of these products is likely to remain robust across different geographic contexts for integrated sugar mill biorefineries, provided similar techno-economic and market conditions apply.

The techno-economic and market analysis indicates that simpler, biologically driven production pathways generally outperform more complex chemical and hybrid routes. The five most attractive products, i.e., scFOS, 2,3-BDO, 1,3-PDO, lactic acid, and glycerol, are all produced via direct fermentation or enzymatic conversion, benefiting from lower energy requirements, fewer chemical inputs, and reduced process complexity. These advantages translate into lower CAPEX and OPEX. In contrast, more complex routes, including bioplastics, ethyl lactate, and SAF, are typically less competitive due to higher processing intensity and cost.

Market dynamics play a critical role in determining feasibility. Ethanol, despite achieving one of the lowest MSPs, ranks poorly overall due to strong global competition and market saturation, particularly from low-cost producers in Brazil and the USA. This highlights a key constraint for emerging bioeconomies, as well as for regions such as the USA, Europe, and China, where the domestic cost of sugar exceeds the global market price. Under these conditions, competitiveness in bulk commodity markets is limited, making lower-volume, higher-value products such as 1,3-PDO and 2,3-BDO more attractive options for biorefinery investment. However, high-value products alone do not provide a complete solution for feedstock utilisation. scFOS, 2,3-BDO, and 1,3-PDO cannot absorb the full A-molasses stream without exceeding realistic market penetration limits. As a result, single-product biorefineries are inherently inefficient in the context of these products as viable diversification options for the sugar industry. Multiproduct biorefineries therefore emerge as a critical strategy, enabling full feedstock utilisation while prioritising the production of high-margin products up to their market limits before allocating remaining capacity to secondary products.

From an environmental perspective, A-molasses-based biorefineries demonstrate consistently low non-biogenic GHG emissions, particularly when biomass-derived energy is utilised. Biological pathways achieve the lowest emissions on average (1.04 kg CO_2eq/kg), owing to milder operating conditions and reduced reliance on solvents and catalysts, compared to chemical and hybrid routes (1.47 kg CO_2eq/kg). While no single pathway is universally superior, processes with strong economic performance generally also perform well environmentally, driven by higher yields, lower energy demand, and reduced material inputs.

The multi-criteria assessment confirms that economic performance remains the dominant driver of overall attractiveness, although environmental performance reinforces these trends. Products achieving strong economic rankings largely retain their positions when GHG emissions are incorporated, reflecting both the relatively low variability in emissions and the alignment between efficient process design and lower environmental impact.

The inclusion of carbon pricing further strengthens the competitiveness of certain pathways but does not fundamentally alter the ranking structure up to 2040. Large-scale products with high GHG saving potential, such as ADA benefit the most, while top-performing products remain largely unchanged. Based on the integrated assessment combining MSP, market potential, and GHG performance, scFOS, 2,3-BDO, 1,3-PDO, glycerol, and lactic acid represent the most attractive near-term opportunities. These products all achieved MSPs below their current market prices, and GHG emissions below 1 kg CO_2eq/kg, positioning them as sustainable options for sugarcane diversification from an economic and environmental perspective. Surprisingly, except for the five products mentioned above and levulinic acid, no other products achieved selling price-to-market price ratios below 100% when accounting for carbon credits at projected prices up to 2040, highlighting that carbon credits alone are insufficient to offset the price premiums of most bioproducts in this study.

For single-mill implementations, glycerol and lactic acid are particularly favourable due to their ability to utilise the full A-molasses stream, followed by 1,3-PDO, which utilised a significant portion (60.1%) of the available feedstock. In contrast, scFOS and 2,3-BDO, and to a lesser extent 1,3-PDO, are better suited to multiproduct biorefinery configurations, reinforcing the importance of integrated system design for maximising both economic and resource efficiency. In future research, the benchmarking analysis presented in this study could be applied to different biorefinery configurations as follows:

Previous studies have shown that multiproduct biorefineries are generally less profitable than single-product biorefineries utilising the same amount of feedstock, due to economies-of-scale disadvantages (Dogbe et al., 2020; Nieder-Heitmann et al., 2019). However, multiproduct biorefineries are less risky with respect to market fluctuations and the trade-offs should be assessed for individual cases. Future benchmarking studies should particularly consider multiproduct configurations for the co-production of high-value, low-volume chemicals such as scFOS, 2,3-BDO, 1,3-PDO, and levulinic acid, where a single-product biorefinery resulted in either market oversaturation or incomplete utilisation of A-molasses.

An alternative to a multiproduct integrated biorefinery co-producing more than one product when the required capacity requires less A-molasses than is available is a small-scale, decentralised (greenfield) biorefinery which can mechanically process sugarcane (juice for 1G biorefinery) into biobased products with the target capacity (OECD, 2021). These can be set up in rural areas to help to relieve poverty. However, the feasibility of such greenfield facilities needs to be assessed compared to the brownfield options considered in this study, as the former lose the benefits gained from an existing structure in terms of energy integration (Sanford et al., 2016).

On the other hand, for the 11 products that are limited by the available A-molasses in a single sugar mill into which the biorefinery is integrated, A-molasses could be diverted from more than one sugar mills to a single, annexed biorefinery to allow all biorefineries to achieve production rates equivalent to the 10% of the product's market size. This arrangement would have economies-of-scale benefit while incurring additional cost of transportation as well as forfeiting the advantage of mass and energy integrations from the A-molasses 'donor' mills (Sanford et al., 2016).

For instance, the biorefinery cannot utilise surplus steam and electricity resulting from diverting A-molasses away from sugar production in the donor mills. The effects of these dynamics (the interplay between economies of scale versus additional costs of energy and transportation) on the overall economic feasibility of the multi-source A-molasses biorefineries could be investigated further in future research.

Another important consideration for future research is to expand and strengthen the market analysis. Due to fluctuations in product selling prices and the expansion or decline of bioproduct markets as new technologies emerge, it is crucial to obtain up-to-date information and future projections to evaluate all products on a consistent basis. Monte Carlo economic analysis should be incorporated into future benchmarking studies to quantify the impact of financial parameters, such as feedstock costs and product market price volatility, on the economic viability of bioproducts. These fluctuations in future projections should be informed by historical price trends to ensure realistic and robust modelling. Furthermore, there is a significant risk of underestimating the market size of bioproducts, particularly platform chemicals, whose derivatives often have substantial markets. Future benchmarking studies should therefore consider the production of these derivatives in addition to their platform chemicals to provide a more comprehensive assessment.

Overall, this work underscores the importance of integrating process optimisation, feedstock selection, and policy mechanisms when evaluating biorefinery opportunities. While biological pathways show strong and consistent advantages in both environmental and economic performance, the results indicate that the optimal pathway is product-specific, and future competitiveness will increasingly depend on the ability to minimise inputs, maximise yields, and capture value from GHG reductions in a carbon-constrained economy.

NOTE

1 European carbon credit prices used: 70.6 US$/t (2024), 146.3 US$/t (2040), and 566 US$/t (2050) (ICAP, 2022; Enerdata, 2025; Intereconomics, 2021).

REFERENCES

Amanful, B., Dogbe, E.S., Bosman, C.E., Görgens, J.F., 2024. Stochastic Techno-economic Analysis for the Co-production of Alternative Sweeteners in Sugarcane Biorefineries. *Food Bioprod. Process.* 143, 9–20. https://doi.org/10.1016/j.fbp.2023.10.006

Armstrong, R.D., Kariuki, B.M., Knight, D.W., Hutchings, G.J., 2017. How to Synthesise High Purity, Crystalline d-Glucaric Acid Selectively. *European J. Org. Chem.* 2017, 6811–6814. https://doi.org/10.1002/ejoc.201701343

Avantium, 2020. Avantium to Build FDCA Flagship Plant at Chemie Park Delfzijl, Netherlands [WWW Document]. URL https://www.avantium.com/2020/avantium-to-build-fdca-flagship-plant-at-chemie-park-delfzijl-netherlands/ (accessed 4.27.23).

Axelsson, L., Franzén, M., Ostwald, M., Berndes, G., Lakshmi, G., Ravindranath, N.H., 2012. Perspective: Jatropha Cultivation in Southern India: Assessing Farmers' Experiences. *Biofuels, Bioprod. Biorefining* 6, 246–256.

Bedzo, O.K.K., Mandegari, M., Görgens, J.F., 2019. Comparison of Immobilized and Free Enzyme Systems in Industrial Production of Short-Chain Fructooligosaccharides from Sucrose Using a Techno-economic Approach. *Biofuels, Bioprod. Biorefining* 13, 1274–1288. https://doi.org/10.1002/bbb.2025

Biddy, M.J., Scarlata, C., Biddy, M.J., Scarlata, C., 2016. *Chemicals from Biomass: A Market Assessment of Bioproducts with Near-Term Potential.* National Renewable Energy Laboratory (NREL), Denver, CO.

Blaga, A.C., Gal, D.G., Tucaliuc, A., 2023. Recent Advances in Muconic Acid Extraction Process. *Appl. Sci.* 13(21), 11691. https://doi.org/10.3390/app132111691

Blois, M., 2024, April 2. World's Biggest Spandex Maker Is Going Biobased. *Chemical and Engineering News*. https://cen.acs.org/business/biobased-chemicals/Worlds-biggest-spandex-maker-biobased/102/web/2024/04

Brobbey, M.S., Louw, J., Görgens, J.F., 2025. A Techno-economic and Life Cycle Assessment of Multiproduct Sugarcane Biorefinery: Lactic Acid as a Platform Chemical. *Biomass and Bioenergy* 193, 107605. https://doi.org/10.1016/j.biombioe.2025.107605

Brobbey, M.S., Louw, J.P., Louw, J., Görgens, J.F., 2024. Effect of Production Scale on the Techno-economic Viability and Environmental Life Cycle Analysis of Lactic Acid Production in a Sugarcane Biorefinery. *Food Bioprod. Process.* 148, 269–284. https://doi.org/10.1016/j.fbp.2024.09.020

Businesswire, 2021. Global Glycerol Industry (2020 to 2027) - Market Trends and Drivers - ResearchAndMarkets.com [WWW Document]. URLhttps://www.businesswire.com/news/home/20210115005399/en/Global-Glycerol-Industry-2020-to-2027---Market-Trends-and-Drivers---ResearchAndMarkets.com (accessed 6.11.21).

Celińska, E., Grajek, W., 2009. Biotechnological Production of 2,3-Butanediol-Current State and Prospects. *Biotechnol. Adv.* 27, 715–725. https://doi.org/10.1016/j.biotechadv.2009.05.002

ChemAnalyst, 2025a. Global Succinic Acid Market Report (2025–2036) [WWW Document]. URL https://www.chemanalyst.com/industry-report/succinic-acid-market-2899

ChemAnalyst, 2025b. Levulinic Acid Price Trends and Forecast [WWW Document]. URL https://www.chemanalyst.com/Pricing-data/levulinic-acid-2352

Climate Catalyst, 2026. Alternative Aviation Fuel Policy in the European Union [WWW Document]. URL https://climatecatalyst.org/learning-hub/sustainable-aviation-fuel-policy-in-the-eu/#:~:text=Crop%2Dbased%20fuels%20(1st%20Generation,the%20subsequent%20reporting%20period5 (accessed 4.9.26).

Costa, M.A.M., Schiavon, N.C.B., Felizardo, M.P., Souza, A.J.D., Dussán, K.J., 2023. Emission Analysis of Sugarcane Bagasse Combustion in a Burner Pilot. *Sustain. Chem. Pharm.* 32, 101028. https://doi.org/10.1016/j.scp.2023.101028

Crago, C.L., Khanna, M., Barton, J., Giuliani, E., 2010. Competitiveness of Brazilian Sugarcane Ethanol Compared to US Corn Ethanol. *Energy Policy* 38, 7404–7415. https://doi.org/10.1016/j.enpol.2010.08.016

Cronjé, Y., Farzad, S., Mandegari, M., Görgens, J.F., 2023. A Critical Review of Multiple Alternative Pathways for the Production of a High-Value Bioproduct from Sugarcane Mill Byproducts: The Case of Adipic Acid. *Biofuel Res. J.* 10, 1933–1947. https://doi.org/10.18331/BRJ2023.10.3.5

D'Este, M., Alvarado-Morales, M., Angelidaki, I., 2018. Amino Acids Production Focusing on Fermentation Technologies - A Review. *Biotechnol. Adv.* 36, 14–25. https://doi.org/10.1016/j.biotechadv.2017.09.001

Dani, M., Ruggiero, G., Perini, D., Bianchi, A., 2017. Process for the Production of 1,4 Butanediol. US 2017/0029852 A1.

Dagle, R.A., Winkelman, A.D., Ramasamy, K.K., Lebarbier Dagle, V., Weber, R.S., 2020. Ethanol as a Renewable Building Block for Fuels and Chemicals. *Ind. Eng. Chem. Res.* 59, 4843–4853. https://doi.org/10.1021/acs.iecr.9b05729

de Jong, E., Stichnothe, H., Bell, G., Henning Jørgensen, M., de Bari, I., Jacco van Haveren, E., Lindorfer, J., an der Johannes Kepler, E., 2020. Biobased Chemicals: A 2020 Update. IEA Bioenergy: Paris, France.

De Jong, S., Antonissen, K., Hoefnagels, R., Lonza, L., Wang, M., Faaij, A., Junginger, M., 2017. Life-cycle Analysis of Greenhouse Gas Emissions from Renewable Jet Fuel Production. *Biotechnol. Biofuels* 10, 64.

Dogbe, E.S., Mandegari, M., Görgens, J.F., 2020. Revitalizing the Sugarcane Industry by Adding Value to A-molasses in Biorefineries. *Biofuels, Bioprod. Biorefining* 14, 1089–1104. https://doi.org/10.1002/bbb.2122

Dogbe, E.S., Mandegari, M.A., Görgens, J.F., 2018. Exergetic Diagnosis and Performance Analysis of a Typical Sugar Mill Based on Aspen Plus® Simulation of the Process. *Energy* 145, 614–625. https://doi.org/10.1016/j.energy.2017.12.134

Dominguez, A.L., Rodrigues, L.R., Lima, N.M., Teixeira, J.A., 2014. An Overview of the Recent Developments on Fructooligosaccharide Production and Applications. *Food Bioprocess Technol.* 7, 324–337. https://doi.org/10.1007/s11947-013-1221-6

Eerhart, A.J.J.E., Faaij, A.P.C., Patel, M.K., 2012. Replacing Fossil Based PET with Biobased PEF; Process Analysis, Energy and GHG Balance. *Energy Environ. Sci.* 5, 6407. https://doi.org/10.1039/c2ee02480b

Eggleston, G., Lima, I., 2015. Sustainability Issues and Opportunities in the Sugar and Sugar-Bioproduct Industries. *Sustainability* 7, 12209–12235. https://doi.org/10.3390/su70912209

Enerdata, 2025. Carbon Price Forecast under the EU ETS2 [WWW Document]. URL https://www.enerdata.net/publications/executive-briefing/carbon-price-projections-eu-ets.html?utm_source=chatgpt.com

European Bioplastics, nova-I., 2025. Bioplastics Market Development Update 2025 [WWW Document]. URL https://www.european-bioplastics.org/bioplastics-market-development-update-2025/

Facts and Factors, 2022. Polyethylene Terephthalate (PET) Market Size, Share Global Analysis Report, 2022–2030 [WWW Document]. URL https://www.fnfresearch.com/polyethylene-terephthalate-pet-market

Farzad, S., Mandegari, M.A., Guo, M., Haigh, K.F., Shah, N., Görgens, J.F., 2017. Multi-product Biorefineries from Lignocelluloses: A Pathway to Revitalisation of the Sugar Industry? *Biotechnol. Biofuels* 10, 1–24. https://doi.org/10.1186/s13068-017-0761–9

Future Market Insights (FMI), 2026. Custom Market Reports for scFOS, 1,3-PDO, PBS, Lactic Acid, 1,4-BDO, Ethyl Lactate, and Glucaric Acid.

Gaffey, J., McMahon, H., Marsh, E., Vehmas, K., Kymäläinen, T., Vos, J., 2021. Understanding Consumer Perspectives of Bio-based Products—A Comparative Case Study from Ireland and the Netherlands. *Sustainability* 13, 6062. https://doi.org/10.3390/su13116062

Geleynse, S., Brandt, K., Garcia-Perez, M., Wolcott, M., Zhang, X., 2018. The Alcohol-to-Jet Conversion Pathway for Drop-in Biofuels: Techno-economic Evaluation. *ChemSusChem* 11, 3728–3741. https://doi.org/10.1002/cssc.201801690

Gérardy, R., Debecker, D.P., Estager, J., Luis, P., Monbaliu, J.-C.M., 2020. Continuous Flow Upgrading of Selected C2–C6 Platform Chemicals Derived from Biomass. *Chem. Rev.* 120, 7219–7347. https://doi.org/10.1021/acs.chemrev.9b00846

Global Market Insights (GMI), 2021. Glutamic Acid and Monosodium Glutamate (MSG) Market Size, Potential, Industry Outlook, Regional Analysis, Application Development, Competitive Landscape & Forecast, 2021–2027 [WWW Document]. URL https://www.gminsights.com/industry-analysis/glutamic-acid-and-monosodium-glutamate-msg-market (accessed 6.14.21).

Godavari, n.d. About Us [WWW Document]. URL https://godavaribiorefineries.com/about-us (accessed 4.8.26).

Gorden, J., Zeiner, T., Sadowski, G., Brandenbusch, C., 2016. Recovery of Cis,Cis-Muconic Acid from Organic Phase after Reactive Extraction. *Sep. Purif. Technol.* 169, 1–8. https://doi.org/10.1016/j.seppur.2016.05.032

Grand View Research (GVR), 2021a. Adipic Acid Market Analysis and Segment Forecast to 2025.

Grand View Research (GVR), 2021b. Biobased Platform Chemicals Market Analysis and Segment Forecast to 2025.

Grand View Research (GVR), 2021c. Global Glycerol Market Analysis and Segment Forecast to 2027.

Grand View Research (GVR), 2021d. Ethyl Acetate Market Analysis & Segment Forecasts to 2028.

Grand View Research (GVR), 2021e. Polylactic Acid Market Estimates and Forecast to 2028.

Greenwood, A., 2017. Consultants Sound Overcapacity Warning for Polyethylene [WWW Document]. ICIS Chemical Business. URL https://www.icis.com/explore/resources/news/2017/11/17/10164197/consultants-sound-overcapacity-warning-for-polyethylene/ (accessed 4.26.23).

Guest, K., Stark, A., Starzak, M., 2019. Development and Verification of an ASPEN Plus ® Model of a Sugarcane Biorefinery. *Proc. S. Afr. Sug. Technol. Ass.* 92, 254–273.

Humbird, D., Davis, R., Tao, L., Kinchin, C., Hsu, D., Aden, A., Schoen, P., Lukas, J., Olthof, B., Worley, M., Sexton, D., Dudgeon, D., 2011. Process Design and Economics for Conversion of Lignocellulosic Biomass to Ethanol. NREL Tech. Rep. NREL/TP-5100–51400 303, 275–3000.

IAR, 2015. International Overview of Biobased Chemical Intermediates: Analysis of the Last 10 Years.

Imarc Group, 2026a. Polyethylene Terephthalate (PET) Prices, Trend, Chart, Demand, Market Analysis, News, Historical and Forecast Data Report 2026 Edition [WWW Document]. URL https://www.imarcgroup.com/polyethylene-terephthalate-pricing-report

Imarc Group, 2026b. Bio-ethanol Prices, Trend, Chart, Demand, Market Analysis, News, Historical and Forecast Data Report 2026 Edition [WWW Document]. URL https://www.imarcgroup.com/bio-ethanol-trend

Imarc Group, 2026c. Polylactic Acid Prices, Trend, Chart, Demand, Market Analysis, News, Historical and Forecast Data Report 2026 Edition [WWW Document]. URL https://www.imarcgroup.com/polylactic-acid-pricing-report

Imarc Group, 2026d. Ethyl Acetate Prices, Trend, Chart, Demand, Market Analysis, News, Historical and Forecast Data Report 2026 Edition [WWW Document]. URL https://www.imarcgroup.com/ethyl-acetate-pricing-report

Intereconomics, 2021. Europe's Climate Target for 2050: An Assessment [WWW Document]. URL https://www.intereconomics.eu/contents/year/2021/number/6/article/europe-s-climate-target-for-2050-an-assessment.html

International Air Transport Association (IATA), 2024. Disappointingly Slow Growth in SAF Production [WWW Document]. URL https://www.iata.org/en/pressroom/2024-releases/2024-12-10-03/#:~:text=In%202024%2C%20SAF%20production%20volumes,of%20global%20renewable%20fuel%20capacity*

International Carbon Action Partnership (ICAP), 2022. EU Emissions Trading System (EU ETS) [WWW Document]. URL https://icapcarbonaction.com/en/ets/eu-emissions-trading-system-eu-ets

Intratec, 2026. Glycerol Prices Worldwide [WWW Document]. URL https://www.intratec.us/solutions/primary-commodity-prices/commodity/glycerol-prices

Ishii, J., Morita, K., Ida, K., Kato, H., Kinoshita, S., Hataya, S., Shimizu, H., Kondo, A., Matsuda, F., 2018. A Pyruvate Carbon Flux Tugging Strategy for Increasing 2,3-Butanediol Production and Reducing Ethanol Subgeneration in the Yeast Saccharomyces cerevisiae. *Biotechnol. Biofuels* 11, 80. https://doi.org/10.1186/s13068-018-1176-y

Kato, S., Ueda, T., Kosaka, N., Nitta, S., 2023. BioPBSTM (Polybutylene Succinate), in: A. Kunkel, G. Battagliarin, M. Winnacker, B. Rieger, & G. Coates (Eds.), *Synthetic Biodegradable and Biobased Polymers*. Springer, Cham, pp. 269–304. https://doi.org/10.1007/12_2023_159

Kendall, A., Yuan, J., 2013. Comparing Life Cycle Assessments of Different Biofuel Options. *Curr. Opin. Chem. Biol.* 17, 439–443. https://doi.org/10.1016/j.cbpa.2013.02.020

Khalil, I., Quintens, G., Junkers, T., Dusselier, M., 2020. Muconic Acid Isomers as Platform Chemicals and Monomers in the Biobased Economy. *Green Chem.* 22, 1517–1541. https://doi.org/10.1039/C9GC04161C

Khatiwada, D., Silveira, S., 2011. Greenhouse Gas Balances of Molasses Based Ethanol in Nepal. *J. Clean. Prod.* 19, 1471–1485. https://doi.org/10.1016/j.jclepro.2011.04.012

Kinoshita, S., 1959. The Production of Amino Acids by Fermentation Processes, in: Umbreit, W. W. (Ed.), *Advances in Applied Microbiology*. Academic Press. Netherlands, pp. 201–214. https://doi.org/10.1016/S0065-2164(08)70480-5

KPMG, 2022. Sustainable Aviation Fuel: Ready for Lift Off? Aviation 2030 Series [WWW Document]. KPMG Ireland. URL https://kpmg.com/ie/en/home/insights/2022/11/sustainable-aviation-fuel.html (accessed 4.27.23).

Lane, J., 2018. The Digest [WWW Document]. Article.

LanzaJet, 2024. LanzaJet Celebrates Grand Opening of Freedom Pines Fuels Plant, the World's First Ethanol to Sustainable Aviation Fuel Production Facility [WWW Document]. URL https://www.lanzajet.com/news-insights/lanzajet-celebrates-grand-opening-of-freedom-pines-fuels-plant-the-worlds-first-ethanol-to-sustainable-aviation-fuel-production-facility

Lees, A.M., 2019. The Challencges of Initial Market Entry [WWW Document]. Toptal. URL https://www.toptal.com/product-managers/market-strategy-research/market-entry-strategy-challenges (accessed 4.28.23).

Li, X., Mupondwa, E., 2021. Empirical Analysis of Large-Scale Bio-succinic Acid Commercialization from a Technoeconomic and Innovation Value Chain Perspective: BioAmber Biorefinery Case Study in Canada. *Renew. Sustain. Energy Rev.* 137, 110587. https://doi.org/10.1016/J.RSER.2020.110587

Louw, J., Louw, J.P., Brobbey, M. S., Foxon, K.M., Görgens, J.F., 2025. An analysis of the financial viability and environmental impact of food products, biochemicals, biofuels and biomaterials as potential diversification strategies for the sugar industry. Proceed. South African Sugar Technol. Assoc. 97, 54–62.

Louw, J., Farzad, S., Görgens, J.F., 2023. Polyethylene Furanoate: Technoeconomic Analysis of Biobased Production. *Biofuels, Bioprod. Biorefining* 17, 135–152. https://doi.org/10.1002/bbb.2430

Louw, J., Farzad, S., Görgens, J.F., 2024. Production of Biobased Polyethylene Terephthalate and Its Precursors for Diversification in the Global Sugarcane Industry. *Biochem. Eng. J.* 210, 109404. https://doi.org/10.1016/j.bej.2024.109404

Löwe, H., Schmauder, L., Hobmeier, K., Kremling, A., Pflüger-Grau, K., 2017. Metabolic Engineering to Expand the Substrate Spectrum of Pseudomonas putida toward Sucrose. *Microbiologyopen* 6, e00473. https://doi.org/10.1002/mbo3.473

Lynd, L.R., Beckham, G.T., Guss, A.M., Jayakody, L.N., Karp, E.M., Maranas, C., McCormick, R.L., Amador-Noguez, D., Bomble, Y.J., Davison, B.H., Foster, C., Himmel, M.E., Holwerda, E.K., Laser, M.S., Ng, C.Y., Olson, D.G., Román-Leshkov, Y., Trinh, C.T., Tuskan, G.A., Upadhayay, V., Vardon, D.R., Wang, L., Wyman, C.E., 2022. Toward Low-Cost Biological and Hybrid Biological/Catalytic Conversion of Cellulosic Biomass to Fuels. *Energy Environ. Sci.* 15, 938–990. https://doi.org/10.1039/d1ee02540f

Maia Research, 2024. Customized Global Chemicals Industry Market Research Report.

Maina, S., Mallouchos, A., Nychas, G.E., Freire, D.M.G., Castro, M.D., Papanikolaou, S., Kookos, I.K., Koutinas, A., 2019. Bioprocess Development for (2R, 3R)-Butanediol and Acetoin Production Using Very High Polarity Cane Sugar and Sugarcane Molasses by a Bacillus amyloliquefaciens Strain. *J. Chem. Technol. Biotechnol.* 94, 2167–2177. https://doi.org/10.1002/jctb.5997

Mancini, E., Dickson, R., Fabbri, S., Udugama, I.A., Ullah, H.I., Vishwanath, S., Gernaey, K.V., Luo, J., Pinelo, M., Mansouri, S.S., 2022. Economic and Environmental Analysis of Bio-succinic Acid Production: From Established Processes to a New Continuous Fermentation Approach with In-Situ Electrolytic Extraction. *Chem. Eng. Res. Des.* 179, 401–414. https://doi.org/10.1016/j.cherd.2022.01.040

Mandegari, M.A., Farzad, S., Görgens, J.F., 2017a. Recent Trends on Techno-economic Assessment (TEA) of Sugarcane Biorefineries. *Biofuel Res. J.* 4, 704–712. https://doi.org/10.18331/BRJ2017.4.3.7

Mandegari, M., Farzad, S., Görgens, J.F., 2018. A New Insight into Sugarcane Biorefineries with Fossil Fuel Co-combustion: Techno-economic Analysis and Life Cycle Assessment. *Energy Convers. Manag.* 165, 76–91. https://doi.org/10.1016/j.enconman.2018.03.057

Mandegari, M.A., Farzad, S., van Rensburg, E., Görgens, J.F., 2017b. Multi-criteria Analysis of a Biorefinery for Co-production of Lactic Acid and Ethanol from Sugarcane Lignocellulose. *Biofuels, Bioprod. Biorefining* 11, 971–990. https://doi.org/10.1002/bbb.1801

MAResearch, 2022. 2022–2030 Report on Global Glutamic Acid and Monosodium Glutamate Market by Player, Region, Type, Application and Sales Channel.

Market Reports World, 2026a. Lactic Acids Market Size, Share, Growth, and Industry Analysis, By Type (DL-Lactic Acid, D-Lactic Acid, L-Lactic Acid), By Application (Food & Beverage, Polylactic Acid, Industrial, Pharmaceutical, Personal Care, Others), Regional Insights and Forecast to 2035 [WWW Document]. URL https://www.marketreportsworld.com/market-reports/lactic-acids-market-14724454

Market Reports World, 2026b. Ethyl Lactate Market Size, Share, Growth, and Industry Analysis, By Type (Food Grade, Pharmaceutical Grade, Industrial Grade, Electronic Grade), By Application (Food & Beverage, Industrial Application, Pharmaceutical, Microelectronics, Other), Regional Insights and Forecast to 2033 [WWW Document]. URL https://www.marketreportsworld.com/market-reports/ethyl-lactate-market-14717884

Market Reserach Future, 2021. Ethyl Acetate Market Research Report: Global Forecast Till 2027 [WWW Document]. URL https://www.marketresearchfuture.com/reports/ethyl-acetate-market-945 (accessed 6.11.21).

market.us, 2026. Ethanol Statistics by Purity, Properties, Consumption [WWW Document]. URL https://media.market.us/ethanol-statistics/

Marr, A.C., 2024. 1,3-Propanediol, an Exemplary Bio-Renewable Organic Platform Chemical. *Adv. Synth. Catal.* 366, 4835–4845. https://doi.org/10.1002/adsc.202401066

Mashoko, L., Mbohwa, C., Thomas, V.M., 2010. LCA of the South African Sugar Industry. *J. Environ. Plan. Manag.* 53, 793–807. https://doi.org/10.1080/09640568.2010.488120

Mazière, A., Prinsen, P., García, A., Luque, R., Len, C., 2017. A Review of Progress in (Bio) catalytic Routes from/to Renewable Succinic acid. *Biofuels, Bioprod. Biorefining,* 11, 908–931. https://doi.org/10.1002/bbb.1785

Mbothu, J., Mutwiwa, U., Eshton, B., Abubakar, L., Kenyatta, J., 2019. Lifecycle Greenhouse Gas Emissions and Energy Balances of Sugarcane Molasses-Based Bioethanol in Kenya. *J. Agric. Sci. Technol.* 19, 118–142.

Morakile, T., Mandegari, M., Farzad, S., Görgens, J.F., 2022. Comparative Techno-economic Assessment of Sugarcane Biorefineries Producing Glutamic Acid, Levulinic Acid and Xylitol from Sugarcane. *Ind. Crops Prod.* 184, 115053. https://doi.org/10.1016/j.indcrop.2022.115053

Nieder-Heitmann, M., Haigh, K., Görgens, J.F., 2019. Process Design and Economic Evaluation of Integrated, Multi-product Biorefineries for the Co-production of Bio-energy, Succinic acid, and Polyhydroxybutyrate (PHB) from Sugarcane Bagasse and Trash Lignocelluloses. *Biofuels, Bioprod. Biorefining* 13, 599–617. https://doi.org/10.1002/BBB.1972

nova-Institut GmbH, 2020. Biobased Building Blocks and Polymers - Global Capacities, Production and Trends 2019–2024.

Novamont, n.d. Novamont in Italy: Our Network of Production and Research Facilities in Italy [WWW Document]. URL https://www.novamont.com/eng/novamont-in-italy (accessed 4.8.26).

OECD, 2021. Guidance for a Biorefining Roadmap for Thailand.

Papageorgiou, G.Z., Papageorgiou, D.G., Terzopoulou, Z., Bikiaris, D.N., 2016. Production of Biobased 2,5-Furan Dicarboxylate Polyesters: Recent Progress and Critical Aspects in Their Synthesis and Thermal Properties. *Eur. Polym. J.* 83, 202–229. https://doi.org/10.1016/j.eurpolymj.2016.08.004

Pereira, C.S.M., Zabka, M., Silva, V.M.T.M., Rodrigues, A.E., 2009. A Novel Process for the Ethyl Lactate Synthesis in a Simulated Moving Bed Reactor (SMBR). *Chem. Eng. Sci.* 64, 3301–3310. https://doi.org/10.1016/j.ces.2009.04.003

Petersen, A.M., Chireshe, F., Okoro, O., Gorgens, J., Dyk, J.V., 2021. Evaluating Refinery Configurations for Deriving Sustainable Aviation Fuel from Ethanol or Syncrude. *Fuel Process. Technol.* 219, 106879.

Phasha, M.J., 2025. Conversion of Sugar Cane A-Molasses and Lignocellulosic Biomass to Value-Added Products: Techno-economics and Greenhouse Gas Emissions (PhD Dissertation). Stellenbosch University.

Phasha, M.J., Petersen, A.M., Görgens, J.F., 2025. Sustainable and Cost-effective Process Design for Energy-Self-Sufficient Ethyl Lactate Production from Sugar Mill Feedstocks: A Techno-economic Approach. *Biofuels, Bioprod. Biorefining* 20, 487–507. https://doi.org/10.1002/bbb.70091

Pothakos, V., Debeer, N., Debonne, I., Rodriguez, A., Starr, J.N., Anderson, T., 2018. Fermentation Titer Optimization and Impact on Energy and Water Consumption during Downstream Processing. *Chem. Eng. Technol.* 41, 2358–2365. https://doi.org/10.1002/ceat.201800279

Ratshoshi, B.K., Farzad, S., Görgens, J.F., 2021. Techno-economic Assessment of Polylactic Acid and Polybutylene Succinate Production in an Integrated Sugarcane Biorefinery. *Biofuels, Bioprod. Biorefining* 15, 1871–1887. https://doi.org/10.1002/bbb.2287

Rosales-Calderon, O., Arantes, V., 2019. A Review on Commercial-Scale High-Value Products That Can Be Produced Alongside Cellulosic Ethanol. *Biotechnol. Biofuels* 12, 240. https://doi.org/10.1186/s13068-019-1529-1

S&P Global, 2020. As Jet Fuel Market Craters, Sustainable Aviation Fuel Prepares for Takeoff [WWW Document]. URL https://www.spglobal.com/platts/en/market-insights/blogs/agriculture/092320-as-jet-fuel-market-craters-sustainable-aviation-fuel-prepares-for-takeoff (accessed 6.14.21).

SA Cane Growers, 2025. Diversification, Not Decline Can Define South Africa's Sugar Industry [WWW Document]. URL https://sacanegrowers.co.za/2025/07/28/diversification-not-decline-can-define-south-africas-sugar-industry/

Sanford, K., Chotani, G., Danielson, N., Zahn, J.A., 2016. Scaling Up of Renewable Chemicals. *Curr. Opin. Biotechnol.* 38, 112–122. https://doi.org/10.1016/j.copbio.2016.01.008

Sano, C., 2009. History of Glutamate Production. *Am. J. Clin. Nutr.* 90, 728S–732S. https://doi.org/10.3945/ajcn.2009.27462F

Santos, C.I., Silva, C.C., Mussatto, S.I., Osseweijer, P., van der Wielen, L.A.M., Posada, J.A., 2017. Integrated 1st and 2nd Generation Sugarcane Bio-refinery for Jet Fuel Production in Brazil: Techno-economic and Greenhouse Gas Emissions Assessment. *Renew. Energy* 129, 733–747. https://doi.org/10.1016/j.renene.2017.05.011

Satam, C.C., Daub, M., Realff, M.J., 2019. Techno-economic Analysis of 1,4-Butanediol Production by a Single-Step Bioconversion Process. *Biofuels, Bioprod. Biorefining* 13, 1261–1273. https://doi.org/10.1002/bbb.2016

Signoretto, M., Taghavi, S., Ghedini, E., Menegazzo, F., 2019. Catalytic Production of Levulinic Acid (LA) from Actual Biomass. *Molecules* 24, 2760. https://doi.org/10.3390/molecules24152760

Sikazwe, M.K., Louw, J., Görgens, J.F., 2025. Process Development Priorities for Scaling Up Direct Microbial Conversion of Sugars into Value-Added Biochemicals. *Biofuels, Bioprod. Biorefining* 20, 747–765. https://doi.org/10.1002/bbb.70095

Skoczinski, P., Chinthapalli, R., Carus, M., Baltus, W., de Guzman, D., Kab, H., Raschka, A., Ravenstijn, J., 2020. Biobased Building Blocks and Polymers - Global Capacities, Production and Trends 2019–2024. URL https://renewable-carbon.eu/publications/product/bio-based-building-blocks-and-polymers-global-capacities-production-and-trends-2019-2024/

Skoog, E., Shin, J.H., Saez-jimenez, V., Mapelli, V., Olsson, L., 2018. Biobased Adipic Acid – The Challenge of Developing the Production Host. *Biotechnol. Adv.* 36, 2248–2263. https://doi.org/10.1016/j.biotechadv.2018.10.012

Statsnex, 2026. Glutamic Acid Market Size by 2032 [WWW Document]. URL https://statsnex-marketinsights.com/reports/glutamic-acid-market

Taylor, R., Nattrass, L., Alberts, G., Robson, P., Chudziak, C., Bauen, A., Libelli, I.M., Lotti, G., Prussi, M., Nistri, R., Chiaramonti, D., López-Contreras, A.M., Bos, H.L., Eggink, G., Springer, J., Bakker, R., Ree, R.v., 2015. From the Sugar Platform to Biofuels and Biochemicals: Final Report for the European Commission Directorate-General Energy. E4tech/Re-CORD/Wageningen UR, 714, FBR BP Biorefinery & Sustainable Value Chains.

Thaore, V.B., Armstrong, R.D., Hutchings, G.J., Knight, D.W., Chadwick, D., Shah, N., 2020. Sustainable Production of Glucaric Acid from Corn Stover via Glucose Oxidation: An Assessment of Homogeneous and Heterogeneous Catalytic Oxidation Production Routes. *Chem. Eng. Res. Des.* 153, 337–349. https://doi.org/10.1016/j.cherd.2019.10.042

The Lycra Company, 2025. Qore Launches in Eddyville, Iowa, Ushering in Large-Scale Corn Processing for Bio-Based Solutions [WWW Document]. URL https://www.thelycracompany.com/en/lycra-company-press-center/press-center/qore-plant-opening

Transparency Market Research (TMR), 2024. Global 2,3-Butanediol Market Outlook 2031 [WWW Document]. URL https://www.transparencymarketresearch.com/2-3-butanediol-market.html

Transport & Environment, 2016. Aviation ETS – Gaining Altitude (An Analysis of the Aviation EU ETS 2013–2015). Brussels.

Ur Rehman, A., Kovacs, Z., Quitmann, H., Ebrahimi, M., Czermak, P., 2016. Enzymatic Production of Fructo-oligosaccharides from Inexpensive and Abundant Substrates Using a Membrane Reactor System. *Sep. Sci. Technol.* 51, 1537–1545. https://doi.org/10.1080/01496395.2016.1167740

Vaswani, S., 2010. *Biobased Succinic Acid.* SRI Consulting Review.

Wang, Y., Cui, J., Liu, Z., Zhou, Y., Chen, W., Gao, S., Ruchala, J., Yu, B., Wang, L., 2026. Industrial Perspective Validation of a Neutralizer-Free L-Lactic Acid Production Process by Engineered Acid-Tolerant Yeast Pichia kudriavzevii. *Trends Biotechnol* 44, 1660–1682. https://doi.org/10.1016/j.tibtech.2025.12.030

Williams, S.M., Sullivan, R.K.P., Scott, H.L., Finkelstein, D.I., Colditz, P.B., Lingwood, B.E., Dodd, P.R., Pow, D.V., 2005. Glial Glutamate Transporter Expression Patterns in Brains from Multiple Mammalian Species. *Glia* 49, 520–541. https://doi.org/10.1002/glia.20139

Zhang, D., Feng, X., Zhou, Z., Zhang, Y., Xu, H., 2012. Economical Production of Poly(γ-glutamic acid) Using Untreated Cane Molasses and Monosodium Glutamate Waste Liquor by Bacillus subtilis NX-2. *Bioresour. Technol.* 114, 583–588.https://doi.org/10.1016/j.biortech.2012.02.114

Zhao, M., Huang, D., Zhang, X., Koffas, M.A.G., Zhou, J., Deng, Y., 2018. Metabolic Engineering of Escherichia coli for Producing Adipic Acid through the Reverse Adipate-Degradation Pathway. *Metab. Eng.* 47, 254–262. https://doi.org/10.1016/j.ymben.2018.04.002

Zion Market Research, 2024. Glucaric Acid Market Size, Share, Trends, Growth 2032 [WWW Document]. URL https://www.zionmarketresearch.com/report/glucaric-acid-market

Appendix to Chapter 6

Multi-Criteria Benchmarking of Bioproducts from Sugarcane Molasses: Perspectives on Production Costs, Market Prices, and Greenhouse Gas Emissions

SIMPLIFIED PROCESS FLOW DIAGRAMS

Food Ingredients

Figures S6.1–S6.20 show the process steps involved for biobased production of the 20 products for the benchmarking study from A-molasses feedstock.

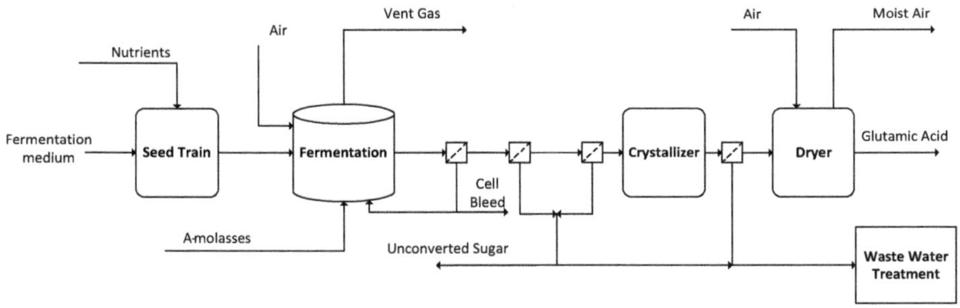

FIGURE S6.1 Flowsheet for glutamic acid production from A-molasses through fermentation and product recovery.

DOI: 10.1201/9781003745150-7

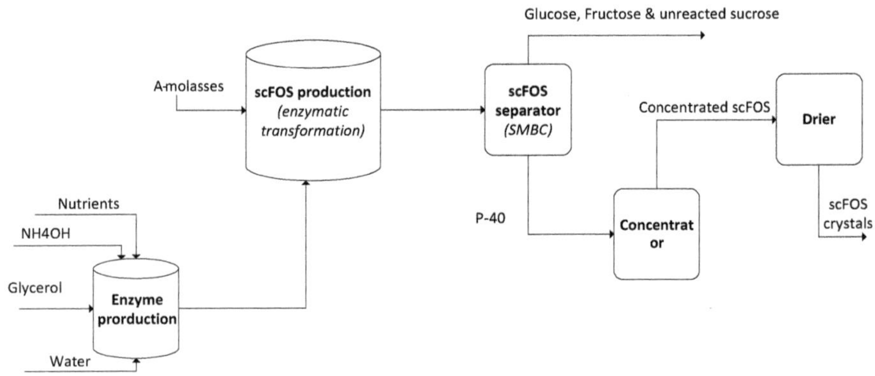

FIGURE S6.2 Flowsheet for the production of β-D-fructofuranosidase (FFase) and scFOS.

Industrial Chemicals and Solvents

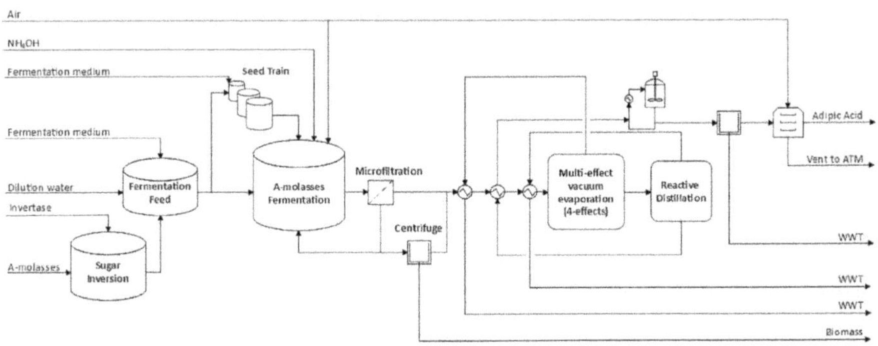

FIGURE S6.3 Flowsheet for bio-adipic acid production from A-molasses through fermentation and reactive distillation assuming that both fructose and glucose can be consumed by microbes.

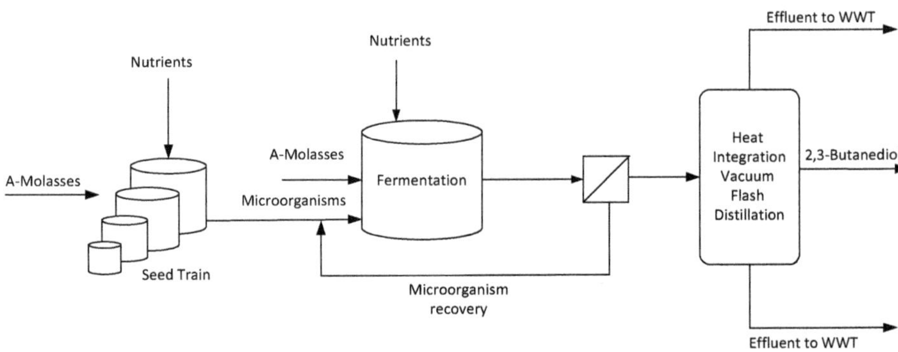

FIGURE S6.4 Flowsheet for 2,3-butanediol production from A-molasses through fermentation and heat-integrated vacuum flash distillation.

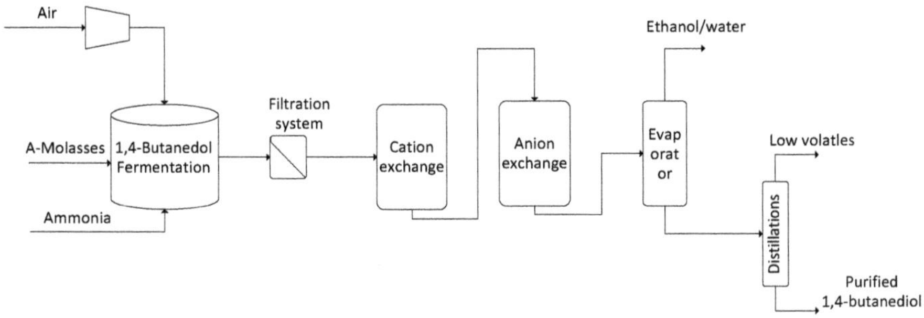

FIGURE S6.5 1 4-BDO fermentation and recovery.

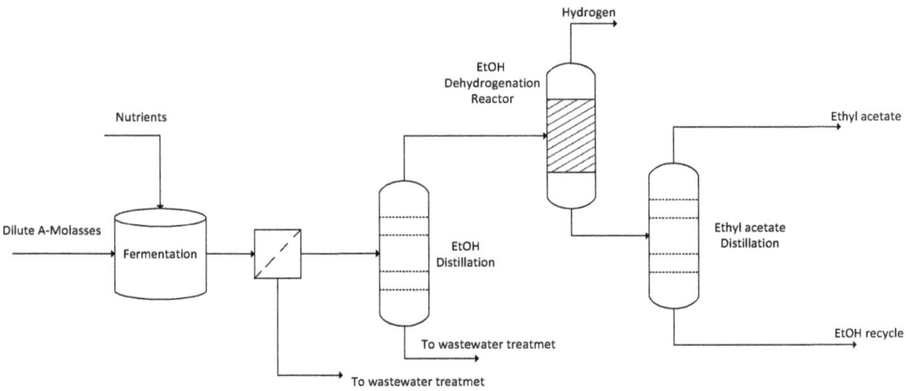

FIGURE S6.6 Flowsheet for bio-ethyl acetate production from A-molasses via fermentation and ethanol dehydrogenation.

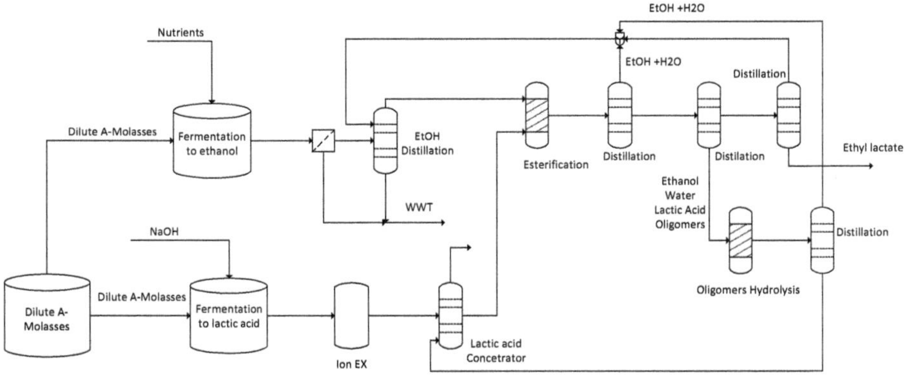

FIGURE S6.7 Flowsheet for bio-ethyl lactate production from A-molasses via fermentation and esterification of ethanol and lactic acid.

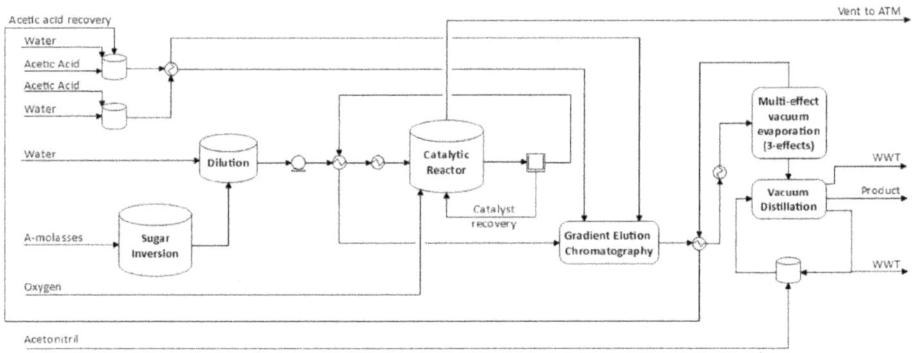

FIGURE S6.8 Flowsheet for bio-glucaric acid production from A-molasses through catalytic conversion.

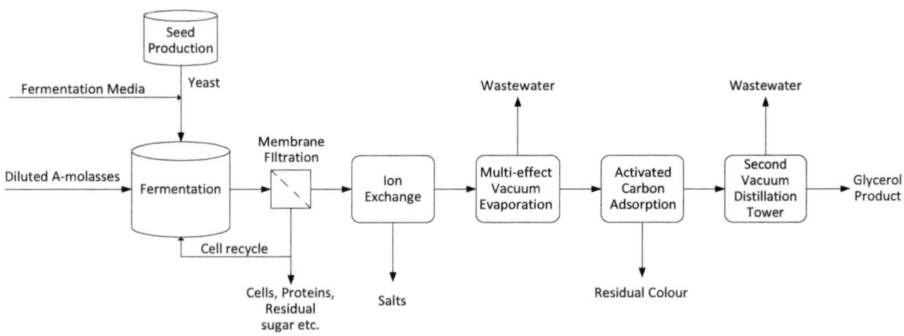

FIGURE S6.9 Flowsheet for glycerol production from A-molasses through batch and evaporation distillation.

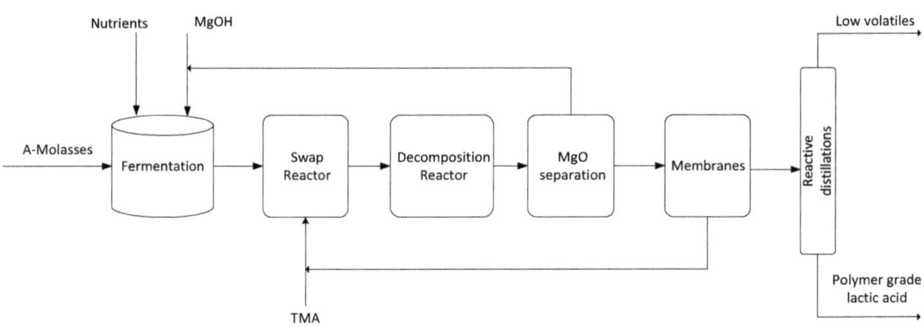

FIGURE S6.10 Lactic acid fermentation and recovery using the gypsum-free process.

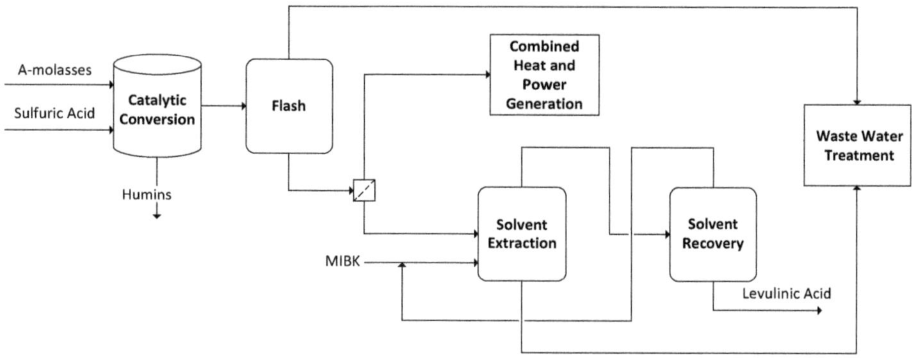

FIGURE S6.11 Flowsheet for levulinic acid production from A-molasses through catalytic conversion and solvent extraction.

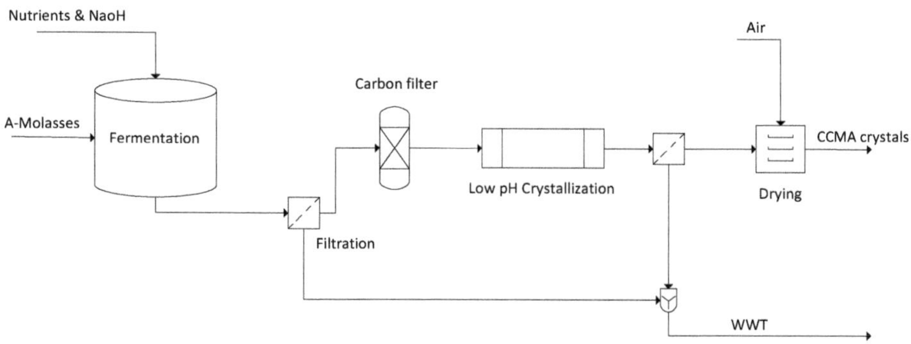

FIGURE S6.12 Flowsheet for bio-muconic acid (CCMA) production from A-molasses via fermentation and low-pH crystallization.

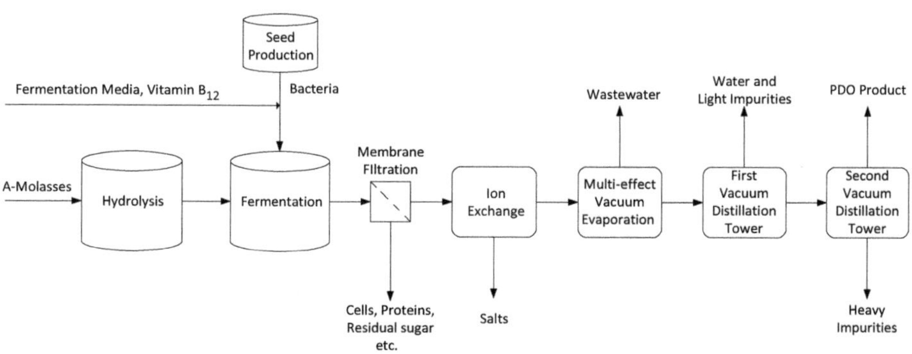

FIGURE S6.13 Flowsheet for 1,3-propanediol production from A-molasses through fed-batch fermentation and evaporation distillation.

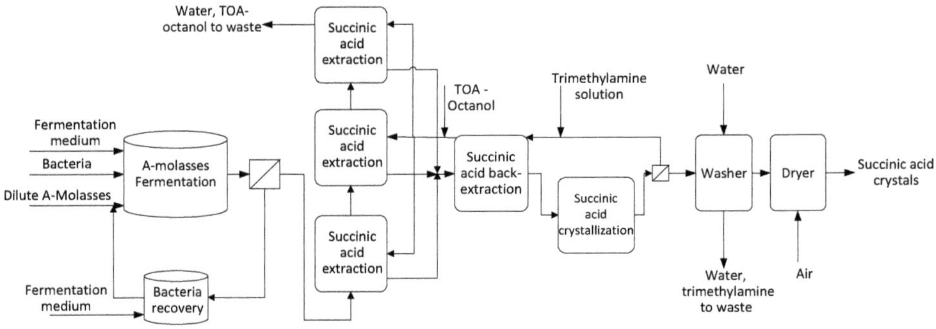

FIGURE S6.14 Flowsheet for bio-succinic acid production from A-molasses through acidic fermentation and reactive extractive.

Biobased Polymers

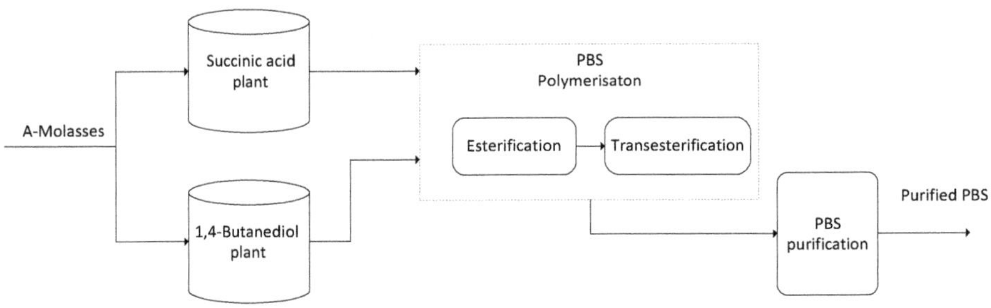

FIGURE S6.15 Bio-derived poly(sulfobetaine methacrylate) (PBS) polymerization.

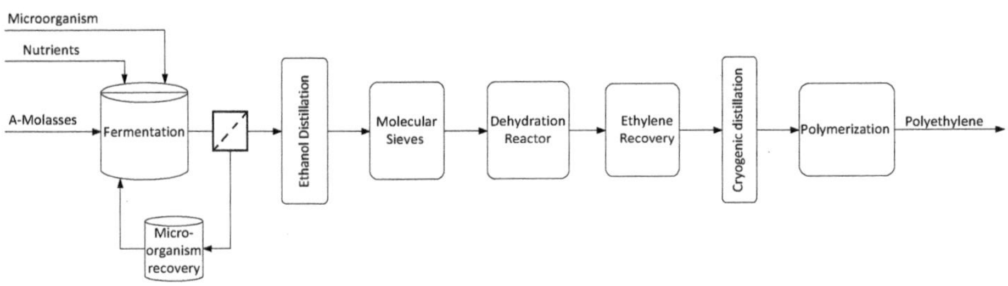

FIGURE S6.16 Flowsheet for bio-polyethylene production from A-molasses through ethanol fermentation, dehydration, and ethylene polymerization.

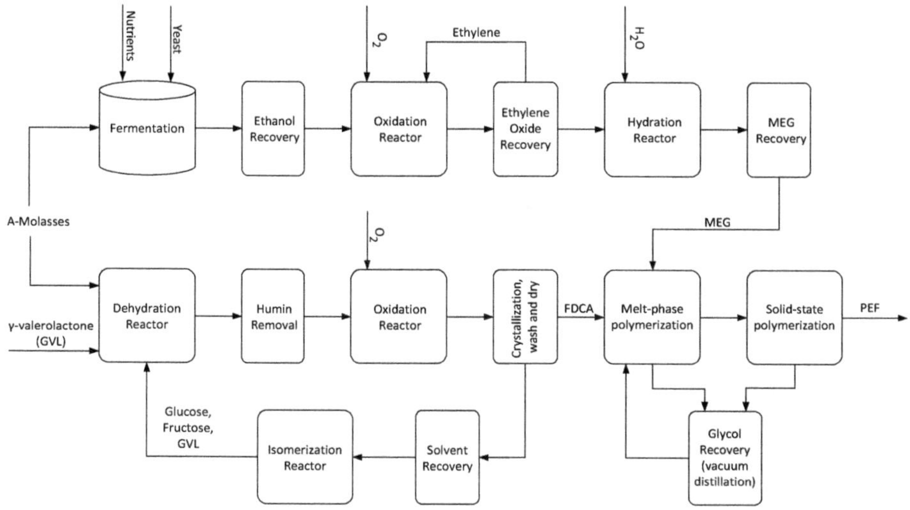

FIGURE S6.17 Flowsheet for bio-polyethylene furanoate production from A-molasses.

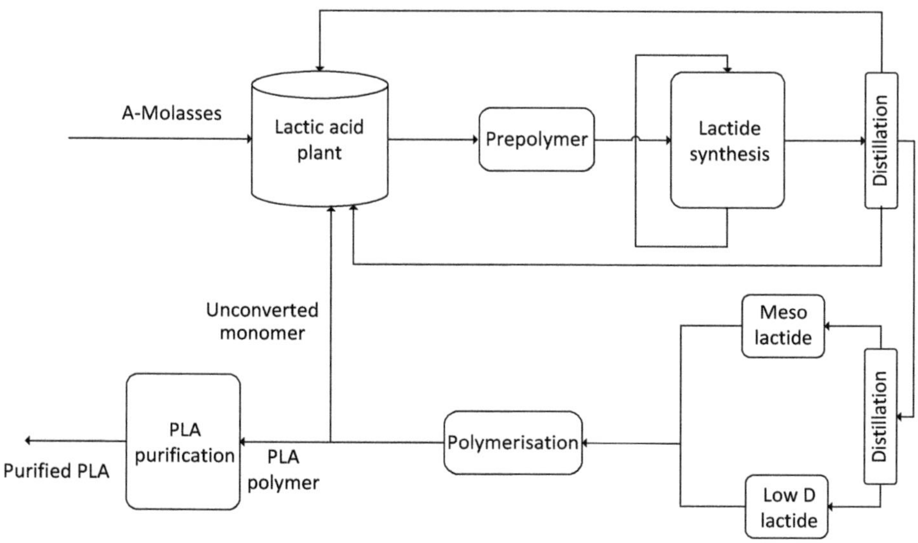

FIGURE S6.18 Biobased polylactic acid production pathway via ring opening polymerization.

Biofuels

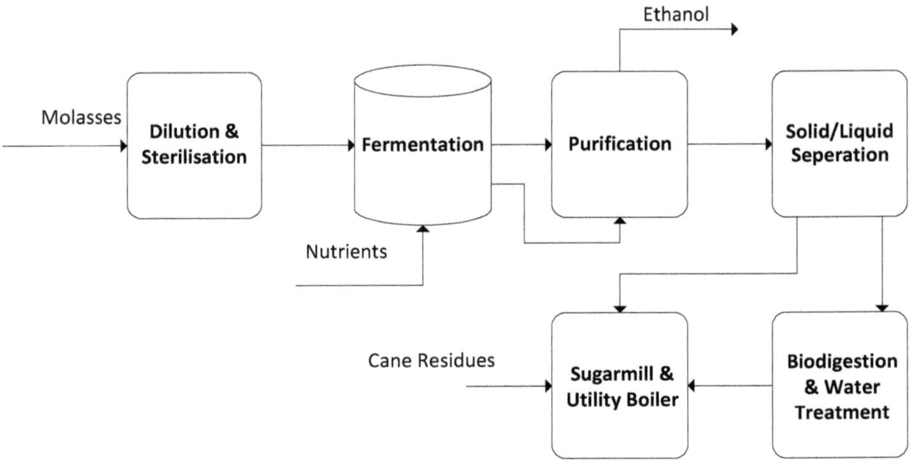

FIGURE S6.19 Flowsheet for ethanol production from A-molasses through fermentation.

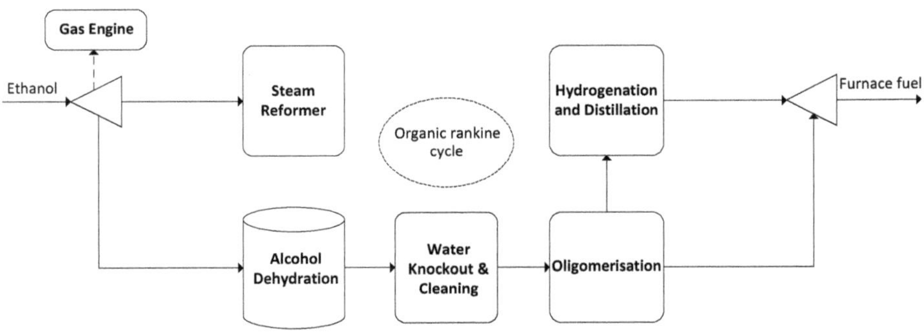

FIGURE S6.20 Ethanol conversion to sustainable aviation fuel (SAF).

MASS AND ENERGY BALANCE RESULTS

Tables S6.1–S6.4 provide the details about the mass and energy balances for the 20 biobased products produced from A-molasses feedstock in the categories of chemicals, food, polymers, and fuels.

TABLE S6.1
Mass balance results of chemicals

Parameter	Adipic acid	2,3-BDO	1,4-BDO	Ethyl acetate	Ethyl lactate	Glucaric acid	Glutamic acid	Glycerol	Lactic acid	Levulinic acid	Muconic acid	1,3-PDO	Succinic acid
Feed flow:													
A-molasses flowrate (t/h)	25.4	25.4	25.4	25.4	21.2	25.4	25.4	25.4	25.4	25.4	21.2	25.4	25.4
Bagasse flowrate for biorefinery steam requirement (t/h)	33.2	6.8	33.3	47.5	50.3	29.2	-	6.7	62.5	-	45.2	9.7	0.8
Product flows:													
Final product flow rate (t/h)	7.5	8.5	9.0	9.014	10.5	10.1	11.2	10.2	17.3	7.2	5.2	9.1	11.6
Product recovery (%)	97.6	90.0	92.7	97.9	99.8	90.4	79.7	99.4	97.1	96.0	92.6	96.0	99.7
Product purity (%)	99.0	98.0	99.9	99.7	99.8	78.0	98.0	99.4	99.9	98.8	91.0	99.9	99.9

TABLE S6.2
Energy balance results of chemicals

Parameter	Adipic acid	2,3-BDO	1,4-BDO	Ethyl acetate	Ethyl lactate	Glucaric acid	Glutamic acid	Glycerol	Lactic acid	Levulinic acid	Muconic acid	1,3-PDO	Succinic acid
Heating requirements:													
High Pressure (HP) steam demand (t/h)	-	15	-	-	-	-	-	3	37	-	-	9	-
Medium Pressure (MP) steam demand (t/h)	94	-	90	0	31	0	-	0	22	-	-	11	-
Low Pressure (LP) steam demand (t/h)	48	1	26	78	157	42	8	15	-	-	18	16	17
Total heating demand (MW)	81.70	9.55	67.40	47.20	113.80	25.20	4.40	10.60	29.10	25.50	11.10	21.54	10.60
Cooling requirements:													
Cooling water demand (t/h)	7 063	3	2 782	2 632	9 100	911	490	631	1 805	-	48 980	1 589	129
Cooling water demand (MW)	73.60	0.03	29.00	27.50	96.13	9.50	5.10	6.58	18.90	-	6.00	16.59	2.24
Chilled water demand (t/h)	29	1 059	7	936	410	-	2 239	82	1 257	-	570	264	2 554
Chilled water demand (MW)	0.37	13.53	0.08	12.00	5.26	-	28.60	1.04	16.10	-	7.30	3.37	38.65
Refrigerant (t/h)	-	-	-	-	-	18	-	-	-	-	-	-	-
Refrigerant (MW)	-	-	-	-	-	18.30	-	-	-	-	-	-	-
Total cooling demand (MW)	74.00	13.56	29.10	39.50	101.38	27.80	33.70	7.62	35.00	17.90	13.30	19.96	40.89
Power requirements:										-			
Electricity demand (MW)	9.24	4.58	0.70	0.85	0.84	3.35	0.44	0.57	0.50	1.30	1.93	16.16	1.13

TABLE S6.3
Mass balance results of food products, polymers, and fuels

Parameter	Food products		Polymers				Fuels	
	Glutamic acid	ScFOS	PBS	PE	PEF	PLA	Ethanol	Jet fuel
Feed flow:								
A-molasses flowrate (t/h)	25.4	25.4	25.4	25.4	25.4	25.4	25.4	-
Bagasse flowrate for biorefinery steam requirement (t/h)	-	5.4	24.4	9.8	3.0	71.9	31.3	-
Ethanol (t/h)	-	-	-	-	-	-	-	30.0
Product flows:								
Final product flow rate (t/h)	11.2	7.4	9.5	5.1	8.9	9.6	10.0	8.3
Product recovery (%)	79.7	97.0	93.1	-	-	91.4	99.7	-
Product purity (%)	98.0	90.0	100.0	100.0	100.0	100.0	99.9	100.0

TABLE S6.4

Energy balance results of food products, polymers, and fuels

Parameter	Food products		Polymers				Fuels	
	Glutamic acid	ScFOS	PBS	PE	PEF	PLA	Ethanol	Jet fuel
Heating requirements:								
HHP steam demand (t/h)	-	11.8	-	-	-	47.1	0	N/A
HP steam demand (t/h)	-	-	68.5	-	11.5		16.3	N/A
MP steam demand (t/h)	-	-	-	47.34	17.21	31.638	16.3	N/A
LP steam demand (t/h)	8	20	31	-	18	-	32.6	N/A
Total heating demand (MW)	**4.40**	**18.37**	**65.00**	**30.56**	**27.92**	**39.30**	**19.31**	N/A
Cooling requirements:								
Cooling water demand (t/h)	490	444	2933	2383	4774	2555	1598	1781
Cooling water demand (MW)	5.1	7.71	30.51	24.82	49.70	26.70	18.5	21.72
Chilled water demand (t/h)	2 239	1 183	1 199	407	267	1 386	Not calculated	N/A
Chilled water demand (MW)	28.60	17.90	15.40	5.20	3.41	17.80	1.9	N/A
Refrigerant demand (t/h)	-		-	2484	-	-	-	-
Refrigerant demand (MW)	-		-	0.93	-	-	-	-
Cooling air demand (t/h)	-		-	-	146	-	-	-
Cooling air demand (MW)	-		-	-	0.20	-	-	-
Total cooling demand (MW)	**33.70**	**25.62**	**45.90**	**30.95**	**53.35**	**44.43**	**20.4**	**21.72**
Power requirements:								
Electricity demand (MW)	**0.44**	**3.28**	**1.20**	**1.06**	**2.02**	**1.80**	**2.7**	**4.4**

GREENHOUSE GAS EMISSION FACTORS

See Table S6.5.

TABLE S6.5
CO_2-equivalent emission factors (EF) for all inputs across scenarios (obtained through Simapro) reported in kg CO_2eq emissions per kg input

Input	EF	Input	EF
1-Hexene	2.00	Iron chloride ($FeCl_3.6H_2O$)	0.83
1-Octanol	2.59	Iron sulfate ($FeSO_4*7H_2O$)	0.28
Acetic acid	1.58	Magnesium hydroxide ($Mg(OH)_2$)	0.80
Acetonitril	4.85	Magnesium sulfate ($MgSO_4*7H_2O$)	0.85
Activated carbon	3.05	Manganese sulfate ($MnO_4S.2H_2O$)	0.80
Aluminum	2.51	Manganese sulfate ($MnSO_4*H_2O$)	0.80
Aluminum oxide	2.26	Methanol	0.65
Ammonia	3.28	Methyl isobutyl ketone (MIBK)	4.43
Ammonium hydroxide (NH_4OH)	1.59	n-Pentane	0.79
A-Molasses	0.19	Oxygen (O_2)	1.07
Antimony trioxide (Sb_2O_3)	3.51	Phosphoric acid (H_3PO_4) (85%)	1.37
Ash	0.28753	Platinum (Pt catalyst)	69123.5
Bagasse	0.03	Polyglycerol-10	2.81
Boiler chemicals (hydrazine)	14.82	Potassium carbonate (K_2CO_3)	3.04
Brine water disposal	0.00038	Potassium chloride (KCl)	0.44
Calcium chloride ($CaCl_2*2H_2O$)	0.71	Potassium dihydrogen phosphate	8.36
Calcium sulfate (gypsum)	0.01	Potassium hydroxide (KOH)	2.61
Chloroform (trichloromethane)	3.62	Potassium sulfate (K_2SO_4)	1.04
Citric acid monohydrate	5.96	Silicon (Si)	3.19
Copper sulfate ($CuSO_4.5H_2O$)	4.43	Silicon dioxide (SiO_2)	1.64
Corn oil	2.05	Silver (Ag)	500.41
Corn steep liquor	0.01	Sodium chloride (NaCl)	0.27
Diammonium phosphate (DAP)	1.66	Sodium hydroxide (NaOH)	1.27
Dipotassium hydrogen phosphate	10.71	Solid waste	0.29
Dowex 77 (anionic IX resin)	3.53	Sulfuric acid (H_2SO_4)	0.16
Dowex 88 (cation IX resin)	1.74	Titanium	47.12
Electricity from grid (kW)	0.06	Titanium isopropoxide ($Ti(Obu)_4$)	6.73
Ethanol	0.53	Titratrant ($NaHCO_3$)	1.25
Gamma valerolactone (GVL)	0.66	Triethylamine	3.85
Glucose	1.29	Trymetylamine	2.73
Glycerol	2.81	Tryoctylamine	3.01
Hydrogen	2.38	Urea	1.69
Hydrogen chloride (HCl) (30%)	0.89	Zinc (Sn)	2.70
Invertase enzyme	10.52	ZnO	0.84

7 From Black Boxes to Blueprints

Engineering of Microbial Consortia for Multiple Product Biorefineries

Olivier Habimana

7.1 INTRODUCTION: A NEW ERA IN MICROBIAL TEAMS IN THE CIRCULAR BIOECONOMY

The circular bioeconomy, which is founded on the innovative conversion of waste into valuable resources (often achieving >95% turnover rates), relies fundamentally on microbiological processes and interactions across different environments (Delvigne et al., 2017; Stikane et al., 2022; Vea et al., 2018; Zaman et al., 2020). Pure cultures have been crucial for industrial biotechnology, and for a century, they have been the basis for the bioprocesses and discoveries in industry (Singh et al., 2022). While effective with well-defined substrates, this technique encounters significant constraints when applied to the complex, diverse matrices inherent in agricultural, industrial, and municipal refuse (Havukainen & Dace, 2023). Resistant polymers such as lignin and cellulose are often maximally broken down when exposed to a single microbial species that consistently cannot produce an extensive suite of enzymes needed for efficient depolymerization and consume the broad spectrum of hydrolysis products for valuable end products, while at the same time, tolerating waste-inhibiting compounds (Bajpai, 2022; Guzmán-Pedraza et al., 2023; Kumar, 2019). This bottleneck in waste valorization stems from the low microbial diversity that can digest the target substrates in practice and often requires expensive physical or chemical pretreatments, yielding underutilized substrates (Devi et al., 2022; Wong et al., 2023) (Figure 7.1, conceptual shift).

On the other hand, complex organic compounds are effectively recycled in nature because of the cooperation of diverse microbial consortia in natural ecosystems (Havukainen & Dace, 2023; Rapp et al., 2020). These communities are more stable and efficient at the metabolic and functional levels than single isolates (Yuan et al., 2022). This perception has prompted a transition in bioprocessing toward a comprehensive examination of 'microbial ecosystem factories' rather than solely 'microbial

 DOI: 10.1201/9781003745150-8

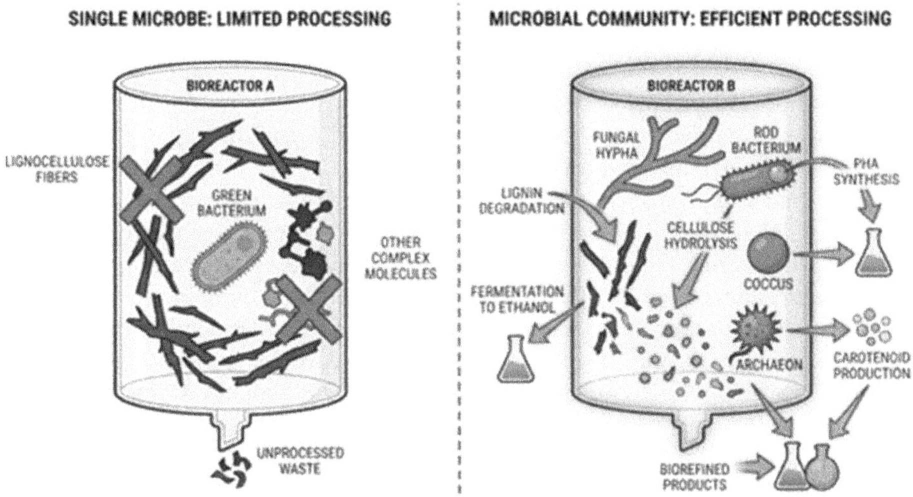

FIGURE 7.1 The conceptual shift: evolving from traditional cultivation to programmable ecosystem factories. The shift from the limitations of single-species bioprocessing to the advantages of customized microbial groups. Left: A single microbial species cannot broadly degrade complex waste streams or produce diverse products. Right: A synthetic microbial consortium enables metabolic division of labor, allowing for simultaneous degradation and the co-production of multiple valuable compounds from mixed feedstocks.

cell factories' (Delvigne et al., 2018). This perspective is founded on the notion that a microbial consortium constitutes a programmable multicellular framework with its own experts that execute particular functions (Atkinson et al., 2022; Moon et al., 2024). For example, one member could produce hydrolytic enzyme, another could convert free sugars, and one could produce high-value products, all linked through synergistic relations such as cross-feeding and cooperation.

In this segment, we propose that monitoring and managing these microbial assemblages is crucial for advancing next-generation, genuine, resilient, and operational biorefineries (Delvigne et al., 2018). Beyond the pure culture paradigm, the valorization of all prospective refuse streams will lead to the efficient use of constraints, as evidenced in natural ecological systems. We will explore the ecological principles underlying these interactions and outline systematic methods for constructing synthetic microbial consortia, including advanced computational and molecular strategies for their engineering (Badr et al., 2022; Chen et al., 2024). We will illustrate how specialized instruments crafted for alliances enhance lignocellulosic breakdown, the simultaneous production of significant compounds, and overall operational efficiency through actual case studies (Troiano & Studer, 2025). We will then discuss some of the critical open questions, including the design for stability and scalability, emphasizing that deliberately engineering microbial synergies is not an alternative – it is a requirement for a carbon-short, sustainable, and circular bioeconomy.

7.2 LEARNING FROM NATURE-ECOLOGICAL PRINCIPLES FOR INDUSTRIAL SYNERGY

Most natural ecosystems provide ideal conditions for the evolution of conversion mechanisms over time, as complex microbial communities evolve to create effective nutrient recycling processes (Ruby, 2020). Anaerobic digesters, landfills, and compost heaps can be considered homologous waste-decomposing ecosystems that depend on interspecies bacterial interaction for waste treatment and nutrient cycling (Jenkins et al., 2021). The stability and operation of these activities are buttressed by symbiotic life forms whose metabolic processes are directly affected and often dependent upon one another. Dissecting these naturally occurring consortia provides a fundamental set of principles for developing synthetic consortia for biorefining (Rapp et al., 2020; Troiano & Studer, 2025). Three major types of mutualistic relations are, in fact, common among species in ecology: syntrophy (a relationship in which a given species is dependent on the metabolic products of another), commensalism (one partner benefits and the other is not affected), and exchange (cross-feeding cooperation among organisms) (Ruby, 2020).

7.2.1 Syntrophy: The Key to a Metabolic Collaboration

'Consuming together' represents a highly specialized type of mutualism, wherein the combined metabolic processes of two or more different species can facilitate a chemical reaction that is nearly unattainable for any single organism to accomplish independently (Schink, 1997; Sieber et al., 2012). This is exemplified by the anaerobic degradation of lipids and ethanol, a crucial process in methanogenic environments (McInerney et al., 2009; Schink, 1997). In this theoretical framework, proton-reducing acetogens convert substrates, including butyrate and ethanol, to generate hydrogen (H_2) and acetate. This reaction is endergonic under standard conditions and becomes exergonic only when the partial pressure of hydrogen is maintained at very low levels by hydrogen-consuming partners such as hydrogenotrophic methanogens (*Methanobrevibacter* spp.) or sulfate-reducing bacteria (Stams et al., 2012). This interspecies hydrogen transfer is a good illustration of the obligatory metabolic cooperation involved in the complete mineralization of organic compounds to methane and CO_2 (Sieber et al., 2012). The recent discovery of direct interspecies electron transfer (DIET), namely, the transmission of electrons through microbial nanowires or conductive materials linking organisms, offers a further example of the intricate mechanisms that have emerged to improve syntrophic metabolism (Nobu et al., 2020; Zhao et al., 2016). The syntrophic relationships described above, which occur as interesting ecological examples, transcend a simple example among the innumerable ecological relationships; rather, they are limiting, obligate, and mandatory conditions for the establishment of stable and efficient anaerobic bioreactors, which are major devices for various biotechnological processes (Nobu et al., 2020; Stams et al., 2012).

7.2.2 Commensalism and Cross-Feeding: Sharing the Load

While syntrophy is obligatory, commensalism and the cross-feeding interactions are more facultative but equally important interactions for the functioning of a community (Wang et al., 2023). Commensalism is a fascinating natural process in which, for example, one organism, via its metabolic activity in complex biochemical reactions, generates a substrate which, as a positive byproduct, is beneficial to another organism but without having any interference or consequence for its own welfare and survival (Wang et al., 2023). A good illustration is the disintegration of complex polymers. Cellulolytic *Clostridium* or filamentous fungi which usually act as primary degraders, secrete enzymes that convert cellulose and hemicellulose to soluble sugars such as cellobiose and xylose (Zambello et al., 2024). These sugars are subsequently accessible to a spectrum of secondary fermenters (e.g., *Escherichia* and *Lactobacillus*), which are unable to degrade the polymer but readily use the hydrolysis end products (Zambello et al., 2024). This stepwise degradation distributes the enzymes over the whole microbial community and makes substrate degradation more efficient. This concept is broadened by the cross-feeding phenomenon that allows for the exchange of essential nutrients, that is, essential vitamins, amino acids (proteins and cofactors of several enzymatic reactions) (Liu et al., 2024a). For example, the population may offset a partner species' auxotrophy by secreting the required growth factor, thereby stabilizing the consortium and preventing the collapse of any member required for a shared metabolic pathway (Kundu et al., 2022).

7.2.3 Functional Redundancy and Diversity as Sources of Robustness

The resilience of waste breakdown assemblages is primarily regulated by complex relationships and specific characteristics of community participants, such as functional redundancy and taxonomic variety (Carvalho et al., 2018; Liu et al., 2025). These attributes bolster the system's resilience and redundancy. Functional redundancy is also critical for robustness, as it ensures that if a primary cellulolytic species experiences harsh conditions, a secondary species with the same functional capabilities can still achieve the appropriate hydrolysis rates for this crucial biochemistry (Carvalho et al., 2018). This richness of the community renders it capable of a prompt response to changes in substrate composition, pH fluctuations, temperature differences, and the presence of contaminants, and a stringent selection of different groups of microbiota in accordance with the prevailing ecological conditions (Carballa et al., 2015). An effective community can maintain reliable metabolic performance over time despite variation in its environment, whereas that of low-efficiency monotypic systems cannot (Wu et al., 2024). Collective resilience is crucial for creating and executing reliable wide-area systems (Carvalho et al., 2018). Through a strategic application of the universal laws of ecology, particularly obligatory syntrophy, mutualistic commensalism, and emergent robustness, we can not only observe but also engineer microbiomes that mimic the most successful tactics invented by nature over billions of years (Regueiro et al., 2015) (Figure 7.2).

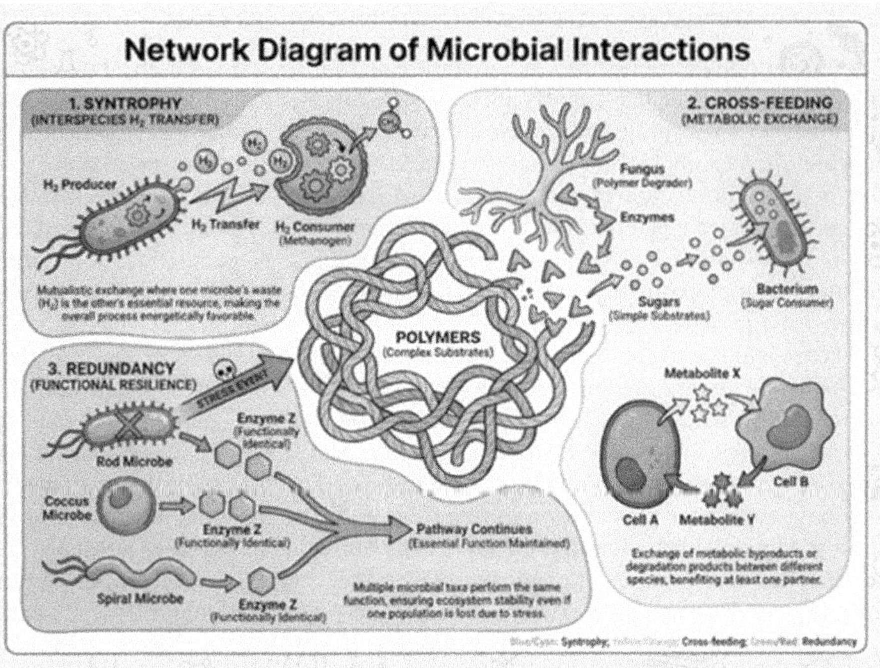

FIGURE 7.2 Ecological principles underpinning consortium design. Core ecological principles for engineering synergistic microbial consortia. (1) Syntrophy, characterized as an obligatory mutualistic relationship, encompasses the metabolic interchange among species where interspecies hydrogen transfer (IHT) and/or DIET are significant factors. (2) Metabolic sharing often occurs in commensalism and cross-feeding where the degradation product from one organism is used as substrate by another one. (3) Functional redundancy, where multiple species perform the same critical function, provides robustness and resilience against environmental perturbations.

7.3 DESIGN STRATEGIES FOR CONSORTIA: TOP-DOWN ENRICHMENT AND BOTTOM-UP ENGINEERING

The application of bioprocesses from ecological ideas necessitates methodical approaches for developing consortia (Chen et al., 2024). Enrichment from the top-down and intervention from the bottom-up are two opposite complementary methods in this discipline (Li et al., 2022). The trade-off requires assessing the advantages of natural selection against the consistency of synthesis (Kang et al., 2022). While the hierarchical method takes advantage of the natural functionality of more complex microbial communities, grassroots as well construct artificial ecosystems consisting of a set of pure strains or isolates (Figure 7.3).

7.3.1 Top-Down Enrichment: Harnessing Natural Selection

The top-down approach starts as the spectrum of organisms to be designed is the one present in a native complex microbial community found in an adequate ecological

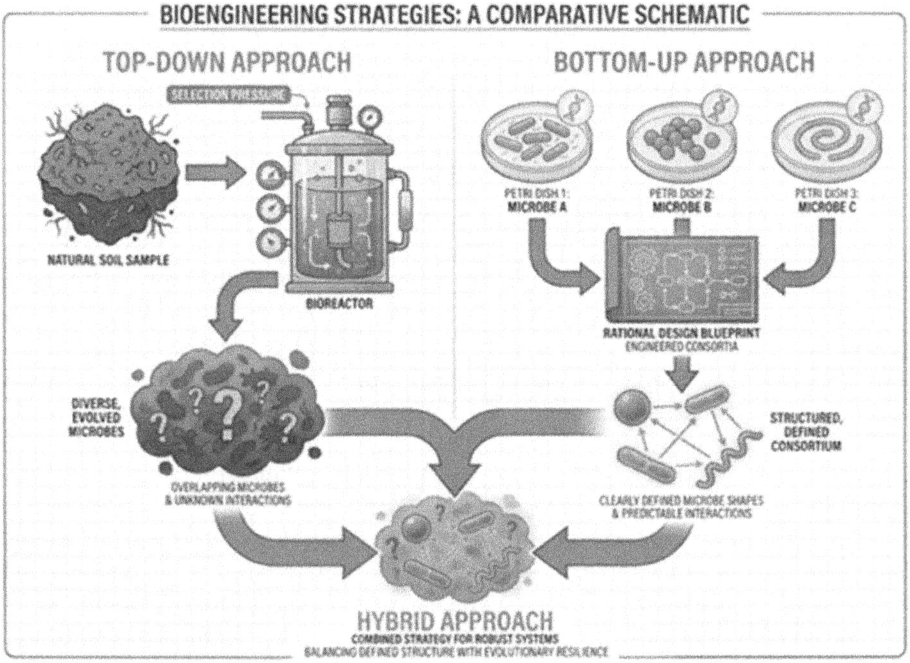

FIGURE 7.3 Top-down vs. bottom-up consortium design strategies. Complementary strategies for consortium development. The top-down method selectively enhances complex, functionally adapted communities from environmental sources, leading to strong yet often unpredictable 'black box' communities. The bottom-up approach rationally assembles defined isolates based on known metabolic traits to create predictable, customizable consortia. An integrated hybrid approach integrates the benefits of the two strategies.

niche, such as anaerobic digestion effluent, organic matter-decomposing systems, or terrestrial sources and substrates (Li et al., 2022). This inoculum possesses the potential (genetic and metabolic capabilities) whose raw materials are established in the metabolism and breakdown of intricate organic compounds (Grimalt-Alemany et al., 2018). The designer then exerts specific forces in order to guide the community toward a desired functional state (Kong et al., 2018). This typically involves a system in which the inoculum is grown in a chemostat or serial batch system on the waste feedstock as the sole or main source of carbon and energy (Gilmore et al., 2019). The enhancement culture throughout successive generations leads to this microbial assembly, which is ideally suited to the physicochemical characteristics of the feed material (Kong et al., 2018). Performance is improved through the stepwise increase of stressors, like substrate concentration, inhibitor levels, or salt that pushes the community toward evolving an increased tolerance or a more efficient metabolism (Adaptive Laboratory Evolution, ALE) (Grimalt-Alemany et al., 2018). The main advantage of the top-down approach is that it has the potential to generate robust, self-organized, and optimally functioning consortia capable of degrading recalcitrant and recalcitrant-containing feedstocks without requiring a priori identification of all microbes involved (Gilmore et al., 2019). However, its major limitation is that the resulting consortium is often a 'black box' with a complex and unpredictable dynamics, not always easy to control, predict or reproduce in another setting.

7.3.2 Bottom-Up Engineering: Predictable Construction from Pure Cultures

In contrast to the top-down approach, the bottom-up strategy works from a more reductionist approach, by assembling consortia in a mimicked fashion by combining two or more distinct microbial isolates in a systematic manner (Lyu et al., 2024). This approach requires extensive knowledge of the physiological attributes and metabolic properties of each member. The system design is intentional, often with a desired allocation of tasks in mind, in which a particular metabolic burden of a complex action is divided into parts among specialists (Dinjaski & Kaplan, 2016; Li et al., 2022). Some illustrative design motifs encompass: Public benefactors that coexist alongside those offering public goods, which include; cellulolytic fungi that release extracellular enzymes and fermenting bacteria that exploit the sugars released; auxotrophic or metabolic exchange-based cross-feeding: for instance, a lactate-producing bacterium in conjunction with a lactate-consuming bacterium (Noto Guillen et al., 2021); and the integration of biocontrol agents that outcompete contaminants while sparing core functionality. It is also worth noting that the bottom-up method produces superior visibility, predictability, and adaptability, as each individual contribution's duties and relationships are well specified (Dinjaski & Kaplan, 2016). This makes it very well-suited for precise coverage determination of specific chemicals and for fundamental studies of microbial interactions (Noto Guillen et al., 2021). Yet, its applicability is confined by the requirement for culturable, matched isolates, a narrow functional spectrum compared with a complex, enriched community, and innate fragility, since the elimination of a single stumbling block may trigger the collapse of the entire compared consortium (Li et al., 2022). The high level of sophistication of consortium engineering is most probably to be found in the combination of rational bottom-up design and top-down omics-based knowledge on successful natural cooperation (Ben Said & Or, 2017).

7.4 ENABLERS FOR ENGINEERING SYNERGY: COMPUTATIONAL AND EXPERIMENTATION PLATFORMS

The transition from observation to engineering of microbial interactions toward better biorefineries is driven by powerful computational and experimental tools (Dukovski et al., 2021; García-Jiménez et al., 2018; García-Jiménez et al., 2021). These tools enable researchers to approach the creation of synthetic communities in a manner distinct from the traditional trial-and-error approach, and to be based on predictions, systematic analysis, and physical principles (Cakir et al., 2023; Ouzounis et al., 2023). By integrating concepts from systems biology, molecular ecology, and bioprocess engineering, these technologies enable the disassembly of complex community behaviors and their reconstruction into engineered bioprocessing platforms that are both optimized and stable (Li et al., 2022; Peng et al., 2016; Roell et al., 2019).

7.4.1 Modeling Metabolism: Predicting Interactions, Not Competition

Computational metabolic models aid in the efficient *in silico* scaffolding for predicting the emergent properties of microbial communities before their physical assembly (Biggs et al., 2015). Constraint-based reconstruction and analysis (COBRA) methods, especially those that involve genome-scale metabolic models (GEMs), are indispensable for this purpose (Esvap & Ulgen, 2021; Tarzi et al., 2024). GEMs are personally curated for each potential consortium participant before being added to the multistep model (Ibrahim et al., 2021). The presented coupled model can be used to simulate metabolite flow in the community network, which allows the prediction of potential synergistic interactions, such as the cross-feeding of amino acids, vitamins, or intermediates, which may be engineered (Götz et al., 2020). Importantly, such frameworks also enable the detection and minimize antagonistic interactions, especially competition for essential resources, such as carbon, nitrogen, or phosphate, that can disrupt a consortium (Gallo et al., 2019; Price et al., 2017). These are either based on OptCom or COMETS (Computation of Microbial Ecosystems in Time and Space) methods, and are capable of dynamically simulating growth and metabolite flow, and can be used to mimic specific environmental conditions and inoculation ratios, allowing testing for stable equilibria that maximize the product yield (Albornoz et al., 2024; Nielsen et al., 2019). Model-informed design significantly minimizes the empirical experimental space, thereby shortening consortium development time (Jansma & El Aidy, 2021; Zuñiga et al., 2020).

7.4.2 Multi-Omics Integration: Deconstructing Community Function

Although frameworks can provide predictions, multi-omics approaches are critical to generate baseline knowledge of community dynamics and to validate model predictions (Hanemaaijer et al., 2015). Metagenomics further enables a full-scale profiling of taxonomic composition and reconstruction of metagenome-assembled genomes (MAGs), and assignment of genetic potential for an enriched top-down community (Zorrilla et al., 2021; Parks et al., 2017). In the progression from potential to actuality, metatranscriptomics reveals which genes are being expressed under particular process conditions, showing which members are actively expressing hydrolytic enzymes or stress responses, for example (Franzosa et al., 2015; Shakya et al., 2019). The presence of the resulting functional proteins is confirmed by metaproteomics and metabolomics reports the metabolic landscape while giving a direct readout of the cross-fed metabolites and final products (Choe et al., 2024). The integration of complementary datasets using advanced bioinformatics techniques further sheds light on complex community dynamics by identifying keystone species and disclosing molecular mechanisms associated with the resilience and the functioning of the consortium (Banerjee et al., 2018). Understanding natural degradative consortia at a systems level is essential for reverse engineering, analyzing effective consortia, and resolving issues in synthetic systems (Röttjers & Faust, 2018; Widder et al., 2016).

7.4.3 Spatial Structuring: Engineering the Microenvironment for Interaction

Microorganism spatial distribution is one of the determinants for the performance of the consortium, because proximity is related to the effectiveness of metabolite exchange as well as to the nature of interspecies relationships (Sing et al., 2014). Stirred-tank reactors ensure a uniform environment that can remove generated metabolites and disrupt symbiotic relationships (Brenner et al., 2008; Großkopf & Soyer, 2014). Advanced design of bioreactors uses spatial arrangement for a favorable interplay (Driscoll & Delmotte, 2019). This can be facilitated by biofilm formation on inert surfaces or granules within reactors, leading to the arrival of the gradient generators and to restriction of the microenvironment, which promotes syntrophy (Flemming et al., 2016; Stewart & Franklin, 2008). It has been applied deliberately to entrap a variety of bacteria in the natural or common hydrogels or alginate beads and to physically connect their different metabolic functions (Augst et al., 2006; Lopez et al., 2018). Membrane bioreactors provide an intricate solution as they compartmentalize numerous species or consortia in different reactors connected by semi-permeable membranes, thus enabling selective exchange of metabolites and signals and avoiding physical mixing and competition (Drews et al., 2005).

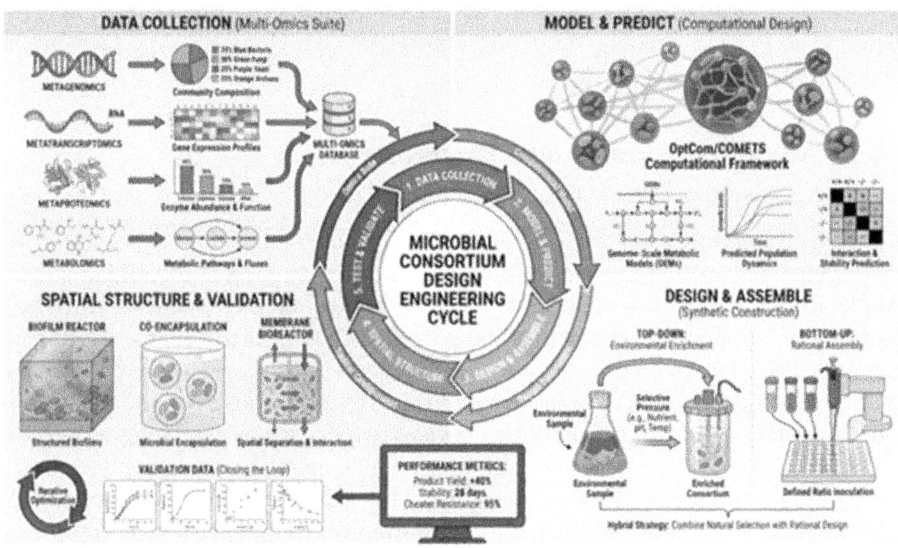

FIGURE 7.4 Enabling technologies for consortium design and control. A comprehensive toolset for designing, analyzing, and spatially managing microbial consortia. Left: Multi-omics cycle, the integrated use of metagenomics, metatranscriptomics, metaproteomics and metabolomics that enables the systems-level analysis of community structure (the potential functions present in a microbiota), function (how it operates to achieve its functions), and metabolite exchange. Right: Spatial engineering strategies, such as biofilm cultivation, co-encapsulation in hydrogels or beads, and membrane-based compartmentalization, structure the microenvironment to enhance cooperative interactions (e.g., syntrophy and cross-feeding), improve process stability, and limit invasion by cheater strains.

This spatial control is then expressed as a mechanism to sequester public goods (hydrolytic enzymes and cross-fed nutrients) within a cooperating consortium, leading to increased overall process efficiency, stability, and resistance to invasions by exploitative species (Ghoul & Mitri, 2016; Price et al., 2017) (Figure 7.4).

7.5 CASE STUDIES IN WASTE VALORIZATION: APPLIED CONSORTIUM ENGINEERING

The rational guidelines and engineering strategies for designing microbial consortia are finally verified by practice (Prabhu et al., 2020). Co-culture and defined community processes for waste valorization have specific advantages over pure culture bioprocesses, particularly for converting complex and diverse feedstocks (Prabhu et al., 2020). The potential of this technology is demonstrated in two key examples: (1) achieving maximum conversion of lignocellulosic biomass into products via direct bioprocessing, normalized to cellulose, and (2) the unique capacity to produce several high-value chemicals from the one waste stream simultaneously, enhancing the economic attractiveness of the associated biorefinery (Singhania et al., 2022).

7.5.1 Designer Consortia for Lignocellulose Consolidated Bioprocessing

Lignocellulosic biomass, such as crop straw, wood chips, and bamboo, is the most abundant and often ignored byproduct derived from diverse agro-industries and forest exploitation and serves as a crucial fuel in establishing and maintaining the circular bioeconomy (Prado-Acebo et al., 2024). However, the resilient framework necessitates a multi-phase transformation, frequently consisting of distinct enzymatic hydrolysis and fermentation stages, rendering it unwieldy and expensive (Brown et al., 2024). This purported consolidated bioprocessing (CBP) is accomplished through the concurrent production of enzymes, substrate breakdown, and fermentation of metabolites taking place within a singular unit, enabled by engineered consortia (Singhania et al., 2022). A common model is to mix a cellulolytic specialist and a robust fermentative bacterium (Shene et al., 2020), illustrating a clear division of labor that can be visually conceptualized (Figure 7.5). Despite the ability of the mycete *Trichoderma reesei* to secrete a powerful cocktail of cellulases and hemicellulases, it is not capable of producing ethanol in large quantities (Scholz et al., 2017). The oligosaccharides and glucose that are generated by fungal hydrolases are efficiently fermented into ethanol when co-cultured with the ethanologenic yeast *S. cerevisiae* (Garrido-Amador et al., 2021). Such a division of work imitates that of natural decomposer communities and can enhance overall conversion yields and rates as compared to sequential processing (Prabhu et al., 2020). More complex consortia add additional members to overcome other challenges (for instance, a lignin-degrading bacterium, e.g., *Pseudomonas putida*, to partly delignify the biomass, providing accessibility for hydrolytic enzymes, and a methanotrophic bacterium to consume inhibitory short-chain fatty acids) (Chen & Wan, 2017; Zong et al., 2022). The results from these joint activities clearly demonstrate how consortia can streamline operations establishment and improve on feedstock utilization (Prabhu et al., 2020; Scholz et al., 2017).

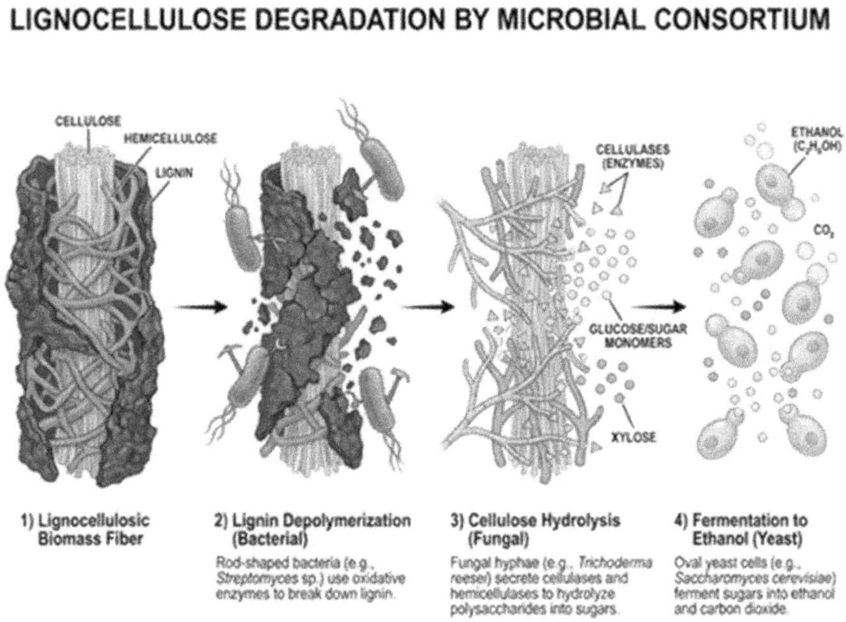

FIGURE 7.5 Case study: consolidated bioprocessing of lignocellulose. Case study of engineered consortia for bioprocessing lignocellulosic biomass. A lignin-decomposing specialist like *Pseudomonas putida* initiates the breakdown, allowing a cellulose-hydrolyzing fungus such as *Trichoderma reesei* to further digest cellulose. Sugars are fermented in one vessel by a strong ethanologen like *Saccharomyces cerevisiae*, which facilitates enzyme production and hydrolysis.

7.5.2 Co-production of Fine Chemicals for Multicomponent Biofactories

The economic sustainability of biorefinery processes that handle waste as feedstock is more likely to be based on the ability to produce a range of marketable products, rather than on single-product production paradigms (Prado-Acebo et al., 2024). Microbial consortia excel in this multiproduct model by enabling simultaneous biosynthetic pathways that would be too costly for a single cell (Katagi et al., 2024). One of the most important research studies dealt with the co-generation of polyhydroxyalkanoates (PHAs), biodegradable biopolymers for environmental applications and carotenoids, organic pigments having significant roles as antioxidants and colorants in both industrial and nutritional applications (Katagi et al., 2024; Prado-Acebo et al., 2024). This method may involve growing a bacterium species that produces PHAs (e.g., *Cupriavidus necator*) along with a yeast species that produces carotenoids (e.g., *Rhodotorula toruloides*) (Możejko-Ciesielska et al., 2023). *C. necator* preferentially consumes glucose and fructose to accumulate intracellular PHA granules; *R. toruloides* can produce carotenoids such as β-carotene and torulene from using the aromatic compounds, organic acids, or even competing for certain sugars (Arreola-Vargas et al., 2022). This process expands product options and enables greater substrate usage as consortium members break down various hydrolysate components (Prabhu et al., 2020). Analogous strategies can be designed for co-producing biosurfactants, enzymes, and biofuels in multi-layered biocatalytic consortia, leading to a microbial analog of a multiproduct industrial complex (Katagi et al., 2024). The success of these systems is critically reliant upon mediating resource competition through careful strain selection, process optimization, and repeated spatial structuring in order to enable the sustainable coexistence of both populations and maximize the volumetric productivity and economic yield from niche habitats such as low-value waste streams (Garrido-Amador et al., 2021).

7.6 OPPORTUNITIES AND CHALLENGES: FROM ENGINEERED CONSORTIA TO THE FACTORY FLOOR

Despite impressive potential and convincing proof-of-concept demonstrations on microbial consortia in experimental systems, their successful application under industrial biorefining conditions is still facing substantial challenges (Jiang et al., 2017). The shift from a regulated, confined system to a widespread, transparent

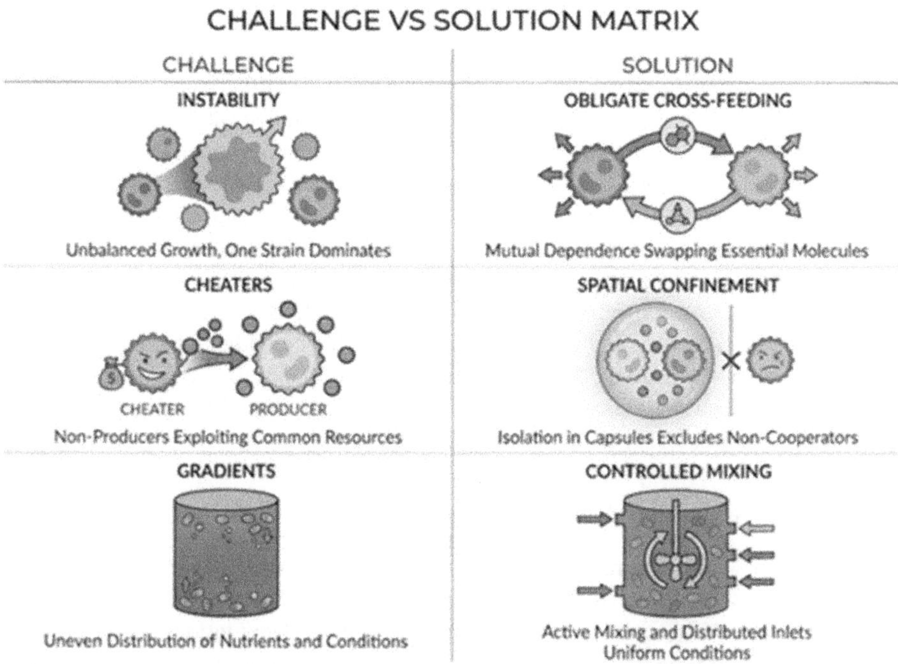

FIGURE 7.6 Challenges and stabilization strategies in industrial scale-up. Key challenges in scaling engineered consortia and corresponding stabilization strategies. Population instability is countered by engineering obligate cross-feeding dependencies. The emergence of non-cooperative 'cheater' strains is mitigated through spatial structuring that localizes public goods. Large-scale bioreactor heterogeneities (e.g., nutrient/pH gradients) are addressed through advanced reactor design and dynamic process control.

industrial operation underscores the intrinsic constraints of artificial microbial ecosystems (Ben Said & Or, 2017; Padmaperuma et al., 2017). Overcoming limitations is essential for improving consortium-based bioprocessing performance (Liu et al., 2024b). The main challenges involve attaining long-term stability, reducing cheater strains as much as possible, and implementing economically scalable bioreactor/process control strategies (Cavaliere et al., 2017; Delvigne et al., 2017) (Figure 7.6).

7.6.1 Achieving Long-Time Stability and Functional Recovery

The difficulty of consortium engineering rests on maintaining a stable and efficient demographic balance over a durable period of exploitation (Losoi et al., 2019). A variety of factors influence the configurations of interactions within a multispecies assemblage, including growth rate fluctuations, resource competition, and the effects of byproducts (Ghoul & Mitri, 2016). If the growth rate distribution is high at the population level for one member, it will competitively exclude the other and take over its function, leading to loss of the intended community function and process failure (Cremer et al., 2019; Zhang et al., 2018). Furthermore, one must also appreciate that there are a number of degrees of environmental perturbations, including sudden shifts in pH, fluctuations in temperature, changes in substrate composition, and, indeed, the inevitable introduction of pollutants into a given habitat that can potentially destabilize the complex and delicate set of synergistic interactions that exist between organisms and their environment (Jiang et al., 2019; Zhang et al., 2018). It should be a priority of future research to identify endogenous stability mechanisms (Lee & Steel, 2022). This includes the design of synthetic dependency networks, such as cross-feeding based on obligatory auxotrophies, in which participants rely on one another for necessary nutrients, thereby necessitating cooperation (Zhao et al., 2019). Quorum-sensing frameworks regulate genomic expression in response to population scale, augmenting homeostasis within biological systems and ensuring stability amidst environmental variations (Waters & Bassler, 2005). It would be necessary to achieve resilient feedback controllers that can adjust nutrient feed rates or dilution rates in accordance with (e.g., in real time) population dynamics (e.g., via flow cytometry and specialized biosensors) to stabilize industrial-scale processes (de Oliveira Lino et al., 2024; Kinet et al., 2024).

7.6.2 Mitigating Cheater Lineage Emergence

These ‘cheater’ strains express instability, gained at the expense of public goods which they consume without producing, created by the rest of the population (Smith & Schuster, 2019; West et al., 2007). In a lignocellulose-degradation community, for example, a cheater type might not have cellulase synthesis genes but might have a high capacity to consume the monosaccharide pool that is made available by the others that synthesize cellulases (Estrela et al., 2016). This lends the cheater increased freedom to distort its development by siphoning resources away from producing members, putting the whole process at risk (Oliveira et al., 2014; Pfeiffer et al., 2010). So we need mechanisms to balance the costs of our cooperation (Nowak, 2006). Spatial structure, through biofilms or immobilization, physically restricts the benefits of public goods to neighboring cells, thereby selecting for the presence of cooperative neighbors (Xavier, 2011). An alternative approach is to genetically link the production of the public good to a private good that is exclusive to the producer, such as an essential intermediate in a private metabolic pathway (Momeni et al., 2013). Guided succession would also be used to make cheater-impervious collaborators (Minty et al., 2013). Understanding ecological and evolutionary drivers of constructed systems is critical for the development of consortia that are strong and stable (Estrela et al., 2021).

7.6.3 Engineering Consortia for Industrial Correlation

The translation of consortium intrinsic processes (behavior) from the lab bench to commercial fermenters or bioreactors poses distinctive engineering challenges (Gustavsson & Lee, 2016). Chemical and physical heterogeneities, such as variability in nutrient content, pH, and dissolved oxygen in large containers, could create microenvironments that, for some species, may be beneficial over others, and disrupt the open non-dominant relationships we observed in the less heterogeneous, well-mixed cultures (Wang et al., 2024). Mass-transfer limits affect metabolite exchange in co-dependent systems. The method of sterilizing, inoculating, and subsequently harvesting multi-strain cocultures according to these procedures is considerably more complex and time-consuming than the generally applied methodology for growing and controlling single, pure cultures (Jiang et al., 2017; Padmaperuma et al., 2017). Bioprocess engineering and microbial ecology should be integral to future visions (Ben Said & Or, 2017; Delvigne et al., 2017). These include generating novel bioreactor topologies that enhance mixing or generate beneficial gradients, tailoring fed-batch and continuous fermentation techniques to control population dynamics, and inventing low-cost methods for co-immobilizing multiple strains (Chang et al., 2022; Losoi et al., 2022). Looking ahead, the future requires a holistic approach that treats the constructed consortium as an integrated bioprocess line, where biological and process designs are optimized simultaneously to enable reliable, scalable, and economically feasible waste valorization in a circular bioeconomy (Delvigne et al., 2017; Liu et al., 2024b).

7.7 CONCLUSION: ENGINEERING CONSORTIA AS THE BUILDING BLOCK FOR NEXT-GENERATION BIOREFINERIES

Progressing toward a sustainable circular bioeconomy is inherently linked to our ability to add value efficiently and effectively to a number of strong waste streams. This chapter has revealed that reliance on pure culture fermentations in the past imposed natural constraints on this effort, defined by the metabolism of those microorganisms. Natural ecosystems showcase how decentralized networks and cooperative microbes enable efficient nutrient cycling and complex transformations. The design of microbial consortia hopes to harness and manipulate these concepts to move beyond the single-cell factory ideal to a new concept of a programmable microbial ecosystem factory. This enhancement constitutes a quintessential advancement and a considerable progression toward the evolution of next-generation, expandable, and adaptable biorefineries.

7.7.1 Creating a Pathway between Concept and Application

Such cooperative engineering begins with intrinsic senses and is informed by endosymbiotic, commensalistic, or cross-feeding interactions that enhance ecosystem stability and efficiency. The environmental principles explain two main design approaches: hierarchical assembly of complex communities for operational resilience, and grassroots methodical assembly for predictability and governance. A potentially clumsy amalgamation of enabling tools has significantly simplified and homogenized the overall design methodology, enabling it to be learned and applied by working professionals in the field. Computational modeling enables *in silico* prediction of interactions and preemptive reduction or mitigation of competition, and multi-omics technologies offer unprecedented understanding of the characteristics of successful consortia in terms of structure and function, impacting both design and troubleshooting. In addition, immobilization and membrane technologies favor good relations between the communities based on the spatial structure. Application examples on practical cases, including integrated bioprocesses for lignocellulose and high-value compounds co-production, demonstrate the advantage in functionality and economy of this method over traditional monoculture.

7.7.2 Developing and Implementing an Integrated Research Agenda

Despite the strong potential, numerous obstacles along the road to industrialization can largely determine future research directions. A unified approach is needed to ensure long-term stability and prevent the emergence of cheater strains, particularly during scale-up. The answer will be the combination of several biological processes. Further progress will rely on establishing evolutionarily resilient cooperation, perhaps through engineered obligate dependencies and private benefits, and constructing dynamic process control systems that can maintain community-level homeostasis despite industrial-level variability. Ultimately, this should pave the way for platforms in which the inherent complexity of waste streams is met not by an ever more complex set of physical pretreatments, but by the sophisticated, self-regulating catalytic performances of engineered microbial communities. The fundamental shift from pure culture to consortium-based biorefining can be summarized across several key operational and strategic dimensions (Table 7.1).

Finally, it is imperative to acknowledge that the innovative advancement and subsequent cultivation of microbial consortia possess the potential to fundamentally transform and significantly enhance the landscape of industrial biotechnology in ways that have not yet been fully realized or explored. Using smart distribution of metabolic work across specialized microorganisms, we can develop next-generation bioprocesses that are not only more efficient but also produce higher-value products from inexpensive waste streams while remaining resilient to multiple stresses. This notion highlights the importance of 'benefit from waste' as it builds a strong foundation that can evolve to support the growth of a circular bioeconomy, promoting efficient resource use. Microbial consortia will, without doubt, be the future of biorefining, and this future will be accomplished by refining the design principles summarized in this chapter.

TABLE 7.1
Comparative framework: pure culture vs. engineered consortium biorefineries

Characteristic	Pure culture bioprocessing	Engineered microbial consortium	Implication for circular bioeconomy
Substrate spectrum	Narrow; optimized for defined, simple substrates (e.g., glucose, sucrose).	Broad; capable of decomposing diverse and intricate substances (like lignocellulose and assorted organic waste).	Facilitates the utilization of various inexpensive waste sources without significant pre-processing.
Metabolic capacity & product diversity	Limited to the native or engineered pathways of a single organism.	High; enables **division of labor** and simultaneous, complementary pathways for co-production of multiple products (e.g., biofuels, bioplastics, fine chemicals).	Improves economic viability by generating multiple revenue streams from a single feedstock.
Process design	Typically involves a series of multistep processes, such as distinct hydrolysis and fermentation.	Streamlines CBP (**Consolidated Bioprocessing**) by integrating enzyme production, hydrolysis of pretreated plant materials, and fermentation into a single vessel.	Reduces capital and operational costs, simplifying biorefinery architecture.
Robustness & stability	Fragile; susceptible to contamination, inhibition, and process perturbations. Susceptible to genetic variation.	**Intrinsically robust** via functional redundancy, cross-feeding, and a distributed response to stress. More resilient to fluctuations in feedstock and environment.	Enhances operational reliability and reduces failure risk in prolonged industrial use.
Design & engineering approach	Well-established: genetic modification of a single chassis organism. Predictable but limited in scope.	Complex: It combines **top-down enrichment** for functional resilience with **bottom-up assembly** for predictable control. Requires ecological principles and multi-omics tools.	Demands interdisciplinary integration (microbial ecology, systems biology, bioprocess engineering) but offers programmable ecosystem factories.
Scalability challenges	Standardized scale-up protocols exist for homogeneous fermentations.	Novel challenges: maintaining population balance, preventing **cheater strain** dominance, managing metabolite gradients in large reactors.	Requires innovative design of bioreactor, process control, and spatial organization for industrial applications.

(Continued)

TABLE 7.1
(Continued)

Characteristic	Pure culture bioprocessing	Engineered microbial consortium	Implication for circular bioeconomy
Primary development cost	High upfront cost in strain engineering and optimization.	High upfront cost in **consortium design, modeling, and stabilization.**	Initial investment mitigated by enhanced feedstock usage, premium product ranges, and operational robustness.
Long-term vision	'Superbug' paradigm: a single, highly engineered universal cell factory.	**Programmable microbial ecosystems** are self-sustaining communities with specialized organisms that replicate natural nutrient cycles.	Indicates a shift from unicellular to multicellular systemic engineering for sustainable biomanufacturing.

REFERENCES

Albornoz, R. V., Oyarzún, D., & Burgess, K. (2024). Optimisation of surfactin yield in Bacillus using data-efficient active learning and high-throughput mass spectrometry. *Computational and Structural Biotechnology Journal, 23*, 1226–1233. https://doi.org/10.1016/j.csbj.2024.02.012

Arreola-Vargas, J., Xu, B., Hu, C., Meng, X., Wang, Y.-y., Ragauskas, A. J., Méndez-Acosta, H. O., Perez-Pimienta, J. A., Dai, S. Y., & Yuan, J. S. (2022). Bioconversion of agave bagasse lignin to medium-chain-length polyhydroxyalkanoates by Pseudomonas putida. *ACS Sustainable Chemistry & Engineering*, *10*(48), 15670–15679. https://doi.org/10.1021/acssuschemeng.2c03677

Atkinson, E., Boo, A., Peng, H., Stan, G. B., & Ledesma-Amaro, R. (2022). Principles, tools, and applications of synthetic consortia toward microbiome engineering. In Chang, M. Wiley (Ed.), *Principles in Microbiome Engineering* (pp. 195–218). Wiley. https://doi.org/10.1002/9783527825462.ch7

Augst, A. D., Kong, H. J., & Mooney, D. J. (2006). Alginate hydrogels as biomaterials. *Macromolecular Bioscience*, *6*(8), 623–633. https://doi.org/10.1002/mabi.200600069

Badr, K., Peter He, Q., & Wang, J. (2022). Understanding the evolution of interspecies metabolic interactions using dynamic genome-scale metabolic modeling. In *2022 American Control Conference (ACC)*.Atlanta, GA, USA.

Bajpai, P. (2022). Enzymatic degradation of lignocellulosic biomass. In Bajpai, P. (Ed.) *Lignocellulosic Biomass in Biotechnology* (pp. 49–74). Elsevier. https://doi.org/10.1016/b978-0-12-821889-1.00009-6

Banerjee, S., Schlaeppi, K., & van der Heijden, M. G. A. (2018). Keystone taxa as drivers of microbiome structure and functioning. *Nature Reviews Microbiology*, *16*(9), 567–576. https://doi.org/10.1038/s41579-018-0024-1

Ben Said, S., & Or, D. (2017). Synthetic microbial ecology: Engineering habitats for modular consortia. *Frontiers in Microbiology*, *8,1125*. https://doi.org/10.3389/fmicb.2017.01125

Biggs, M. B., Medlock, G. L., Kolling, G. L., & Papin, J. A. (2015). Metabolic network modeling of microbial communities. *WIREs Systems Biology and Medicine*, *7*(5), 317–334. https://doi.org/10.1002/wsbm.1308

Brenner, K., You, L., & Arnold, F. H. (2008). Engineering microbial consortia: A new frontier in synthetic biology. *Trends in Biotechnology*, *26*(9), 483–489. https://doi.org/10.1016/j.tibtech.2008.05.004

Brown, J., Lindstrom, J. K., Ghosh, A., Rollag, S. A., & Brown, R. C. (2024). Production of sugars from lignocellulosic biomass via biochemical and thermochemical routes. *Frontiers in Energy Research, 12,* 1347373. https://doi.org/10.3389/fenrg.2024.1347373

Cakir, T., Scott, W. T., Benito-Vaquerizo, S., Zimmermann, J., Bajić, D., Heinken, A., Suarez-Diez, M., & Schaap, P. J. (2023). A structured evaluation of genome-scale constraint-based modeling tools for microbial consortia. *PLOS Computational Biology*, *19*(8), e1011363. https://doi.org/10.1371/journal.pcbi.1011363

Carballa, M., Regueiro, L., & Lema, J. M. (2015). Microbial management of anaerobic digestion: Exploiting the microbiome-functionality nexus. *Current Opinion in Biotechnology*, *33*, 103–111. https://doi.org/10.1016/j.copbio.2015.01.008

Carvalho, G., Pedras, I., Karst, S. M., Oliveira, C. S. S., Duque, A. F., Nielsen, P. H., & Reis, M. A. M. (2018). Functional redundancy ensures performance robustness in 3-stage PHA-producing mixed cultures under variable feed operation. *New Biotechnology*, *40*, 207–217. https://doi.org/10.1016/j.nbt.2017.08.007

Cavaliere, M., Feng, S., Soyer, O. S., & Jiménez, J. I. (2017). Cooperation in microbial communities and their biotechnological applications. *Environmental Microbiology*, *19*(8), 2949–2963. https://doi.org/10.1111/1462-2920.13767

Chang, W. L., Hou, W., Xu, M., & Yang, S. T. (2022). High-rate continuous n-butanol production by Clostridium acetobutylicum from glucose and butyric acid in a single-pass fibrous-bed bioreactor. *Biotechnology and Bioengineering*, *119*(12), 3474–3486. https://doi.org/10.1002/bit.28223

Chen, X., He, C., Zhang, Q., Bayakmetov, S., & Wang, X. (2024). Modularized design and construction of tunable microbial consortia with flexible topologies. *ACS Synthetic Biology*, *13*(1), 183–194. https://doi.org/10.1021/acssynbio.3c00420

Chen, Z., & Wan, C. (2017). Biological valorization strategies for converting lignin into fuels and chemicals. *Renewable and Sustainable Energy Reviews*, *73*, 610–621. https://doi.org/10.1016/j.rser.2017.01.166

Choe, D., Olson, C. A., Szubin, R., Yang, H., Sung, J., Feist, A. M., & Palsson, B. O. (2024). Advancing the scale of synthetic biology via cross-species transfer of cellular functions enabled by iModulon engraftment. *Nature Communications*, *15*(1), 2356. https://doi.org/10.1038/s41467-024-46486-3

Cremer, J., Melbinger, A., Wienand, K., Henriquez, T., Jung, H., & Frey, E. (2019). Cooperation in microbial populations: Theory and experimental model systems. *Journal of Molecular Biology*, *431*(23), 4599–4644. https://doi.org/10.1016/j.jmb.2019.09.023

de Oliveira Lino, F. S., Garg, S., Li, S. S., Misiakou, M.-A., Kang, K., Vale da Costa, B. L., Beyer-Pedersen, T. S.-A., Giacon, T. G., Basso, T. O., Panagiotou, G., & Sommer, M. O. A. (2024). Strain dynamics of contaminating bacteria modulate the yield of ethanol biorefineries. *Nature Communications*, *15*(1), 5323. https://doi.org/10.1038/s41467-024-49683-2

Delvigne, F., Takors, R., Mudde, R., van Gulik, W., & Noorman, H. (2017). Bioprocess scale-up/down as integrative enabling technology: From fluid mechanics to systems biology and beyond. *Microbial Biotechnology*, *10*(5), 1267–1274. https://doi.org/10.1111/1751-7915.12803

Delvigne, F., Zacchetti, B., Fickers, P., Fifani, B., Roulling, F., Lefebvre, C., Neubauer, P., & Junne, S. (2018). Improving control in microbial cell factories: From single cell to large-scale bioproduction. *FEMS Microbiology Letters, 365*, fny236. https://doi.org/10.1093/femsle/fny236

Devi, A., Bajar, S., Kour, H., Kothari, R., Pant, D., & Singh, A. (2022). Lignocellulosic biomass valorization for bioethanol production: A circular bioeconomy approach. *BioEnergy Research*, *15*(4), 1820–1841. https://doi.org/10.1007/s12155-022-10401-9

Dinjaski, N., & Kaplan, D. L. (2016). Recombinant protein blends: Silk beyond natural design. *Current Opinion in Biotechnology*, *39*, 1–7. https://doi.org/10.1016/j.copbio.2015.11.002

Drews, A., Vocks, M., Bracklow, U., & Kraume, M. (2005). Nutrients removal in MBRs for municipal wastewater treatment. *Water Science and Technology*, *51*(6–7), 391–402. https://doi.org/10.2166/wst.2005.0661

Driscoll, M., & Delmotte, B. (2019). Leveraging collective effects in externally driven colloidal suspensions: Experiments and simulations. *Current Opinion in Colloid & Interface Science*, *40*, 42–57. https://doi.org/10.1016/j.cocis.2018.10.002

Dukovski, I., Bajić, D., Chacón, J. M., Quintin, M., Vila, J. C. C., Sulheim, S., Pacheco, A. R., Bernstein, D. B., Riehl, W. J., Korolev, K. S., Sanchez, A., Harcombe, W. R., & Segrè, D. (2021). A metabolic modeling platform for the computation of microbial ecosystems in time and space (COMETS). *Nature Protocols*, *16*(11), 5030–5082. https://doi.org/10.1038/s41596-021-00593-3

Estrela, S., Kerr, B., & Morris, J. J. (2016). Transitions in individuality through symbiosis. *Current Opinion in Microbiology*, *31*, 191–198. https://doi.org/10.1016/j.mib.2016.04.007

Estrela, S., Sanchez-Gorostiaga, A., Vila, J. C. C., & Sanchez, A. (2021). Nutrient dominance governs the assembly of microbial communities in mixed nutrient environments. *eLife*, *10,* e65948. https://doi.org/10.7554/eLife.65948

Esvap, E., & Ulgen, K. O. (2021). Advances in genome-scale metabolic modeling toward microbial community analysis of the human microbiome. *ACS Synthetic Biology*, *10*(9), 2121–2137. https://doi.org/10.1021/acssynbio.1c00140

Flemming, H.-C., Wingender, J., Szewzyk, U., Steinberg, P., Rice, S. A., & Kjelleberg, S. (2016). Biofilms: An emergent form of bacterial life. *Nature Reviews Microbiology*, *14*(9), 563–575. https://doi.org/10.1038/nrmicro.2016.94

Franzosa, E. A., Hsu, T., Sirota-Madi, A., Shafquat, A., Abu-Ali, G., Morgan, X. C., & Huttenhower, C. (2015). Sequencing and beyond: Integrating molecular 'omics' for microbial community profiling. *Nature Reviews Microbiology*, *13*(6), 360–372. https://doi.org/10.1038/nrmicro3451

Gallo, J., Forsgren, M. F., Karlsson, M., Dahlqvist Leinhard, O., Dahlström, N., Norén, B., Romu, T., Ignatova, S., Ekstedt, M., Kechagias, S., Lundberg, P., & Cedersund, G. (2019). Model-inferred mechanisms of liver function from magnetic resonance imaging data: Validation and variation across a clinically relevant cohort. *PLOS Computational Biology*, *15*(6), e1007157. https://doi.org/10.1371/journal.pcbi.1007157

García-Jiménez, B., García, J. L., & Nogales, J. (2018). FLYCOP: Metabolic modeling-based analysis and engineering microbial communities. *Bioinformatics*, *34*(17), i954–i963. https://doi.org/10.1093/bioinformatics/bty561

García-Jiménez, B., Torres-Bacete, J., & Nogales, J. (2021). Metabolic modelling approaches for describing and engineering microbial communities. *Computational and Structural Biotechnology Journal*, *19*, 226–246. https://doi.org/10.1016/j.csbj.2020.12.003

Garrido-Amador, P., Kniaziuk, M., Vekeman, B., & Kartal, B. (2021). Learning from microorganisms: Using new insights in microbial physiology for sustainable nitrogen management. *Current Opinion in Biotechnology*, *67*, 42–48. https://doi.org/10.1016/j.copbio.2020.12.017

Ghoul, M., & Mitri, S. (2016). The ecology and evolution of microbial competition. *Trends in Microbiology*, *24*(10), 833–845. https://doi.org/10.1016/j.tim.2016.06.011

Gilmore, S. P., Lankiewicz, T. S., Wilken, S. E., Brown, J. L., Sexton, J. A., Henske, J. K., Theodorou, M. K., Valentine, D. L., & O'Malley, M. A. (2019). Top-down enrichment guides in formation of synthetic microbial consortia for biomass degradation. *ACS Synthetic Biology*, *8*(9), 2174–2185. https://doi.org/10.1021/acssynbio.9b00271

Götz, R., Panzer, S., Trinks, N., Eilts, J., Wagener, J., Turrà, D., Di Pietro, A., Sauer, M., & Terpitz, U. (2020). Expansion microscopy for cell biology analysis in fungi. *Frontiers in Microbiology*, *11,* 574. https://doi.org/10.3389/fmicb.2020.00574

Grimalt-Alemany, A., Łężyk, M., Lange, L., Skiadas, I. V., & Gavala, H. N. (2018). Enrichment of syngas-converting mixed microbial consortia for ethanol production and thermodynamics-based design of enrichment strategies. *Biotechnology for Biofuels*, *11*(1), 198. https://doi.org/10.1186/s13068-018-1189-6

Großkopf, T., & Soyer, O. S. (2014). Synthetic microbial communities. *Current Opinion in Microbiology*, *18*, 72–77. https://doi.org/10.1016/j.mib.2014.02.002

Gustavsson, M., & Lee, S. Y. (2016). Prospects of microbial cell factories developed through systems metabolic engineering. *Microbial Biotechnology*, *9*(5), 610–617. https://doi.org/10.1111/1751-7915.12385

Guzmán-Pedraza, R., Medina Peralta, S., Rojas-Herrera, R. A., & Sánchez-González, M. N. (2023). Metasecretome and biochemical analysis of consortium PM-06 during the degradation of nixtamalized maize pericarp. *Biocatalysis and Agricultural Biotechnology*, *48,*102634. https://doi.org/10.1016/j.bcab.2023.102634

Hanemaaijer, M., Rölling, W. F. M., Olivier, B. G., Khandelwal, R. A., Teusink, B., & Bruggeman, F. J. (2015). Systems modeling approaches for microbial community studies: From metagenomics to inference of the community structure. *Frontiers in Microbiology*, *6,* 213. https://doi.org/10.3389/fmicb.2015.00213

Havukainen, J., & Dace, E. (2023). Waste to energy and circular economy: The case of anaerobic digestion. In Prasad, M.N.V., & Smol, M (Eds.) *Sustainable and Circular Management of Resources and Waste towards a Green Deal* (pp. 105–115). Elsevier. https://doi.org/10.1016/b978-0-323-95278-1.00017-6

Ibrahim, M., Raajaraam, L., & Raman, K. (2021). Modelling microbial communities: Harnessing consortia for biotechnological applications. *Computational and Structural Biotechnology Journal*, *19*, 3892–3907. https://doi.org/10.1016/j.csbj.2021.06.048

Jansma, J., & El Aidy, S. (2021). Understanding the host-microbe interactions using metabolic modeling. *Microbiome*, *9*(1), 16. https://doi.org/10.1186/s40168-020-00955-1

Jenkins, S. N., Weerasekara, M. L. M. A. W., & Zavahir, J. S. (2021). Role of microbial communities in the low-cost, sustainable treatment of pig effluent waste. In Gamini, S., & Zavahir, J.S (Eds.)*Role of Microbial Communities for Sustainability* (pp. 289–316). Springer Nature. https://doi.org/10.1007/978-981-15-9912-5_11

Jiang, L.-L., Zhou, J.-J., Quan, C.-S., & Xiu, Z.-L. (2017). Advances in industrial microbiome based on microbial consortium for biorefinery. *Bioresources and Bioprocessing*, *4*(1), 11. https://doi.org/10.1186/s40643-017-0141-0

Jiang, Y., Wu, R., Zhou, J., He, A., Xu, J., Xin, F., Zhang, W., Ma, J., Jiang, M., & Dong, W. (2019). Recent advances of biofuels and biochemicals production from sustainable resources using co-cultivation systems. *Biotechnology for Biofuels*, *12*(1), 155. https://doi.org/10.1186/s13068-019-1495-7

Kang, C. W., Lim, H. G., Won, J., Cha, S., Shin, G., Yang, J.-S., Sung, J., & Jung, G. Y. (2022). Circuit-guided population acclimation of a synthetic microbial consortium for improved biochemical production. *Nature Communications*, *13*(1), 6506. https://doi.org/10.1038/s41467-022-34190-z

Katagi, V., Vytla, R. M., & Somashekara, D. (2024). Integrated production of microbial biopolymer (PHA) with other value-added bioproducts: An innovative approach for sustainable production. *Green Chemistry Letters and Reviews*, *17*(1), 2289983. https://doi.org/10.1080/17518253.2023.2289983

Kinet, R., Richelle, A., Colle, M., Demaegd, D., von Stosch, M., Sanders, M., Sehrt, H., Delvigne, F., & Goffin, P. (2024). Giving the cells what they need when they need it: Biosensor-based feeding control. *Biotechnology and Bioengineering*, *121*(4), 1270–1282. https://doi.org/10.1002/bit.28657

Kong, W., Meldgin, D. R., Collins, J. J., & Lu, T. (2018). Designing microbial consortia with defined social interactions. *Nature Chemical Biology*, *14*(8), 821–829. https://doi.org/10.1038/s41589-018-0091-7

Kumar, A. (2019). Myco-degradation of lignocellulose: An update on the reaction mechanism and production of lignocellulolytic enzymes by fungi. In Naraian, R. (Ed.) *Mycodegradation of Lignocelluloses* (pp. 81–117). Springer Nature. https://doi.org/10.1007/978-3-030-23834-6_6

Kundu, P., Mondal, S., & Ghosh, A. (2022). Bacterial species metabolic interaction network for deciphering the lignocellulolytic systems in fungal cultivating termite gut microbiota. *SSRN Electronic Journal* 22, 104763. https://doi.org/10.2139/ssrn.4043589

Lee, T. A., & Steel, H. (2022). Cybergenetic control of microbial community composition. *Frontiers in Bioengineering and Biotechnology*, *10*, 957140 https://doi.org/10.3389/fbioe.2022.957140

Li, X., Zhou, Z., Li, W., Yan, Y., Shen, X., Wang, J., Sun, X., & Yuan, Q. (2022). Design of stable and self-regulated microbial consortia for chemical synthesis. *Nature Communications*, *13*(1), 1554. https://doi.org/10.1038/s41467-022-29215-6

Liu, T., Li, J., Hao, X., & Meng, J. (2024a). Efficient caproic acid production from lignocellulosic biomass by bio-augmented mixed microorganisms. *Bioresource Technology*, *399*, 130565. https://doi.org/10.1016/j.biortech.2024.130565

Liu, X., Wang, M., Liu, B., Chen, X., An, L., Nie, Y., & Wu, X.-L. (2025). Keystone taxa mediate the trade-off between microbial community stability and performance in activated sludges. *Nature Water*, *3*(6), 723–733. https://doi.org/10.1038/s44221-025-00451-6

Liu, Y., Xue, B., Liu, H., Wang, S., & Su, H. (2024b). Rational construction of synthetic consortia: Key considerations and model-based methods for guiding the development of a novel biosynthesis platform. *Biotechnology Advances*, *72*, 108348. https://doi.org/10.1016/j.biotechadv.2024.108348

Lopez, T., Chuan, C., Ramirez, A., Chen, K. H. E., Lorenson, M. Y., Benitez, C., Mustafa, Z., Pham, H., Sanchez, R., Walker, A. M., & Ge, X. (2018). Epitope-specific affinity maturation improved stability of potent protease inhibitory antibodies. *Biotechnology and Bioengineering*, *115*(11), 2673–2682. https://doi.org/10.1002/bit.26814

Losoi, P., Konttinen, J., & Santala, V. (2022). Substantial gradient mitigation in simulated large-scale bioreactors by optimally placed multiple feed points. *Biotechnology and Bioengineering*, *119*(12), 3549–3566. https://doi.org/10.1002/bit.28232

Losoi, P. S., Santala, V. P., & Santala, S. M. (2019). Enhanced population control in a synthetic bacterial consortium by interconnected carbon cross-feeding. *ACS Synthetic Biology*, *8*(12), 2642–2650. https://doi.org/10.1021/acssynbio.9b00316

Lyu, X., Nuhu, M., Candry, P., Wolfanger, J., Betenbaugh, M., Saldivar, A., Zuniga, C., Wang, Y., & Shrestha, S. (2024). Top-down and bottom-up microbiome engineering approaches to enable biomanufacturing from waste biomass. *Journal of Industrial Microbiology and Biotechnology*, *51*, kuae025. https://doi.org/10.1093/jimb/kuae025

McInerney, M. J., Sieber, J. R., & Gunsalus, R. P. (2009). Syntrophy in anaerobic global carbon cycles. *Current Opinion in Biotechnology*, *20*(6), 623–632. https://doi.org/10.1016/j.copbio.2009.10.001

Minty, J. J., Singer, M. E., Scholz, S. A., Bae, C.-H., Ahn, J.-H., Foster, C. E., Liao, J. C., & Lin, X. N. (2013). Design and characterization of synthetic fungal-bacterial consortia for direct production of isobutanol from cellulosic biomass. *Proceedings of the National Academy of Sciences*, *110*(36), 14592–14597. https://doi.org/10.1073/pnas.1218447110

Momeni, B., Waite, A. J., & Shou, W. (2013). Spatial self-organization favors heterotypic cooperation over cheating. *eLife*, *2*, e00960. https://doi.org/10.7554/eLife.00960

Moon, S. Y., Son, S.-H., Baek, S.-H., & Lee, J. Y. (2024). Designing microbial cell factories for programmable control of cellular metabolism. *Current Opinion in Systems Biology*, *37*, 100493. https://doi.org/10.1016/j.coisb.2023.100493

Możejko-Ciesielska, J., Ray, S., & Sankhyan, S. (2023). Recent challenges and trends of polyhydroxyalkanoate production by extremophilic bacteria using renewable feedstocks. *Polymers*, *15*(22), 4385 . https://doi.org/10.3390/polym15224385

Nielsen, J., Zampieri, G., Vijayakumar, S., Yaneske, E., & Angione, C. (2019). Machine and deep learning meet genome-scale metabolic modeling. *PLOS Computational Biology*, *15*(7), e1007084. https://doi.org/10.1371/journal.pcbi.1007084

Nobu, M. K., Narihiro, T., Mei, R., Kamagata, Y., Lee, P. K. H., Lee, P.-H., McInerney, M. J., & Liu, W.-T. (2020). Catabolism and interactions of uncultured organisms shaped by ecothermodynamics in methanogenic bioprocesses. *Microbiome*, *8*(1), 111. https://doi.org/10.1186/s40168-020-00885-y

Noto Guillen, M., Rosener, B., Sayin, S., & Mitchell, A. (2021). Assembling stable syntrophic Escherichia coli communities by comprehensively identifying beneficiaries of secreted goods. *Cell Systems*, *12*(11), 1064–1078.e1067. https://doi.org/10.1016/j.cels.2021.08.002

Nowak, M. A. (2006). Five rules for the evolution of cooperation. *Science*, *314*(5805), 1560–1563. https://doi.org/10.1126/science.1133755

Oliveira, N. M., Niehus, R., & Foster, K. R. (2014). Evolutionary limits to cooperation in microbial communities. *Proceedings of the National Academy of Sciences*, *111*(50), 17941–17946. https://doi.org/10.1073/pnas.1412673111

Ouzounis, C. A., Thompson, J. C., Zavala, V. M., & Venturelli, O. S. (2023). Integrating a tailored recurrent neural network with Bayesian experimental design to optimize microbial community functions. *PLOS Computational Biology*, *19*(9), e1011436. https://doi.org/10.1371/journal.pcbi.1011436

Padmaperuma, G., Kapoore, R. V., Gilmour, D. J., & Vaidyanathan, S. (2017). Microbial consortia: A critical look at microalgae co-cultures for enhanced biomanufacturing. *Critical Reviews in Biotechnology*, *38*(5), 690–703. https://doi.org/10.1080/07388551.2017.1390728

Parks, D. H., Rinke, C., Chuvochina, M., Chaumeil, P.-A., Woodcroft, B. J., Evans, P. N., Hugenholtz, P., & Tyson, G. W. (2017). Recovery of nearly 8,000 metagenome-assembled genomes substantially expands the tree of life. *Nature Microbiology*, *2*(11), 1533–1542. https://doi.org/10.1038/s41564-017-0012-7

Peng, X. N., Gilmore, S. P., & O'Malley, M. A. (2016). Microbial communities for bioprocessing: Lessons learned from nature. *Current Opinion in Chemical Engineering*, *14*, 103–109. https://doi.org/10.1016/j.coche.2016.09.003

Pfeiffer, T., Nadell, C. D., Foster, K. R., & Xavier, J. B. (2010). Emergence of spatial structure in cell groups and the evolution of cooperation. *PLOS Computational Biology*, *6*(3), e1000716. https://doi.org/10.1371/journal.pcbi.1000716

Prabhu, A. A., Ledesma-Amaro, R., Lin, C. S. K., Coulon, F., Thakur, V. K., & Kumar, V. (2020). Bioproduction of succinic acid from xylose by engineered Yarrowia lipolytica without pH control. *Biotechnology for Biofuels*, *13*(1), 113. https://doi.org/10.1186/s13068-020-01747-3

Prado-Acebo, I., Cubero-Cardoso, J., Lu-Chau, T. A., & Eibes, G. (2024). Integral multi-valorization of agro-industrial wastes: A review. *Waste Management*, *183*, 42–52. https://doi.org/10.1016/j.wasman.2024.05.001

Price, N. D., Chan, S. H. J., Simons, M. N., & Maranas, C. D. (2017). SteadyCom: Predicting microbial abundances while ensuring community stability. *PLOS Computational Biology*, *13*(5), e1005539. https://doi.org/10.1371/journal.pcbi.1005539

Rapp, K. M., Jenkins, J. P., & Betenbaugh, M. J. (2020). Partners for life: building microbial consortia for the future. *Current Opinion in Biotechnology*, *66*, 292–300. https://doi.org/10.1016/j.copbio.2020.10.001

Regueiro, L., Lema, J. M., & Carballa, M. (2015). Key microbial communities steering the functioning of anaerobic digesters during hydraulic and organic overloading shocks. *Bioresource Technology*, *197*, 208–216. https://doi.org/10.1016/j.biortech.2015.08.076

Roell, G. W., Zha, J., Carr, R. R., Koffas, M. A., Fong, S. S., & Tang, Y. J. (2019). Engineering microbial consortia by division of labor. *Microbial Cell Factories*, *18*(1), 35. https://doi.org/10.1186/s12934-019-1083-3

Röttjers, L., & Faust, K. (2018). From hairballs to hypotheses–biological insights from microbial networks. *FEMS Microbiology Reviews*, *42*(6), 761–780. https://doi.org/10.1093/femsre/fuy030

Ruby, E. (2020). Getting to know our microbial friends by dropping into their neighbourhood. *Environmental Microbiology Reports*, *13*(1), 27–30. https://doi.org/10.1111/1758-2229.12895

Schink, B. (1997). Energetics of syntrophic cooperation in methanogenic degradation. *Microbiology and Molecular Biology Reviews*, *61*(2), 262–280. https://doi.org/10.1128/mmbr.61.2.262-280.1997

Scholz, S. A., Graves, I., Minty, J. J., & Lin, X. N. (2017). Production of cellulosic organic acids via synthetic fungal consortia. *Biotechnology and Bioengineering*, *115*(4), 1096–1100. https://doi.org/10.1002/bit.26509

Shakya, M., Lo, C.-C., & Chain, P. S. G. (2019). Advances and challenges in metatranscriptomic analysis. *Frontiers in Genetics*, *10*, 904. https://doi.org/10.3389/fgene.2019.00904

Shene, C., Paredes, P., Flores, L., Leyton, A., Asenjo, J. A., & Chisti, Y. (2020). Dynamic flux balance analysis of biomass and lipid production by Antarctic thraustochytrid Oblongichytrium sp. RT2316-13. *Biotechnology and Bioengineering*, *117*(10), 3006–3017. https://doi.org/10.1002/bit.27463

Sieber, J. R., McInerney, M. J., & Gunsalus, R. P. (2012). Genomic insights into syntrophy: The paradigm for anaerobic metabolic cooperation. *Annual Review of Microbiology*, *66*(1), 429–452. https://doi.org/10.1146/annurev-micro-090110-102844

Sing, A., Tsatskis, Y., Fabian, L., Hester, I., Rosenfeld, R., Serricchio, M., Yau, N., Bietenhader, M., Shanbhag, R., Jurisicova, A., Brill, Julie A., McQuibban, G. A., & McNeill, H. (2014). The atypical cadherin fat directly regulates mitochondrial function and metabolic state. *Cell*, *158*(6), 1293–1308. https://doi.org/10.1016/j.cell.2014.07.036

Singh, H. M., Raina, S., Tyagi, V. V., & Kothari, R. (2022). Microbial bioprospecting for biorefinery application: Bottlenecks and sustainability. In Verma, P., & Shah, M.P. (Eds.) *Bioprospecting of Microbial Diversity* (pp. 277–296). Elsevier. https://doi.org/10.1016/b978-0-323-90958-7.00004-2

Singhania, R. R., Patel, A. K., Singh, A., Haldar, D., Soam, S., Chen, C.-W., Tsai, M.-L., & Dong, C.-D. (2022). Consolidated bioprocessing of lignocellulosic biomass: Technological advances and challenges. *Bioresource Technology*, *354*, 127153. https://doi.org/10.1016/j.biortech.2022.127153

Smith, P., & Schuster, M. (2019). Public goods and cheating in microbes. *Current Biology*, *29*(11), R442–R447. https://doi.org/10.1016/j.cub.2019.03.001

Stams, A. J. M., Worm, P., Sousa, D. Z., Alves, M. M., & Plugge, C. M. (2012). Syntrophic degradation of fatty acids by methanogenic communities. In Hallenbeck, P.C. (Ed.), *Microbial Technologies in Advanced Biofuels Production* (pp. 127–142). Springer Nature. https://doi.org/10.1007/978-1-4614-1208-3_8

Stewart, P. S., & Franklin, M. J. (2008). Physiological heterogeneity in biofilms. *Nature Reviews Microbiology*, *6*(3), 199–210. https://doi.org/10.1038/nrmicro1838

Stikane, A., Dace, E., & Stalidzans, E. (2022). Closing the loop in bioproduction: Spent microbial biomass as a resource within circular bioeconomy. *New Biotechnology*, *70*, 109–115. https://doi.org/10.1016/j.nbt.2022.06.001

Tarzi, C., Zampieri, G., Sullivan, N., & Angione, C. (2024). Emerging methods for genome-scale metabolic modeling of microbial communities. *Trends in Endocrinology & Metabolism*, *35*(6), 533–548. https://doi.org/10.1016/j.tem.2024.02.018

Troiano, D. T., & Studer, M. H. P. (2025). Microbial consortia for the conversion of biomass into fuels and chemicals. *Nature Communications*, *16*(1), 6712. https://doi.org/10.1038/s41467-025-61957-x

Vea, E. B., Romeo, D., & Thomsen, M. (2018). Biowaste valorisation in a future circular bioeconomy. *Procedia CIRP*, *69*, 591–596. https://doi.org/10.1016/j.procir.2017.11.062

Wang, D., Hunt, K. A., Candry, P., Tao, X., Wofford, N. Q., Zhou, J., McInerney, M. J., Stahl, D. A., Tanner, R. S., Zhou, A., Winkler, M., Pan, C., Zhao, L., & Bello, M. G. D. (2023). Cross-feedings, competition, and positive and negative synergies in a four-species synthetic community for anaerobic degradation of cellulose to methane. *mBio*, *14*(2), e0318922. https://doi.org/10.1128/mbio.03189-22

Wang, X., Yang, Q., Haringa, C., Wang, Z., Chu, J., Zhuang, Y., & Wang, G. (2024). An industrial perspective on metabolic responses of Penicillium chrysogenum to periodic dissolved oxygen feast-famine cycles in a scale-down system. *Biotechnology and Bioengineering*, *121*(10), 3076–3098. https://doi.org/10.1002/bit.28782

Waters, C. M., & Bassler, B. L. (2005). Quorum sensing: Cell-to-cell communication in bacteria. *Annual Review of Cell and Developmental Biology*, *21*(1), 319–346. https://doi.org/10.1146/annurev.cellbio.21.012704.131001

West, S. A., Diggle, S. P., Buckling, A., Gardner, A., & Griffin, A. S. (2007). The social lives of microbes. *Annual Review of Ecology, Evolution, and Systematics*, *38*(1), 53–77. https://doi.org/10.1146/annurev.ecolsys.38.091206.095740

Widder, S., Allen, R. J., Pfeiffer, T., Curtis, T. P., Wiuf, C., Sloan, W. T., Cordero, O. X., Brown, S. P., Momeni, B., Shou, W., Kettle, H., Flint, H. J., Haas, A. F., Laroche, B., Kreft, J.-U., Rainey, P. B., Freilich, S., Schuster, S., Milferstedt, K., … Soyer, O. S. (2016). Challenges in microbial ecology: Building predictive understanding of community function and dynamics. *The ISME Journal*, *10*(11), 2557–2568. https://doi.org/10.1038/ismej.2016.45

Wong, M. T., Nesbø, C. L., Wang, W., Couturier, M., Lombard, V., Lapebie, P., Terrapon, N., Henrissat, B., Edwards, E. A., & Master, E. R. (2023). Taxonomic composition and carbohydrate-active enzyme content in microbial enrichments from pulp mill anaerobic granules after cultivation on lignocellulosic substrates. *Frontiers in Microbiomes*, *2,* 1094865. https://doi.org/10.3389/frmbi.2023.1094865

Wu, Z.-L., Shi, W.-J., Zhang, L., Xia, Z.-Y., Gou, M., Sun, Z.-Y., & Tang, Y.-Q. (2024). Investigating the robustness of microbial communities in municipal sludge anaerobic digestion under organic loading rate disturbance. *Journal of Environmental Management*, *372*, 123326. https://doi.org/10.1016/j.jenvman.2024.123326

Xavier, J. B. (2011). Social interaction in synthetic and natural microbial communities. *Molecular Systems Biology*, *7*(1), MSB201116. https://doi.org/10.1038/msb.2011.16

Yuan, S., Guo, S., Huang, X., & Meng, F. (2022). Time-lagged interspecies interactions prevail during biofilm development in moving bed biofilm reactor. *Biotechnology and Bioengineering*, *119*(10), 2770–2783. https://doi.org/10.1002/bit.28177

Zaman, G., Grădinaru, G. I., & Matei, B. F. (2020). Looking ahead - From fossil fuel to a circular bioeconomy in Europe. Evolution and behaviour. *Proceedings of the International Conference on Applied Statistics*, *2*(1), 296–306.

Zambello, I. U., Holwerda, E. K., & Lynd, L. R. (2024). Characterization of sugarcane bagasse solubilization and utilization by thermophilic cellulolytic and saccharolytic bacteria at increasing solid loadings. *Bioresource Technology*, *406*, 130973. https://doi.org/10.1016/j.biortech.2024.130973

Zhang, S., Merino, N., Okamoto, A., & Gedalanga, P. (2018). Interkingdom microbial consortia mechanisms to guide biotechnological applications. *Microbial Biotechnology*, *11*(5), 833–847. https://doi.org/10.1111/1751-7915.13300

Zhao, K., Liu, L., Chen, X., Huang, T., Du, L., Lin, J., Yuan, Y., Zhou, Y., Yue, B., Wei, K., & Chu, Y. (2019). Behavioral heterogeneity in quorum sensing can stabilize social cooperation in microbial populations. *BMC Biology*, *17*(1), 20. https://doi.org/10.1186/s12915-019-0639-3

Zhao, Z., Zhang, Y., Yu, Q., Dang, Y., Li, Y., & Quan, X. (2016). Communities stimulated with ethanol to perform direct interspecies electron transfer for syntrophic metabolism of propionate and butyrate. *Water Research*, *102*, 475–484. https://doi.org/10.1016/j.watres.2016.07.005

Zong, Q.-J., Xu, T., Liu, H., Xu, L., Zhang, R.-K., Li, B.-Z., Liu, Z.-H., & Yuan, Y.-J. (2022). Microbial valorization of lignin to bioplastic by genome-reduced Pseudomonas putida. *Frontiers in Microbiology*, *13*, 923664. https://doi.org/10.3389/fmicb.2022.923664

Zorrilla, F., Buric, F., Patil, K. R., & Zelezniak, A. (2021). metaGEM: Reconstruction of genome scale metabolic models directly from metagenomes. *Nucleic Acids Research*, *49*(21), e126–e126. https://doi.org/10.1093/nar/gkab815

Zuñiga, C., Li, T., Guarnieri, M. T., Jenkins, J. P., Li, C.-T., Bingol, K., Kim, Y.-M., Betenbaugh, M. J., & Zengler, K. (2020). Synthetic microbial communities of heterotrophs and phototrophs facilitate sustainable growth. *Nature Communications*, *11*(1), 3803. https://doi.org/10.1038/s41467-020-17612-8

Index

Note: **Bold** page numbers refer to tables and *italic* page numbers refer to figures.

For Product Safety Concerns and Information please contact our EU
representative GPSR@taylorandfrancis.com
Taylor & Francis Verlag GmbH, Kaufingerstraße 24, 80331 München, Germany

www.ingramcontent.com/pod-product-compliance
Ingram Content Group UK Ltd.
Pitfield, Milton Keynes, MK11 3LW, UK
UKHW020200200926
475702UK00004B/548

* 9 7 8 1 0 4 1 2 5 1 0 8 8 *